THE SINGULARITY

Could artificial intelligence really out-think us (and would we want it to)?

Edited by Uziel Awret
with keynote author David Chalmers

imprint-academic.com

Published in the UK by
Imprint Academic, PO Box 200, Exeter EX5 5YX, UK

Distributed in the USA by
Ingram Book Company,
One Ingram Blvd., La Vergne, TN 37086, USA

ISBN 9781845409074 (paperback)
9781845409142 (hardback)

A CIP catalogue record for this book is available from the
British Library and US Library of Congress

The commentaries in this volume were originally published as two
special issues of the *Journal of Consciousness Studies*, vols. 19 (1–2) and 19
(7–8). The original target article was published in *JCS*, vol. 17 (9–10)

Contents

About Authors

Igor Aleksander, FREng, is Emeritus Professor and Senior Reserch Investigator at Imperial College. He has researched AI and Neural Systems since the 1970s, published 14 books and is currently working on informational models of mind.

Bryan Appleyard was educated at Bolton School and King's College, Cambridge, graduating with a degree in English. He was Financial News Editor and Deputy Arts Editor at *The Times* from 1976 to 1984. Subsequently he became a freelance journalist and author. He has been Feature Writer of the Year at the British Press Awards three times and once Interviewer of the Year. Currently he is a special feature writer, commentator, reviewer and columnist for *The Sunday Times*. He also writes for the *New Statesman*, *Vanity Fair*, and *Intelligent Life* and *Prospect*. He has also written for, among other publications, *The New York Times*, *The Spectator*, *The Daily Telegraph*, *The Times*, *The Independent*, *The Tablet*, *The Times Literary Supplement*, *Literary Review*, and *The Sun*.

Uziel Awret, a previous contributor to JCS with a life long interest in consciousness studies, studied physics in the Technion (the Israeli Technical Institute), biophysics in Georgetown University, and some continental philosophy in George Mason University. He is currently a faculty member in the College of Arts and Sciences in TrinityDC University in Washington D.C. where he teaches physics. He is also a member of the Inspire Institute and a research associate at Chapman University East where he is involved in theoretical research on the connections between quantum mechanics and biology.

Susan Blackmore is a psychologist and writer researching consciousness, memes, and anomalous experiences, and a Visiting Professor at the University of Plymouth. She blogs for the *Guardian*, and often appears on radio and television. *The Meme Machine* (1999) has been translated into 16 other languages; other books include *Conversations on Consciousness* (2005), *Zen and the Art of Consciousness* (2011), and a textbook, *Consciousness: An Introduction* (2nd ed. 2010).

Nick Bostrom is a professor in the Faculty of Philosophy and the Oxford Martin School at Oxford University, where he directs the Future of Humanity Institute and the Programme on the Impacts of

Future Technology. His research interests intersect areas within philosophy of science, moral philosophy, practical ethics, and technology policy and futures. He is currently writing a book about machine intelligence.

Selmer Bringsjord is Professor of Cognitive Science, Computer Science, and Management & Technology at Rensselaer Polytechnic Institute. He specializes in leading the engineering of logic-based AI systems, and in the logico-mathematical and philosophical foundations of AI.

Damien Broderick holds a PhD from Deakin University and is a Senior Fellow in the School of Culture and Communication at the University of Melbourne. His 50 books comprise novels (such as the post-Singularity *The White Abacus, Transcension*, and *Godplayers/ K-Machines*), critical and theoretical studies (including *Reading by Starlight, The Architecture of Babel, Theory and its Discontents*), popular science (including *The Spike* and *The Last Mortal Generation*), and edited volumes (such as *Warriors of the Tao* and *Skiffy and Mimesis*). He does not expect to live long enough to see the singularity, but you never know.

Richard Brown is a philosopher at the City University of New York. In particular he is an Associate Professor in the philosophy program at LaGuardia Community College. He earned his PhD in Philosophy with a concentration in Cognitive Science from the CUNY Graduate Center in 2008. To learn more visit his website: http://onemorebrown.com.

David J. Chalmers is Professor of Philosophy and co-director of the Center for Mind, Brain, and Consciousness at NYU, and also Professor of Philosophy at ANU. He works in the philosophy of mind and related areas of philosophy and cognitive science, with a special interest in consciousness. He was among the founders of the Association for the Scientific Study of Consciousness and his publications include *The Conscious Mind* (OUP 1996), *Philosophy of Mind: Classical and Contemporary Readings* (ed., OUP 2002), *The Character of Consciousness* (OUP 2010), and *Constructing the World* (OUP 2012).

Joseph Corabi is a philosophy professor at Saint Joseph's University in Philadelphia. He is the author of numerous articles in philosophy of mind and philosophy of religion.

Barry Dainton is a Professor in the Philosophy Department at the University of Liverpool. He has a PhD from the University of Cambridge (2001) and University of Durham (2008). Recent books include *Time and Space* (Acumen, 2nd ed. 2010) and *The Phenomenal Self* (OUP 2008). He has contributed to a number of other books and has published numerous journal articles.

Daniel C. Dennett is the author of *Breaking the Spell* (Viking 2006), *Freedom Evolves* (Viking Penguin 2003), and *Darwin's Dangerous Idea* (Simon & Schuster 1995), and most recently (with Matthew Hurley and Reginald Adams) *Inside Jokes: Using Humor to Reverse Engineer the Mind* (2011). He is University Professor and Austin B. Fletcher Professor of Philosophy, and Co-Director of the Center for Cognitive Studies at Tufts University.

Ben Goertzel is CEO of AI software company Novamente LLC and bioinformatics company Biomind LLC; co-leader of the open-source OpenCog Artificial General Intelligence software project; Vice Chairman of futurist nonprofit Humanity+; Advisor to the Singularity University and Singularity Institute; Research Professor in the Fujian Key Lab for Brain-Like Intelligent Systems at Xiamen University; and general Chair of the Artificial General Intelligence conference series. He has published a dozen scientific books, 100+ technical papers, and numerous journalistic articles.

Susan Greenfield CBE (www.susangreenfield.com) is a Senior Research Fellow at the University Dept. of Pharmacology Oxford where she explores novel neural mechanisms in three scenarios: neurodegeneration, the impact of contemporary cyber-culture, and consciousness. As well as writing a range of neuroscience-based books for the non-specialist she also is active in the print and broadcast media, has 30 honorary degrees from British and foreign universities, and a seat in the House of Lords.

Robin Hanson is an associate professor of economics at George Mason University, a research associate at the Future of Humanity Institute of Oxford University, and chief scientist at Consensus Point. After receiving his PhD in social science from the California Institute of Technology in 1997, Robin was a Robert Wood Johnson Foundation health policy scholar at the University of California at Berkeley. He spent nine years researching artificial intelligence, Bayesian statistics, and hypertext publishing at Lockheed, NASA, and independently. Robin has 70+ publications.

Francis Heylighen is a research professor at the Vrije Universiteit Brussel, where he directs the Evolution, Complexity and Cognition group, and the newly founded Global Brain Institute. He has authored over 100 scientific publications on a broad variety of subjects and disciplines. The main focus of his research is the emergence of intelligent organization in complex systems.

Marcus Hutter is Professor in the RSCS at the Australian National University in Canberra. He received his PhD in physics from the LMU in Munich and a Habilitation in informatics from the TU Munich. Since 2000, his research at IDSIA and ANU is centred around the information-theoretic foundations of inductive reasoning and reinforcement learning, which has resulted in 100+ publications and several awards. He has a book titled *Universal Artificial Intelligence* (Springer, EATCS, 2005). He also runs the Human Knowledge Compression Contest (50,000 Euro H-prize).

Ray Kurzweil is an inventor, author, and futurist. He was the principal developer of the first CCD flat-bed scanner, the first omni-font optical character recognition, the first print-to-speech reading machine for the blind, the first text-to-speech synthesizer, the first music synthesizer capable of recreating the grand piano and other orchestral instruments, and the first commercially marketed large-vocabulary speech recognition. He has written four national best-selling books. His book *The Singularity is Near* was a *New York Times* best seller and has been the #1 book on Amazon in both science and philosophy. His next book, *How to Create a Mind, The Secret of Human Thought Revealed*, was released by Viking in November 2012.

Pamela McCorduck has written extensively about artificial intelligence, including her history of AI, *Machines Who Think*, which was reissued in a 25th anniversary edition. For the past few years, her interests have turned to the sciences of complexity, and she has published two novels of a planned trilogy that explore the human side of complexity: *The Edge of Chaos* (2007) and *Bounded Rationality* (2012). She divides her time between New York City and Santa Fe, New Mexico.

Drew McDermott is Professor of Computer Science at Yale University. He has done work in several areas of artificial intelligence. He did seminal work in the area of 'hierarchical planning' in the 1970s. In the last decade, his focus has switched to regression-based techniques for classical planning. He was instrumental in starting the biannual series of AI Planning Systems (AIPS) conferences. In 1998, he ran the first ever Planning Competition in conjunction with AIPS. Professor McDermott is a Fellow of the American Association for Artificial Intelligence.

Chris Nunn is a retired psychiatrist whose research interests were in mind–body relationships and bipolar disorder. He has been actively involved in the 'consciousness studies' movement since 1992 and has published books on 'free will', mentality, and theories of consciousness.

Arkady Plotnitsky is a professor of English and Theory and Cultural Studies at Purdue University, where he is also a director of the Theory and Cultural Studies Program. He has published extensively on philosophy of physics and mathematics, continental philosophy, British and European Romanticism, Modernism, and the relationships among literature, philosophy, and science. His most recent books are *Epistemology and Probability: Bohr, Heisenberg, Schrödinger and the Nature of Quantum-Theoretical Thinking* and *Niels Bohr and Complementarity: An Inroduction.*

Jesse Prinz is Distinguished Professor of Philosophy at the City University of New York, Graduate Center. He is author of *Furnishing the Mind* (MIT 2002), *Gut Reactions* (OUP 2004), *The Emotional Construction of Morals* (OUP 2007), *Beyond Human Natures* (Penguin/Norton 2012), and *The Conscious Brain* (OUP 2012), with another title, *Works of Wonder* (OUP), in progress.

Jürgen Schmidhuber is Director of the Swiss Artificial Intelligence Lab IDSIA (since 1995), Professor of Artificial Intelligence at the University of Lugano, Switzerland (since 2009), and Professor SUPSI (since 2003). He has published nearly 300 works on topics such as machine learning, neural nets, complexity theory, digital physics, and his formal theory of beauty and creativity and humour. The algorithms developed in his lab won first prizes in many pattern recognition competitions, and led to several best paper awards. In 2008 he was elected member of the European Academy of Sciences and Arts.

Susan Schneider is a philosophy professor at the University of Pennsylvania and the author of several books and articles in metaphysics and philosophy of mind/cognitive science. Recent works include: *Science Fiction and Philosophy* (Wiley-Blackwell 2009); *The*

Language of Thought (MIT Press 2011); and *Blackwell Companion to Consciousness* (with Max Velmans, Wiley-Blackwell 2007).

Murray Shanahan is Professor of Cognitive Robotics at Imperial College London. In the 1980s, he studied computer science as an undergraduate at Imperial College, and then obtained his PhD from Cambridge University. Since then he has carried out work in AI, robotics, and cogitive science, and has numerous peer-reviewed publications in these areas. For the past decade or so he has turned his attention to the brain, and the relationship between cognition and consciousness. His book *Embodiment and the Inner Life* was published by OUP in 2010.

Carl Shulman is a Research Fellow at the Singularity Institute for Artificial Intelligence, where his research focuses on forecasting and policy issues raised by the long-term future of artificial intelligence. He attended New York University School of Law and holds a degree in philosophy from Harvard University.

Eric Steinhart received his BS in Computer Science from the Pennsylvania State University in 1983, after which he worked as a software designer for several years. He earned an MA in Philosophy from Boston College and a PhD in Philosophy from SUNY at Stony Brook in 1996. Since 1997, he has taught in the Philosophy Department at William Paterson University. His past work has concerned Nietzsche as well as metaphor. He has written extensively in metaphysics, and is especially interested in alternative and non-theistic types of Western religion.

Frank J. Tipler is Professor in the Department of Mathematics at Tulane University, Louisiana. His books include *The Anthropic Cosmological Principle* (co-authored with John Barrow) and *The Physics of Immortality* (both Anchor Books/Doubleday 1995). His main area of interest concerns the Omega Point Theory and he has numerous publications in this and other areas. He is a fellow of the International Society for Complexity, Information, and Design.

Burton Voorhees received his PhD from the University of Texas at Austin in 1971, working in general relativity theory where he is known for the Voorhees solutions of the Einstein equations. He later received a postdoc appointment in mathematics at the University of Alberta where he worked in mathematical biology, theoretical psychology, philosophy of science, and consciousness studies. In 1987 he began work in the field of cellular automata, which has continued to be a major area of research. He has published papers on the trade-offs between stability and flexibility, and between quickness and accuracy of response in risk/reward situations. He has recently become interested in evolutionary dynamics.

Roman V. Yampolskiy is the director of the Cybersecurity Research Lab at the CECS Dept, University of Louisville. He holds a PhD degree from the Department of Computer Science and Engineering at the University at Buffalo. There he was a recipient of a four-year NSF fellowship. After completing his PhD Dr. Yampolskiy was an Affiliate Academic at the Centre for Advanced Spatial Analysis, University College London. He is currently an assistant professor in the Department of Computer Engineering and Computer Science.

Bryan Appleyard

Preface

An Unbeliever's Introduction to the Singularity[1]

I first became fully conscious of the technological singularity in 2006 when I was writing a book about immortality. The ultimate machine was then expected — and still is — within about thirty years and the conquest of death was one of its most tantalizing promises. Since many of the thinkers and scientists I encountered were in their fifties, this time frame was awkward. Could they live long enough to live for ever or would they be among the last humans to die?

They were clearly determined — unattractively so — to make it across the line. I considered their rather rabid daily consumption of hundreds of supposedly life-prolonging pills and pursuit of dangerously calorie-restricted diets undignified and I was profoundly sceptical — for, perhaps, dubious reasons — of the singularity itself. At one level it looked suspiciously like another expression of American religiosity. The imminent arrival of transcendence, of uplift into a higher, better state, echoed the 'rapture' which some denominations derive from the Book of Revelations. Adding immortality to the mix was further evidence of the connection.

In addition, there was something all too obviously self-serving about the enthusiasm of Silicon Valley's embrace of the idea. Valley adepts like to believe they and only they own the future; the imminent arrival of a quasi-deity constructed from their own technology would seem to endorse this belief. (I once asked Eric Schmidt, chairman of

---

1 This guest editorial continues in the tradition of an earlier collection of *JCS* essays (*The Volitional Brain*), lauded by Daniel C. Dennett as 'an unsurpassed paragon of open-mindedness, the proof of which is that it includes as its closing essay a trenchant review of itself' (Dennett, 2004, p. 256).

Google, if he believed in the singularity and he answered, unhelpfully, it depended on what I meant. However, since he had just employed Ray Kurzweil, the singularity's most vociferous prophet, I think we can take it that Google, at the very least, is hedging its bets.)

Finally, I knew the undistinguished history of artificial intelligence research since the creation of the term in the 1950s. Wildly publicized promises were routinely followed by rather less publicized failures. This already applies to the singularity. As David Chalmers points out, in 1965 the statistician I.J. Good forecast the arrival of the 'First Ultraintelligent Machine' in 2000. If they always get it so wrong, why should they suddenly get it right in the future? To say they *must* eventually get it right is a statement of the faith that Popper called 'promissory materialism'.

I now accept that my first two reasons for suspicion were, though still pointed, expressions of my own prejudices and had nothing to do with the central issues which are: is the singularity a real possibility and, if so, is it imminent? My third suspicion, however, stands. As Luciano Floridi has pointed out, computers are just Turing machines and 'No conscious, intelligent entity is going to emerge from a Turing Machine' (Floridi, 2016). In other words, it is the height of vanity to think the machines we have created for making our lives more efficient, easy, or entertaining are precisely the sort of machines that will become super-intelligent and/or conscious.

Of course, this is not to say that the singularity will not happen, it is just to say it is unlikely to happen with the technology we now employ. New technology could still emerge within the promised time frame. Furthermore, we may be setting an excessively high bar for a machine to qualify as sufficiently singular. As David Chalmers, and many others, have said, an intelligent machine does not have to be conscious to be a world-transforming event, it merely has to be effective.

This certainly takes the pressure off singularity fans. Consciousness, in spite of sundry claims to the contrary, remains a very hard problem indeed and there is no prospect of it being solved in the near future — that is to say in the future in which the machine is currently expected to be built. Even in the long term, it may be that framing consciousness as a problem requiring a solution may be a meaningless task. If, in terms of the singularity, consciousness is an irrelevant problem then only technical software/hardware issues would seem to remain. Then, of course, there are the human problems.

Machine takeover, whether expressed in terms of the singularity or not, is commonly expressed as a grim prospect in popular culture. In the *Terminator* films humans and machines are in perpetual conflict and the machines are definitely the bad guys. Similarly, in the *Matrix* films the machines have enslaved humanity within a virtual reality of which, for the most part, they are entirely unaware. There was also the psychopathic computer Hal in *2001: A Space Odyssey* whose homicidal behaviour seems to arise from a software conflict — not an improbable outcome in any machine we may devise. There are countless other examples; one way or another the trope of the bad machine has become a commonplace of the contemporary imagination.

There are some counter-examples of good or cute machines. Perhaps the most interesting is the android Data in *Star Trek: The Next Generation*. He is fully competent in his starship officer role, more so than most of his human colleagues, but he is stricken with the desire to understand humanity and become more like us. Data is in pursuit of a reverse singularity, a machine aspiring to salvation by becoming human.

Meanwhile, in advertising and especially in animated movies, the cute robot is now pervasive. Allowing machines to be funny and loveable because of their very machineness is a way of seducing ourselves into a wider acceptance of a future that seems inevitable.

There is a further, closely related, seduction. The machines that accompany us daily — notably phone and tablets — are insinuating themselves ever more effectively into our inner lives. They do this not simply by becoming more reliable, competent, or efficient but by becoming more personal. Machine interactions by voice are, at last, highly effective so that we have become used, for example, to addressing our iPhones as 'Siri' and our Microsoft phones as 'Cortana'. The pervasive obsession with and increasing intimacies of social media, meanwhile, seem to transform the machine itself into an aspect of our consciousness and our bodies. To the young, in particular, the removal of a smartphone seems like a form of solitary confinement or an amputation.[2]

The importance of this latter development is that it represents a reversal. Once our machines were adapted to us, now we adapt to our machines, becoming in some sense their creations. To put it another

---

[2] Sherry Turkle provides copious evidence in her books *Alone Together* (2011) and *Reclaiming Conversation* (2015).

way: a social network is as much a product of machine logic as of human aspiration. This has implications not just for the nature of the singularity but for the probability of our acceptance.

Or, indeed, for its arrival. We can, obviously, will a machine future into existence. To some, the present phase seems like a softening up process. The Turing test is a widely pursued test of machine intelligence. It is, as Floridi has pointed out, a low bar to set since it seems to test not intelligence — and certainly not consciousness — but rather a machine's ability to fake it. Furthermore, Silicon Valley dissident Jaron Lanier pointed out over twenty years ago in the *Journal of Consciousness Studies* that there are, in fact, two ways in which the test may be passed: either we make the machines smarter or we make ourselves dumber (Lanier, 1995; 2011). That is to say we can lower our own criteria for intelligence by reducing ourselves to machine readable entities. In this context, elements of the singularity are already in place. The assorted technologies — the internet, GPS, cloud computing, video on demand, programmatic advertising, robotic surgery, machine trading, and so on — that we now take for granted and which would have seemed miraculous thirty years ago are all ways in which we have handed over agency to machines. They don't seem that smart only because we have adjusted to their astounding competence. Certainly, for all their chummy fakery they display no evidence of consciousness, but, as I say, that is not a problem so long as they are effective.

Thus scepticism about our technical capacity to build the ultimate machine may be misplaced; we may simply decide, perhaps with a consumerist sigh of relief, that some extension of our current machines is close enough. We have already discarded our maps and handed over our senses of direction and place — pretty crucial to human identity I would have thought — to GPS. That said, the crucial part of the singularity hypothesis is that the ultimate machine is able to build machines more intelligent than itself. We shall have proved this is possible by our own construction of the singularity. As it will be able to do this with ever greater speed, it will, conceivably within hours, create machines far beyond our understanding and control. My consumer-driven, good-enough machine might not be able to do this, but, then again, it might. Either way, the point I am making is that the singularity is all too often defined in terms of technology or, indeed, philosophy. It should also be seen as a product of human desires, of political and social forces.

This means we can stop it and some rather surprising people have now said we should be prepared to do just that. This began with Bill Joy, a computer scientist and co-founder of Sun Microsystems, who, in 2000, wrote an essay entitled 'Why the Future Doesn't Need Us' in *Wired* magazine (Joy, 2000). This was a warning that three technologies in particular — robotics, genetic engineering, and nano-technology — had the capacity to render irrelevant or terminate our species. The stature of the author opened the way for a small but impressive wave of celebrity hi-tech neo-luddites: Stephen Hawking, Elon Musk, and Bill Gates among them.

The threat of a (non-fictional) technological termination of humanity is, as a result, out there in the popular imagination and it is now almost entirely centred on the threat of the singularity. In the context of global warming, Vladimir Putin, and Donald Trump, it is probably not an enormous anxiety but it is there ready to rise up and fight a war — which one would expect it to lose — with consumers' carefully nurtured love of their increasingly smart gadgets.

If the war is lost and the ultimate machine takes over, what would be the outcome? A good deal of thought has gone into the possibility of ensuring that the machine is programmed to be nice and choose not to enslave or terminate humanity. To be honest I find almost all of this absurd. First, if the machine is able to create ever more intelligent machines — or to boot itself into ever higher levels of intelligence — then, however nicely we design its initial program, it is certain to be overwritten and abandoned. Secondly, niceness is a slippery concept. Niceness to liberal democratic Westerners will not be the same as niceness to ISIS or even the Chinese communist party. Thirdly, even if the machine is exactly like its makers — i.e. human — we will not have any control over which human traits it may choose to amplify. This came out in a conversation I had with Eliezer Yudkowsky, the leading thinker in the area of friendly artificial intelligence. When I asked the question why the machine should choose this trait or that, he replied that, for example, I would not kill a baby would I? My response was that I wouldn't but there were millions of examples of human beings who have done just that.

Whatever the likelihood of it happening, therefore, the construction of the singularity would mean the creation of a supremely powerful entity over which we would have no control whatsoever. Since there is at least a possibility of this happening, then we should indeed be thinking about whether it should be done at all. Some will say, in spite of our inability to control the machine, the benefits that will flow justify its creation. They might add that even human extinction is a worthwhile price to pay simply because this ever better being represents a higher stage in the evolution of intelligence in the cosmos. But, at that point, one loses interest since there is nothing merely human that is worth saying.

In short, the singularity is a matter that may be urgent and is certainly important. Indeed, I think that it is a crucial concept that forms the contemporary imagination, even if most people aren't familiar with the words 'technological singularity'. The rise of the machines is apparent to all, whether they find themselves boxed in at work by the demands of corporate computers, or at play they find themselves addicted to gadgets that intrude on their privacy and, often, their sanity. We are all aware that we have taken the first steps towards ceding stewardship of the human world to our own creations. If so, then a discussion of a future is also a discussion of the present. The singularity has arrived if only, for the moment, in our imaginations. Now is the time, therefore, to imagine the perils and possibilities.

References

Dennett, D.C. (2004) *Freedom Evolves*, London: Penguin Books.

Floridia, L. (2016) Should we be afraid of AI?, [Online], https://aeon.co/essays/true-ai-is-both-logically-possible-and-utterly-implausible

Joy, B. (2000) Why the future doesn't need us, http://www.wired.com/2000/04/joy-2/

Lanier, J. (1995) Agents of alienation, *Journal of Consciousness Studies*, **2** (1), pp. 76–81.

Lanier, J. (2011) *You Are Not a Gadget: A Manifesto*, London: Penguin Books.

Turkle, S. (2011) *Alone Together: Why We Expect More from Technology and Less from Each Other*, New York: Basic Books.

Turkle, S. (2015) *Reclaiming Conversation: The Power of Talk in a Digital Age*, London: Penguin Books.

Uziel Awret

Introduction

In his 1965 article 'Speculations Concerning the First Ultraintelligent Machine' statistician I.J. Good predicted the coming of a technological singularity:

> Let an ultra-intelligent machine be defined as a machine that can far surpass all the intellectual activities of any man however clever. Since the design of machines is one of these intellectual activities, an ultra-intelligent machine could design even better machines; there would then unquestionably be an 'intelligence explosion', and the intelligence of man would be left far behind. Thus the first ultraintelligent machine is the last invention that man need ever make.

The term 'singularity' was introduced by the science fiction writer Vernor Vinge in a 1983 opinion article. The underlying idea has always captured the imagination of science fiction writers from John Campbell's 1932 short story 'The Last Evolution' to Robert A. Heinlein's 1952[1] essay 'Where To?', Asimov's 1956 'The last question' and many more recent works. There has also been discussion by mathematicians, AI researchers and futurists. Until recently, the subject has not been as popular with philosophers.

This book is based on two special highly interdisciplinary issues of the *Journal of Consciousness Studies* (JCS) on the singularity and the future relationship of humanity and AI that are centred on David Chalmers' 2010 JCS article 'The Singularity: A Philsophical Analysis'. To quote Chalmers:

> One might think that the singularity would be of great interest to academic philosophers, cognitive scientists, and artificial intelligence researchers. In practice, this has not been the case. Good was an eminent academic, but his article was largely unappreciated at the time. The subsequent discussion of the singularity has largely taken place in non-academic circles, including Internet forums, popular media and books, and workshops organized by the independent Singularity Institute. Perhaps

[1] See Damien Broderick's article on the singularity and science fiction in this volume.

the highly speculative flavour of the singularity idea has been responsible for academic resistance to it. I think this resistance is a shame, as the
singularity idea is clearly an important one. The argument for a singularity is one that we should take seriously. And the questions surrounding the singularity are of enormous practical and philosophical
concern.[2]

It is fair to say that Chalmers is the first to provide a detailed comprehensive philosophical analysis of the idea of the singularity that
brings into focus not only questions about the nature of intelligence
and the prospects for an intelligence explosion but also important
philosophical questions about consciousness, identity and the relationship between facts and values.

At the end of the 2010 Tucson consciousness conference during
one of the celebrated 'end of consciousness' parties, Chalmers and I
discussed his plenary talk on the singularity and agreed that it was
well-suited for a special edition of JCS. The idea was to solicit commentaries on his target article from philosophers, AI researchers, science fiction writers, futurists, cognitive scientists, biologists and
others to which Chalmers would respond. This project is also com-
mensurate with the JCS credo of exploring controversies in science
and the humanities.

We received many invited and submitted commentaries, including
from Igor Aleksander, Nick Bostrom and Carl Shulman, Sue
Blackmore, Selmer Bringsjord, Damien Broderick, Richard Brown,
Joseph Corabi and Susan Schneider, Barry Dainton, Dan Dennett,
Ben Goertzel, Susan Greenfield, Robin Hanson, Francis Heylighen,
Marcus Hutter, Ray Kurzweil, Pamela McCorduck, Drew
McDermott, Chris Nunn, Arkady Plotnitsky, Jesse Prinz, Juergen
Schmidhuber, Murray Shanahan, Eric Steinhart, Frank Tipler, Burt
Voorhees and Roman Yampolskiy.

Chalmers' paper is divided into three parts, the likelihood of the singularity, negotiating the singularity, and the place of humans in a post
singularity world with a special emphasis on uploading. I will use a
synopsis of his paper to present short descriptions of the different contributions to this volume.

Chalmers' basic argument for the singularity is:

1. There will be AI (before long, absent defeaters).

2. If there is AI, there will be AI+ (soon after, absent defeaters).

[2] Chalmers mentions some exceptions to this academic neglect including Bostrom (1998; 2003), Hanson (2008), Hofstadter (2005), and Moravec (1988; 1998).

3. If there is AI+, there will be AI++ (soon after, absent defeaters).

---

4. There will be AI++ (before too long, absent defeaters).

AI is human-level intelligence, AI+ is greater than human intelligence and AI++ is much greater than human intelligence (standing to humans as humans stand to ants). 'Before too long' means within centuries while 'soon after' means within decades or years. Defeaters are defined as anything that prevents intelligent systems from realizing their capacities to design intelligent systems.

Chalmers analyses the first three premises separately, describing them accordingly as the equivalence premise, the extension premise and the amplification premise.

The equivalence premise (we will construct AI as intelligent as ourselves) includes the brain emulation argument and the evolutionary argument. The emulation argument claims that:

(i) The human brain is a machine.

(ii) We will have the capacity to emulate this machine (before long).

(iii) If we emulate this machine, there will be AI.

(iv) Absent defeaters, there will be AI (before long).

Neuroscientist Susan Greenfield argues against both premise (i) and (ii) and attempts to provide what she calls a reality check arguing that the brain is non-computational and *that whilst the hypothetical scenario of neuron substitution is conceptually logical and plausible, in reality it's meaningless and unhelpful*. Greenfield also feels that consciousness is crucial for values, understanding and 'wisdom'.

AI researcher Francis Heylighen also seems to reject both premises (i) and (ii) embracing the embedded paradigm in which the brain does not simply crunch symbols and is inseparable from its immediate environment.

Cognitive and computer scientist Selmer Bringsjord holds that a more formal analysis of the singularity suggests that it is logically brittle. Philosopher and cultural theorist Arkady Plotnitsky holds that microphysical processes cannot be simulated arbitrarily closely and that the emulation argument and the evolution argument fail to convince us that we will have AI soon.

The evolutionary argument proceeds as follows:

(i)   Evolution produced human-level intelligence mechanically
      and non-miraculously.

(ii) If evolution produced human-level intelligence, then we can
      produce AI (before long).

───────────

(iii) Absent defeaters, there will be AI (before long).

How difficult is it for evolutionary mechanisms to produce intelligence similar to ours? Carl Shulman and Nick Bostrom address this question by salvaging the evolutionary argument from the 'observation selection effect' objection. They do so by combining arguments which are based on relevant examples of terrestrial convergent evolution with probabilistic arguments that are based on the 'sleeping beauty paradox' concluding that the evolution of human level intelligence on an earth type planet is not exceedingly improbable.

Economist Robin Hanson agrees with all three premises but claims that concluding that human level AI is near is based less on Good's recursive argument with its ensuing intelligence explosion and more on the extrapolation of general historic and economic trends that are clearly exponential.[3] Hanson also holds that the relevant parameters that should be traced in the context of an intelligence explosion are not those of individual systems, whether biological or artificial, but rather more collective 'cognitive' feats. This leads us to the extension premise leading from AI to AI+.

(i)   If there is AI, AI will be produced by an extendible method.

(ii) If AI is produced by an extendible method, we will have the
      capacity to extend the method (soon after).

(iii) Extending the method that produces an AI will yield an
      AI+.

───────────

(iv) Absent defeaters, if there is AI, there will (soon after) be AI+.

Three extendible methods are put forward: direct programming, machine learning, and artificial evolution. AI researcher Drew McDermott argues against all three forms of extendibility considering both the extendibility of hardware and software and, with Schmidhuber, he holds that direct programming may not be extend-

───────────

[3] See his article in this volume.

ible.[4] McDermott also holds that the extendibility of hardware is not guaranteed because of the lack of a smooth manifold as breakthroughs in hardware design are discontinuous and unpredictable.

Among the routes to extendibility, Chalmers also considers brains embedded in a rapidly improving environment that result in an extended mind (*à la* Clark and Chalmers) similar to the scenario considered by Heylighen. The section also considers extendibility and brain enhancement, elaborated on by Ray Kurzweil.

The third premise, the amplification premise, claims that:

> Premise 3: If there is AI+, there will be AI++ (soon after, absent defeaters)

The premise relies crucially on assuming that increases in intelligence always lead to proportionate increases in the capacity to design intelligent systems. AI researcher Igor Aleksander argues that designing AI that can design machines as well as itself is much harder than Chalmers imagines and that increases in intelligence may lead to diminishing returns in design capacity. He holds that we will not be able to design machines that design machines as well as us in the foreseeable future.

Frank Tipler, the mathematical physicist and cosmologist (*The Anthropic Principle*), gives an alternative argument for the singularity, based on considerations from physics. Tipler argues that the entropy in a contracting universe cannot grow indefinitely and that the needed entropic cooling can only be supplied by an intelligence explosion. On Tipler's view, biological life forms will not survive the heat and pressure generated by a contracting universe and the only way to prevent an entropy explosion is for biological life forms to either be uploaded or to design more robust AI that will be able to survive these extreme conditions. This means that the inevitability of the singularity is a direct outcome of our natural laws.

The second major part of Chalmers' article, 'Negotiating the Singularity', is concerned with maximizing the expected value of a post-singularity world.

> In the near term, the question that matters is: how (if at all) should we go about designing AI, in order to maximize the expected value of the resulting outcome? Are there some policies or strategies that we might adopt? In particular, are there certain constraints on design of AI and AI+ that we might impose, in order to increase the chances of a good outcome?

---

[4] McDermott also holds that artificial evolution is not extendible: it is interesting to compare some of his arguments with those of Shulman and Bostrom.

Here Chalmers divides these constraints into external and internal constraints.

Section 6, 'Internal Constraints: Constraining Values', analyses ways in which we can maximize a positive outcome, for us humans, by designing AI with the right kinds of values. Chalmers distinguishes Humean approaches to AI, on which values are largely independent of intelligence (being built into a fixed utility function, for example), from Kantian approaches on which values are themselves rationally revisable.

Schmidhuber's Gödel machines rewrite their value functions and are Kantian in the sense of connecting morality and rationality even if they decide at some stage to rid the planet of sentient biological systems. Tipler's view also has a Kantian element in that he holds that an intelligence explosion must be based on honest agents and that if AI+ is to produce good science it must be honest. While in Schmidhuber's case constraining the evolving value system of his self-referential machines will significantly diminish their capacity, Tipler's insistence on scientific honesty can only improve AI+ and AI++. However most AI researchers (and Chalmers) are more inclined to the Humean view that separates values and rationality.

Philosopher Pamela McCorduck holds that the human value system is too heterogeneous to lend itself to simplistic internal constraint scenarios. Like McDermott, and unlike AI researcher Murray Shanahan who entertains motivational defeater scenarios, McCorduck believes that structural defeaters are more likely.

AI researcher and mathematician Ben Goertzel who feels that the design of AI and AI+ must be constrained both internally and externally proposes an original solution:

> ... the deliberate human creation of an 'AI Nanny' with mildly superhuman intelligence and surveillance powers, designed either to forestall Singularity eternally, or to delay the Singularity until humanity more fully understands how to execute a Singularity in a positive way. It is suggested that as technology progresses, humanity may find the creation of an AI Nanny desirable as a means of protecting against the destructive potential of various advanced technologies such as AI, nanotechnology and synthetic biology.

Section 7 titled 'External Constraints: The Leakproof Singularity' explores ways of externally constraining the AI designs that might lead towards a singularity, especially constraining such AI to a virtual world from which it cannot leak into the real world.

AI researcher Roman Yampolskiy's article, 'Leakproofing the Singularity: Artificial Intelligence Confinement Problem', provides us with a detailed and well-reasoned analysis of this possibility.

Another external type of constraint mitigating unwanted outcomes is Robin Hanson's suggestion to create legally binding contracts that AI+, for example, will have to abide by, minimizing intergenerational conflicts and guaranteeing our continued existence. In this scenario AI ++ will be legally obligated to upload us.

Francis Heylighen, who advances an embedded approach, rejects 'brain in a vat' scenarios and holds that confining AI to a virtual environment will result in greatly diminished capacity. Heylighen also holds that our sensory capacities honed by hundreds of millions of years of evolution cannot be successfully simulated, unlike AI researcher Burt Voorhees who explores the consequences of exponential advances in artificial sensory capacity.

In his article, 'Can Intelligence Explode?' AI researcher Marcus Hutter, who believes that the singularity is near, explores what it means to be inside and outside a singularity whose default state consists of interacting super-intelligent systems in a virtual world. Hutter believes that some aspects of this singularitarian society might be theoretically studied with current scientific tools (for example, superintelligent machine sociology) and that entering a singularity might be similar to crossing the event horizon of a black hole where we don't know that we have entered a singularity. However unlike crossing a black hole event horizon it is the outside which slows down to a crawl. Another reason that an outsider may miss the singularity altogether is that maximally compressed information is indistinguishable from random noise[5]. Arguing for a speed explosion, Hutter holds that what is meant by an intelligence explosion needs to be clarified by a better definition of universal intelligence. However psychiatrist Chris Nunn holds that improving the definition of intelligence is complicated by the intrinsically contextual nature of information.

The last major part of Chalmers' article concerns uploading and the questions that it raises about consciousness and identity. We are introduced to destructive uploading as in 'serial sectioning', gradual uploading as in 'nano-transfer' and reconstructive uploading as a virtual resurrection. Will we survive uploading? Chalmers holds that the most agreeable form of uploading is probably gradual uploading in conjunction with a 'continuity of consciousness' approach to identity.

---

[5]   In line with John Smart's transcention scenario.

In his contribution 'On Singularities and Simulations' philosopher Barry Dainton, who like Chalmers believes that the singularity scenario is certainly not out of the question, explores the mechanics of uploading and its relationship to identity. Much of his chapter is devoted to an analysis of the possibility that we are already uploaded inhabitants in a virtual world, concluding that such a possibility may be higher than it seems. Dainton bases his argument on his interpretation of Bostrom's 'simulation argument'. His simulation based approach towards distinguishing 'cartesian scepticism' from 'simulation scepticism' is another example of the relevance of the singularity scenario to some of our deepest philosophical questions about the nature of identity, reality and intentionality.

Philosopher and cognitive scientist Jesse Prinz holds that either we already are uploads and the singularity is here or we are not uploads and the singularity will not materialize, arguing that in both cases we are doomed but adding that we have nothing to worry about(!)

Section nine, 'Uploading and Consciousness', asks whether an upload can be conscious. Chalmers holds a 'further fact' view of consciousness that leaves the question wide open. He suggests that an analysis of the gradual uploading scenario tends to support the functionalist approach.

Here philosopher Dan Dennett sets aside issues about the singularity and discusses Chalmers' 'further fact' view of consciousness. Dennett suggests that Chalmers' own 1996 work concerning gradual replacement shows that the 'further fact' view is unfounded, and offers some speculation about why Chalmers himself holds the view. The nature of this ongoing disagreement itself raises some interesting questions about the nature of philosophical truth and the philosophical endeavour. Ray Kurzweil also discusses issues about consciousness in his contribution.

Philosopher Richard Brown argues against the principle of organizational invariance and holds that uploading may force us to modify the conclusion of Chalmers' conceivability argument.

Section ten, 'Uploading and Personal Identity', asks whether uploading preserves our identity. In a comprehensive analysis of the questions that relate identity, survival and uploading, Chalmers gives an argument based on destructive uploading that supports a pessimistic view and an argument based on gradual uploading that supports an optimistic view. While these arguments lead to diametrically opposed conclusions and cannot both be right we are not sure which view is correct. Chalmers reaches the conclusion that while holding a further fact view on consciousness is justified, holding a 'further fact' view on identity is probably not.[6]

In her short contribution 'She Won't Be Me', psychologist and memeticist Susan Blackmore, who is also sympathetic to the 'singularity soon' scenario, explains why contrary to (her take on) a pessimistic approach to the deflationary position, *that we never survive from moment to moment, or from day to day*, she finds this position to be exhilarating and liberating. Dainton also discusses issues about personal identity, holding that a continuity of consciousness approach to identity can resolve some of the problems encountered by the more orthodox 'Parfitian view'.

Philosophers Susan Schneider and Joseph Corabi explore the way in which the very idea of the singularity forces us to reconsider identity, especially due to enhancement.

To borrow from Plotnitsky, *the debate concerning the possibility of artificial intelligence goes back at least to Descartes and is, thus, coextensive with the history of modern philosophy.* As this symposium on the singularity shows, and as this collection of responses shows, this debate is gaining a sense of urgency.

Acknowledgments

I enjoyed being the guest editor of this collection. I would like to thank David Chalmers for contributing the target paper, responding to the authors, and helping in many ways. I was privileged to collaborate on another JCS issue with Anthony Freeman. I would also like to thank Ben Goertzel for sound advice, Arkady Plotnitsky, Bernard Baars, Hava Siegelmann and Yotam Hoffman for useful discussions, and Minerva San Juan, Ron Chrisley and TrinityDC University for their support.

---

[6] Joe Levin separates the hard problem into the problem of phenomenal content and the 'puzzle' of subjectivity. Perhaps it's possible to engage the latter while suspending the former.

Journal of Consciousness Studies

David J. Chalmers

The Singularity

A Philosophical Analysis

1. Introduction

What happens when machines become more intelligent than humans? One view is that this event will be followed by an explosion to ever-greater levels of intelligence, as each generation of machines creates more intelligent machines in turn. This intelligence explosion is now often known as the 'singularity'.[1]

The basic argument here was set out by the statistician I.J. Good in his 1965 article 'Speculations Concerning the First Ultraintelligent Machine':

> Let an ultraintelligent machine be defined as a machine that can far sur-pass all the intellectual activities of any man however clever. Since the design of machines is one of these intellectual activities, an ultra-intelligent machine could design even better machines; there would then unquestionably be an 'intelligence explosion', and the intelligence of man would be left far behind. Thus the first ultraintelligent machine is the last invention that man need ever make.

The key idea is that a machine that is more intelligent than humans will be better than humans at designing machines. So it will be capable of designing a machine more intelligent than the most intelligent machine that humans can design. So if it is itself designed by humans,

[1] I first became interested in this cluster of ideas as a student, before first hearing explicitly of the 'singularity' in 1997. I was spurred to think further about these issues by an invita-tion to speak at the 2009 Singularity Summit in New York City. I thank many people at that event for discussion, as well as many at later talks and discussions at West Point, CUNY, NYU, Delhi, ANU, Tucson, Oxford, and UNSW. Thanks also to Doug Hofstadter, Marcus Hutter, Ole Koksvik, Drew McDermott, Carl Shulman, and Michael Vassar for comments on this paper.

it will be capable of designing a machine more intelligent than itself. By similar reasoning, this next machine will also be capable of designing a machine more intelligent than itself. If every machine in turn does what it is capable of, we should expect a sequence of ever more intelligent machines.[2]

This intelligence explosion is sometimes combined with another idea, which we might call the 'speed explosion'. The argument for a speed explosion starts from the familiar observation that computer processing speed doubles at regular intervals. Suppose that speed doubles every two years and will do so indefinitely. Now suppose that we have human-level artificial intelligence designing new processors. Then faster processing will lead to faster designers and an ever-faster design cycle, leading to a limit point soon afterwards.

The argument for a speed explosion was set out by the artificial intelligence researcher Ray Solomonoff in his 1985 article 'The Time Scale of Artificial Intelligence'.[3] Eliezer Yudkowsky gives a succinct version of the argument in his 1996 article 'Staring at the Singularity':

> Computing speed doubles every two subjective years of work. Two years after Artificial Intelligences reach human equivalence, their speed doubles. One year later, their speed doubles again. Six months — three months — 1.5 months ... Singularity.

The intelligence explosion and the speed explosion are logically independent of each other. In principle there could be an intelligence explosion without a speed explosion and a speed explosion without an intelligence explosion. But the two ideas work particularly well together. Suppose that within two subjective years, a greater-than-human machine can produce another machine that is not only twice as fast but 10% more intelligent, and suppose that this principle is indefinitely extensible. Then within four objective years there will have been an infinite number of generations, with both speed and intelligence increasing beyond any finite level within a finite time. This process would truly deserve the name 'singularity'.

Of course the laws of physics impose limitations here. If the currently accepted laws of relativity and quantum mechanics are correct — or even if energy is finite in a classical universe — then we cannot expect the principles above to be indefinitely extensible. But even with these physical limitations in place, the arguments give some

---

[2] Scenarios of this sort have antecedents to the argument in science fiction, perhaps most notably in John Campbell's 1932 short story 'The Last Evolution'.

[3] Solomonoff also discusses the effects of what we might call the 'population explosion': a rapidly increasing population of artificial AI researchers.

reason to think that both speed and intelligence might be pushed to the limits of what is physically possible. And on the face of it, it is unlikely that human processing is even close to the limits of what is physically possible. So the arguments suggest that both speed and intelligence might be pushed far beyond human capacity in a relatively short time. This process might not qualify as a 'singularity' in the strict sense from mathematics and physics, but it would be similar enough that the name is not altogether inappropriate.

The term 'singularity' was introduced by the science fiction writer Vernor Vinge in a 1983 opinion article.[4] It was brought into wider circulation by Vinge's influential 1993 article 'The Coming Technological Singularity', and by the inventor and futurist Ray Kurzweil's popular 2005 book *The Singularity is Near*. In practice, the term is used in a number of different ways. A loose sense refers to phenomena whereby ever-more-rapid technological change leads to unpredictable consequences.[5] A very strict sense refers to a point where speed and intelligence go to infinity, as in the hypothetical speed/intelligence explosion above. Perhaps the core sense of the term, though, is a moderate sense in which it refers to an intelligence explosion through the recursive mechanism set out by I.J. Good, whether or not this intelligence explosion goes along with a speed explosion or with divergence to infinity. I will always use the term 'singularity' in this core sense in what follows.

One might think that the singularity would be of great interest to academic philosophers, cognitive scientists, and artificial intelligence researchers. In practice, this has not been the case.[6] Good was an eminent academic, but his article was largely unappreciated at the time. The subsequent discussion of the singularity has largely taken place in nonacademic circles, including Internet forums, popular media and

---

[4] As Vinge (1993) notes, Stanislaw Ulam (1958) describes a conversation with John von Neumann in which the term is used in a related way: 'One conversation centered on the ever accelerating progress of technology and changes in the mode of human life, which gives the appearance of approaching some essential singularity in the history of the race beyond which human affairs, as we know them, could not continue.'

[5] A useful taxonomy of uses of 'singularity' is set out by Yudkowsky (2007). He distinguishes an 'accelerating change' school, associated with Kurzweil, an 'event horizon' school, associated with Vinge, and an 'intelligence explosion' school, associated with Good. Smart (1999–2008) gives a detailed history of associated ideas, focusing especially on accelerating change.

[6] With some exceptions: discussions by academics include Bostrom (1998; 2003), Hanson (2008), Hofstadter (2005), and Moravec (1988; 1998). Hofstadter organized symposia on the prospect of superintelligent machines at Indiana University in 1999 and at Stanford University in 2000, and more recently, Bostrom's Future of Humanity Institute at the University of Oxford has organized a number of relevant activities.

books, and workshops organized by the independent Singularity Institute. Perhaps the highly speculative flavour of the singularity idea has been responsible for academic resistance to it.

I think this resistance is a shame, as the singularity idea is clearly an important one. The argument for a singularity is one that we should take seriously. And the questions surrounding the singularity are of enormous practical and philosophical concern.

Practically: If there is a singularity, it will be one of the most important events in the history of the planet. An intelligence explosion has enormous potential benefits: a cure for all known diseases, an end to poverty, extraordinary scientific advances, and much more. It also has enormous potential dangers: an end to the human race, an arms race of warring machines, the power to destroy the planet. So if there is even a small chance that there will be a singularity, we would do well to think about what forms it might take and whether there is anything we can do to influence the outcomes in a positive direction.

Philosophically: The singularity raises many important philosophical questions. The basic argument for an intelligence explosion is philosophically interesting in itself, and forces us to think hard about the nature of intelligence and about the mental capacities of artificial machines. The potential consequences of an intelligence explosion force us to think hard about values and morality and about consciousness and personal identity. In effect, the singularity brings up some of the hardest traditional questions in philosophy and raises some new philosophical questions as well.

Furthermore, the philosophical and practical questions intersect. To determine whether there might be an intelligence explosion, we need to better understand what intelligence is and whether machines might have it. To determine whether an intelligence explosion will be a good or a bad thing, we need to think about the relationship between intelligence and value. To determine whether we can play a significant role in a post-singularity world, we need to know whether human identity can survive the enhancing of our cognitive systems, perhaps through uploading onto new technology. These are life-or-death questions that may confront us in coming decades or centuries. To have any hope of answering them, we need to think clearly about the philosophical issues.

In what follows, I address some of these philosophical and practical questions. I start with the argument for a singularity: is there good reason to believe that there will be an intelligence explosion? Next, I consider how to negotiate the singularity: if it is possible that there will be a singularity, how can we maximize the chances of a good outcome? Finally, I consider the place of humans in a post-singularity world, with special attention to questions about uploading: can an uploaded human be conscious, and will uploading preserve personal identity?

My discussion will necessarily be speculative, but I think it is possible to reason about speculative outcomes with at least a modicum of rigour. For example, by formalizing arguments for a speculative thesis with premises and conclusions, one can see just what opponents need to deny in order to deny the thesis, and one can then assess the costs of doing so. I will not try to give knockdown arguments in this paper, and I will not try to give final and definitive answers to the questions above, but I hope to encourage others to think about these issues further.[7]

2. The Argument for a Singularity

To analyse the argument for a singularity in a more rigorous form, it is helpful to introduce some terminology. Let us say that AI is artificial intelligence of human level or greater (that is, at least as intelligent as an average human). Let us say that AI+ is artificial intelligence of greater than human level (that is, more intelligent than the most intelligent human). Let us say that AI++ (or superintelligence) is AI of far greater than human level (say, at least as far beyond the most intelligent human as the most intelligent human is beyond a mouse).[8] Then we can put the argument for an intelligence explosion as follows:

1. There will be AI+.
2. If there is AI+, there will be AI++.

———————

3. There will be AI++.

Here, premise 1 needs independent support (on which more soon), but is often taken to be plausible. Premise 2 is the key claim of the intelligence explosion, and is supported by Good's reasoning set out above. The conclusion says that there will be superintelligence.

———————

[7] The main themes in this article have been discussed many times before by others, especially in the nonacademic circles mentioned earlier. My main aims in writing the article are to subject some of these themes (especially the claim that there will be an intelligence explosion and claims about uploading) to a philosophical analysis, with the aim of exploring and perhaps strengthening the foundations on which these ideas rest, and also to help bring these themes to the attention of philosophers and scientists.

[8] Following common practice, I use 'AI' and relatives as a general term ('An AI exists'), an adjective ('An AI system exists'), and as a mass term ('AI exists').

The argument depends on the assumption that there is such a thing as intelligence and that it can be compared between systems: otherwise the notion of an AI+ and an AI++ does not even make sense. Of course these assumptions might be questioned. Someone might hold that there is no single property that deserves to be called 'intelligence', or that the relevant properties cannot be measured and compared. For now, however, I will proceed with under the simplifying assumption that there is an intelligence measure that assigns an intelligence value to arbitrary systems. Later I will consider the question of how one might formulate the argument without this assumption. I will also assume that intelligence and speed are conceptually independent, so that increases in speed with no other relevant changes do not count as increases in intelligence.

We can refine the argument a little by breaking the support for premise 1 into two steps. We can also add qualifications about timeframe, and about potential defeaters for the singularity.

1. There will be AI (before long, absent defeaters).
2. If there is AI, there will be AI+ (soon after, absent defeaters).
3. If there is AI+, there will be AI++ (soon after, absent defeaters).

_____

4. There will be AI++ (before too long, absent defeaters).

Precise values for the timeframe variables are not too important. But we might stipulate that 'before long' means 'within centuries'. This estimate is conservative compared to those of many advocates of the singularity, who suggest decades rather than centuries. For example, Good (1965) predicts an ultraintelligent machine by 2000, Vinge (1993) predicts greater-than-human intelligence between 2005 and 2030, Yudkowsky (1996) predicts a singularity by 2021, and Kurzweil (2005) predicts human-level artificial intelligence by 2030.

Some of these estimates (e.g. Yudkowsky's) rely on extrapolating hardware trends.[9] My own view is that the history of artificial intelligence suggests that the biggest bottleneck on the path to AI is software, not hardware: we have to find the right algorithms, and no-one has come close to finding them yet. So I think that hardware extrapolation is not a good guide here. Other estimates (e.g. Kurzweil's) rely on

_____

[9] Yudkowsky's web-based article is now marked 'obsolete', and in later work he does not endorse the estimate or the argument from hardware trends. See Hofstadter (2005) for scepticism about the role of hardware extrapolation here and more generally for scepticism about timeframe estimates on the order of decades.

estimates for when we will be able to artificially emulate an entire human brain. My sense is that most neuroscientists think these estimates are overoptimistic. Speaking for myself, I would be surprised if there were human-level AI within the next three decades. Nevertheless, my credence that there will be human-level AI before 2100 is somewhere over one-half. In any case, I think the move from decades to centuries renders the prediction conservative rather than radical, while still keeping the timeframe close enough to the present for the conclusion to be interesting.

By contrast, we might stipulate that 'soon after' means 'within decades'. Given the way that computer technology always advances, it is natural enough to think that once there is AI, AI+ will be just around the corner. And the argument for the intelligence explosion suggests a rapid step from AI+ to AI++ soon after that. I think it would not be unreasonable to suggest 'within years' here (and some would suggest 'within days' or even sooner for the second step), but as before 'within decades' is conservative while still being interesting. As for 'before too long', we can stipulate that this is the sum of a 'before long' and two 'soon after's. For present purposes, that is close enough to 'within centuries', understood somewhat more loosely than the usage in the first premise to allow an extra century or so.

As for defeaters: I will stipulate that these are anything that prevents intelligent systems (human or artificial) from manifesting their capacities to create intelligent systems. Potential defeaters include disasters, disinclination, and active prevention.[10] For example, a nuclear war might set back our technological capacity enormously, or we (or our successors) might decide that a singularity would be a bad thing and prevent research that could bring it about. I do not think considerations internal to artificial intelligence can exclude these possibilities, although we might argue on other grounds about how likely they are. In any case, the notion of a defeater is still highly constrained (importantly, a defeater is *not* defined as anything that would prevent a singularity, which would make the conclusion near-trivial), and the conclusion that absent defeaters there will be superintelligence is strong enough to be interesting.

---

[10] I take it that when someone has the capacity to do something, then if they are sufficiently motivated to do it and are in reasonably favourable circumstances, they will do it. So defeaters can be divided into *motivational defeaters*, involving insufficient motivation, and *situational defeaters*, involving unfavourable circumstances (such as a disaster). There is a blurry line between unfavorable circumstances that prevent a capacity from being manifested and those that entail that the capacity was never present in the first place — for example, resource limitations might be classed on either side of this line — but this will not matter much for our purposes.

We can think of the three premises as an *equivalence* premise (there will be AI at least equivalent to our own intelligence), an *extension* premise (AI will soon be extended to AI+), and an *amplification* premise (AI+ will soon be greatly amplified to AI++). Why believe the premises? I will take them in order.

Premise 1: There will be AI (before long, absent defeaters)

One argument for the first premise is the *emulation argument*, based on the possibility of brain emulation. Here (following the usage of Sandberg and Bostrom, 2008), emulation can be understood as close simulation: in this case, simulation of internal processes in enough detail to replicate approximate patterns of the system's behaviour.

(i)   The human brain is a machine.
(ii)   We will have the capacity to emulate this machine (before long).
(iii) If we emulate this machine, there will be AI.

(iv)   Absent defeaters, there will be AI (before long).

The first premise is suggested by what we know of biology (and indeed by what we know of physics). Every organ of the body appears to be a machine: that is, a complex system comprised of law-governed parts interacting in a law-governed way. The brain is no exception. The second premise follows from the claims that microphysical processes can be simulated arbitrarily closely and that any machine can be emulated by simulating microphysical processes arbitrarily closely. It is also suggested by the progress of science and technology more generally: we are gradually increasing our understanding of biological machines and increasing our capacity to simulate them, and there do not seem to be limits to progress here. The third premise follows from the definitional claim that if we emulate the brain, this will replicate approximate patterns of human behaviour along with the claim that such replication will result in AI. The conclusion follows from the premises along with the definitional claim that absent defeaters, systems will manifest their relevant capacities.

One might resist the argument in various ways. One could argue that the brain is more than a machine; one could argue that we will never have the capacity to emulate it; and one could argue that emulating it need not produce AI. Various existing forms of resistance to AI take each of these forms. For example, J.R. Lucas (1961) has argued that for reasons tied to Gödel's theorem, humans are more sophisticated than any machine. Hubert Dreyfus (1972) and Roger

Penrose (1994) have argued that human cognitive activity can never be emulated by any computational machine. John Searle (1980) and Ned Block (1981) have argued that even if we can emulate the human brain, it does not follow that the emulation itself has a mind or is intelligent.

I have argued elsewhere that all of these objections fail.[11] But for present purposes, we can set many of them to one side. To reply to the Lucas, Penrose, and Dreyfus objections, we can note that nothing in the singularity idea requires that an AI be a *classical* computational system or even that it be a computational system at all. For example, Penrose (like Lucas) holds that the brain is not an algorithmic system in the ordinary sense, but he allows that it is a mechanical system that relies on certain nonalgorithmic quantum processes. Dreyfus holds that the brain is not a rule-following symbolic system, but he allows that it may nevertheless be a mechanical system that relies on subsymbolic processes (for example, connectionist processes). If so, then these arguments give us no reason to deny that we can build artificial systems that exploit the relevant nonalgorithmic quantum processes, or the relevant subsymbolic processes, and that thereby allow us to simulate the human brain.

As for the Searle and Block objections, these rely on the thesis that even if a system duplicates our behaviour, it might be missing important 'internal' aspects of mentality: consciousness, understanding, intentionality, and so on. Later in the paper, I will advocate the view that if a system in our world duplicates not only our outputs but our internal computational structure, then it will duplicate the important internal aspects of mentality too. For present purposes, though, we can set aside these objections by stipulating that for the purposes of the argument, intelligence is to be measured wholly in terms of behaviour and behavioural dispositions, where behaviour is construed operationally in terms of the physical outputs that a system produces. The conclusion that there will be AI++ in this sense is still strong enough to be interesting. If there are systems that produce apparently superintelligent outputs, then whether or not these systems are truly conscious or intelligent, they will have a transformative impact on the rest of the world.

Perhaps the most important remaining form of resistance is the claim that the brain is not a mechanical system at all, or at least that

---

[11] For a general argument for strong artificial intelligence and a response to many different objections, see Chalmers (1996, chapter 9). For a response to Penrose and Lucas, see Chalmers (1995). For a in-depth discussion of the current prospects for whole brain emulation, see Sandberg and Bostrom (2008).

nonmechanical processes play a role in its functioning that cannot be emulated. This view is most naturally combined with a sort of Cartesian dualism holding that some aspects of mentality (such as consciousness) are nonphysical and nevertheless play a substantial role in affecting brain processes and behaviour. If there are nonphysical processes like this, it might be that they could nevertheless be emulated or artificially created, but this is not obvious. If these processes cannot be emulated or artificially created, then it may be that human-level AI is impossible.

Although I am sympathetic with some forms of dualism about consciousness, I do not think that there is much evidence for the strong form of Cartesian dualism that this objection requires. The weight of evidence to date suggests that the brain is mechanical, and I think that even if consciousness plays a causal role in generating behaviour, there is not much reason to think that its role is not emulable. But while we know as little as we do about the brain and about consciousness, I do not think the matter can be regarded as entirely settled. So this form of resistance should at least be registered.

Another argument for premise 1 is the *evolutionary argument*, which runs as follows.

(i)   Evolution produced human-level intelligence.
(ii) If evolution produced human-level intelligence,
      then we can produce AI (before long).

_____

(iii) Absent defeaters, there will be AI (before long).

Here, the thought is that since evolution produced human-level intelligence, this sort of intelligence is not entirely unattainable. Furthermore, evolution operates without requiring any antecedent intelligence or forethought. If evolution can produce something in this unintelligent manner, then in principle humans should be able to produce it much faster, by using our intelligence.

Again, the argument can be resisted, perhaps by denying that evolution produced intelligence, or perhaps by arguing that evolution produced intelligence by means of processes that we cannot mechanically replicate. The latter line might be taken by holding that evolution needed the help of superintelligent intervention, or needed the aid of other nonmechanical processes along the way, or needed an enormously complex history that we could never artificially duplicate, or needed an enormous amount of luck. Still, I think the argument makes at least a prima facie case for its conclusion.

We can clarify the case against resistance of this sort by changing 'Evolution produced human-level intelligence' to 'Evolution produced human-level intelligence mechanically and nonmiraculously' in both premises of the argument. Then premise (ii) is all the more plausible. Premise (i) will now be denied by those who think evolution involved nonmechanical processes, supernatural intervention, or extraordinary amounts of luck. But the premise remains plausible, and the structure of the argument is clarified.

Of course these arguments do not tell us how AI will first be attained. They suggest at least two possibilities: brain emulation (simulating the brain neuron by neuron) and artificial evolution (evolving a population of AIs through variation and selection). There are other possibilities: direct programming (writing the program for an AI from scratch, perhaps complete with a database of world knowledge), for example, and machine learning (creating an initial system and a learning algorithm that on exposure to the right sort of environment leads to AI). Perhaps there are others still. I doubt that direct programming is likely to be the successful route, but I do not rule out any of the others.

It must be acknowledged that every path to AI has proved surprisingly difficult to date. The history of AI involves a long series of optimistic predictions by those who pioneer a method, followed by a periods of disappointment and reassessment. This is true for a variety of methods involving direct programming, machine learning, and artificial evolution, for example. Many of the optimistic predictions were not obviously unreasonable at the time, so their failure should lead us to reassess our prior beliefs in significant ways. It is not obvious just what moral should be drawn: Alan Perlis has suggested 'A year spent in artificial intelligence is enough to make one believe in God'. So optimism here should be leavened with caution. Still, my own view is that the balance of considerations still distinctly favours the view that AI will eventually be possible.

Premise 2: If there is AI, then there will be AI+ (soon after, absent defeaters)

One case for the extension premise comes from advances in information technology. Whenever we come up with a computational product, that product is soon afterwards obsolete due to technological advances. We should expect the same to apply to AI. Soon after we have produced a human-level AI, we will produce an even more intelligent AI: an AI+.

We might put the argument as follows.

(i)   If there is AI, AI will be produced by an extendible method.
(ii) If AI is produced by an extendible method, we will have the capacity to extend the method (soon after).
(iii) Extending the method that produces an AI will yield an AI+.
_____
(iv) Absent defeaters, if there is AI, there will (soon after) be AI+.

Here, an extendible method is a method that can easily be improved, yielding more intelligent systems. Given this definition, premises (ii) and (iii) follow immediately. The only question is premise (i).

Not every method of creating human-level intelligence is an extendible method. For example, the currently standard method of creating human-level intelligence is biological reproduction. But biological reproduction is not obviously extendible. If we have better sex, for example, it does not follow that our babies will be geniuses. Perhaps biological reproduction will be extendible using future technologies such as genetic engineering, but in any case the conceptual point is clear.

Another method that is not obviously extendible is brain emulation. Beyond a certain point, it is not the case that if we simply emulate brains better, then we will produce more intelligent systems. So brain emulation on its own is not clearly a path to AI+. It may nevertheless be that brain emulation speeds up the path to AI+. For example, emulated brains running on faster hardware or in large clusters might create AI+ much faster than we could without them. We might also be able to modify emulated brains in significant ways to increase their intelligence. We might use brain simulations to greatly increase our understanding of the human brain and of cognitive processing in general, thereby leading to AI+. But brain emulation will not on its own suffice for AI+: if it plays a role, some other path to AI+ will be required to supplement it.

Other methods for creating AI do seem likely to be extendible, however. For example, if we produce an AI by direct programming, then it is likely that like almost every program that has yet been written, the program will be improvable in multiple respects, leading soon after to AI+. If we produce an AI by machine learning, it is likely that soon after we will be able to improve the learning algorithm and extend the learning process, leading to AI+. If we produce an AI by artificial evolution, it is likely that soon after we will be able to improve the evolutionary algorithm and extend the evolutionary process, leading to AI+.

To make the case for premise (i), it suffices to make the case that either AI will be produced directly by an extendible method, or that if it is produced by a nonextendible method, this method will itself lead soon after to an extendible method. My own view is that both claims are plausible. I think that if AI is possible at all (as the antecedent of this premise assumes), then it should be possible to produce AI through a learning or evolutionary process, for example. I also think that if AI is produced through a nonextendible method such as brain emulation, this method is likely to greatly assist us in the search for an extendible method, along the lines suggested above. So I think there is good reason to believe premise (i).

To resist the premise, an opponent might suggest that we lie at a limit point in intelligence space: perhaps we are as intelligent as a system could be, or perhaps we are at least at a local maximum in that there is no easy path from systems like us to more intelligent systems. An opponent might also suggest that although intelligence space is not limited in this way, there are limits on our capacity to create intelligence, and that as it happens those limits lie at just the point of creating human-level intelligence. I think that there is not a great deal of antecedent plausibility to these claims, but again, the possibility of this form of resistance should at least be registered.

There are also potential paths to greater-than-human intelligence that do not rely on first producing AI and then extending the method. One such path is brain enhancement. We might discover ways to enhance our brains so that the resulting systems are more intelligent than any systems to date. This might be done genetically, pharmacologically, surgically, or even educationally. It might be done through implantation of new computational mechanisms in the brain, either replacing or extending existing brain mechanisms. Or it might be done simply by embedding the brain in an ever more sophisticated environment, producing an 'extended mind' (Clark & Chalmers, 1998) whose capacities far exceed that of an unextended brain.

It is not obvious that enhanced brains should count as AI or AI+. Some potential enhancements will result in a wholly biological system, perhaps with artificially enhanced biological parts (where to be biological is to be based on DNA, let us say). Others will result in a system with both biological and nonbiological parts (where we might use organic DNA-based composition as a rough and ready criterion for being biological). At least in the near-term, all such systems will count as human, so there is a sense in which they do not have greater-than-human intelligence. For present purposes, I will stipulate that the baseline for human intelligence is set at current human standards, and I will stipulate that at least the systems with nonbiological components to their cognitive systems (brain implants and technologically extended minds, for example) count as artificial. So intelligent enough systems of this sort will count as AI+.

Like other AI+ systems, enhanced brains suggest a potential intelligence explosion. An enhanced system may find further methods of enhancement that go beyond what we can find, leading to a series of ever-more-intelligent systems. Insofar as enhanced brains always rely on a biological core, however, there may be limitations. There are likely to be speed limitations on biological processing, and there may well be cognitive limitations imposed by brain architecture in addition. So beyond a certain point, we might expect non-brain-based systems to be faster and more intelligent than brain-based systems. Because of this, I suspect that brain enhancement that preserves a biological core is likely to be at best a first stage in an intelligence explosion. At some point, either the brain will be 'enhanced' in a way that dispenses with the biological core altogether, or wholly new systems will be designed. For this reason I will usually concentrate on non-biological systems in what follows. Still, brain enhancements raise many of the same issues and may well play an important role.

Premise 3: If there is AI+, there will be AI++ (soon after, absent defeaters)

The case for the amplification premise is essentially the argument from I.J. Good given above. We might lay it out as follows. Suppose there exists an AI+. Let us stipulate that AI_1 is the first AI+, and that AI_0 is its (human or artificial) creator. (If there is no sharp borderline between non-AI+ and AI+ systems, we can let AI_1 be any AI+ that is more intelligent than its creator.) Let us stipulate that δ is the difference in intelligence between AI_1 and AI_0, and that one system is significantly more intelligent than another if there is a difference of at least δ between them. Let us stipulate that for $n > 1$, an AI_{n+1} is an AI that is created by an AI_n and is significantly more intelligent than its creator.

(i)   If there exists AI+, then there exists an AI_1.
(ii) For all $n>0$, if an AI_n exists, then absent defeaters, there will be an AI_{n+1}.
(iii) If for all n there exists an AI, there will be AI++.

_____

(iv) If there is AI+, then absent defeaters, there will be AI++.

Here premise (i) is true by definition. Premise (ii) follows from three claims: (a) the definitional claim that if AI_n exists, it is created by AI_{n-1}

and is more intelligent than AI_{n-1}, (b) the definitional claim that if AI_n exists, then absent defeaters it will manifest its capacities to create intelligent systems, and (c) the substantive claim that if AI_n is significantly more intelligent than AI_{n-1}, it has the capacity to create a system significantly more intelligent than any that AI_{n-1} can create. Premise (iii) follows from the claim that if there is a sequence of AI systems each of which is significantly more intelligent than the last, there will eventually be superintelligence. The conclusion follows by logic and mathematical induction from the premises.

The conclusion as stated here omits the temporal claim 'soon after'. One can make the case for the temporal claim by invoking the ancillary premise that AI+ systems will be running on hardware much faster than our own, so that steps from AI+ onward are likely to be much faster than the step from humans to AI+.

There is room in logical space to resist the argument. For a start, one can note that the soundness of the argument depends on the intelligence measure used: if there is an intelligence measure for which the argument succeeds, there will almost certainly be a rescaled intelligence measure (perhaps a logarithmic measure) for which it fails. So for the argument to be interesting, we need to restrict it to intelligence measures that accord sufficiently well with intuitive intelligence measures that the conclusion captures the intuitive claim that there will be AI of far greater than human intelligence.

Relatedly, one could resist premise (iii) by holding that an arbitrary number of increases in intelligence by δ need not add up to the difference between AI+ and AI++. If we stipulate that δ is a ratio of intelligences, and that AI++ requires a certain fixed multiple of human intelligence (100 times, say), then resistance of this sort will be excluded. Of course for the conclusion to be interesting, then as in the previous paragraph, the intelligence measure must be such that this fixed multiple suffices for something reasonably counted as superintelligence.

The most crucial assumption in the argument lies in premise (ii) and the supporting claim (c). We might call this assumption a *proportionality thesis*: it holds that increases in intelligence (or increases of a certain sort) always lead to proportionate increases in the capacity to design intelligent systems. Perhaps the most promising way for an opponent to resist is to suggest that this thesis may fail. It might fail because here are upper limits in intelligence space, as with resistance to the last premise. It might fail because there are points of diminishing returns: perhaps beyond a certain point, a 10% increase in intelligence yields only a 5% increase at the next generation, which yields only a 2.5% increase at the next generation, and so on. It might fail because intelligence does not correlate well with design capacity: systems that are more intelligent need not be better designers. I will return to resistance of these sorts in section 4, under 'structural obstacles'.

One might reasonably doubt that the proportionality thesis will hold across all possible systems and all the way to infinity. To handle such an objection, one can restrict premise (ii) to AI systems in a certain class. We just need some property ϕ such that an AI_n with ϕ can always produce an AI_{n+1} with ϕ, and such that we can produce an AI+ with ϕ. One can also restrict the proportionality thesis to a specific value of δ (rather than all possible values), and one can restrict n to a relatively small range $n < k$ (where $k = 100$, say) as long as k increases of δ suffices for superintelligence.

It is worth noting that in principle the recursive path to AI++ need not start at the human level. If we had a system whose overall intelligence were far lower than human level but which nevertheless had the capacity to improve itself or to design further systems, resulting in a system of significantly higher intelligence (and so on recursively), then the same mechanism as above would lead eventually to AI, AI+, and AI++. So in principle the path to AI++ requires only that we create a certain sort of self-improving system, and does not require that we directly create AI or AI+. In practice, the clearest case of a system with the capacity to amplify intelligence in this way is the human case (via the creation of AI+), and it is not obvious that there will be less intelligent systems with this capacity.[12] But the alternative hypothesis here should at least be noted.

3. The Intelligence Explosion Without Intelligence

The arguments so far have depended on an uncritical acceptance of the assumption that there is such a thing as intelligence and that it can be measured. Many researchers on intelligence accept these assumptions. In particular, it is widely held that there is such a thing as 'gen-

---

[12] The 'Gödel machines' of Schmidhuber (2003) provide a theoretical example of self-improving systems at a level below AI, though they have not yet been implemented and there are large practical obstacles to using them as a path to AI. The process of evolution might count as an indirect example: less intelligent systems have the capacity to create more intelligent systems by reproduction, variation and natural selection. This version would then come to the same thing as an evolutionary path to AI and AI++. For present purposes I am construing 'creation' to involve a more direct mechanism than this.

eral intelligence', often labeled g, that lies at the core of cognitive ability and that correlates with many different cognitive capacities.[13]

Still, many others question these assumptions. Opponents hold that there is no such thing as intelligence, or at least that there is no single thing. On this view, there are many different ways of evaluating cognitive agents, no one of which deserves the canonical status of 'intelligence'. One might also hold that even if there is a canonical notion of intelligence that applies within the human sphere, it is far from clear that this notion can be extended to arbitrary non-human systems, including artificial systems. Or one might hold that the correlations between general intelligence and other cognitive capacities that hold within humans need not hold across arbitrary non-human systems. So it would be good to be able to formulate the key theses and arguments without assuming the notion of intelligence.

I think that this can be done. We can rely instead on the general notion of a cognitive capacity: some specific capacity that can be compared between systems. All we need for the purpose of the argument is (i) a self-amplifying cognitive capacity G: a capacity such that increases in that capacity go along with proportionate (or greater) increases in the ability to create systems with that capacity, (ii) the thesis that we can create systems whose capacity G is greater than our own, and (iii) a correlated cognitive capacity H that we care about, such that certain small increases in H can always be produced by large enough increases in G. Given these assumptions, it follows that absent defeaters, G will explode, and H will explode with it. (A formal analysis that makes the assumptions and the argument more precise follows at the end of the section.)

In the original argument, intelligence played the role of both G and H. But there are various plausible candidates for G and H that do not appeal to intelligence. For example, G might be a measure of programming ability, and H a measure of some specific reasoning ability. Here it is not unreasonable to hold that we can create systems with greater programming ability than our own, and that systems with greater programming ability will be able to create systems with greater programming ability in turn. It is also not unreasonable to hold that programming ability will correlate with increases in various specific reasoning abilities. If so, we should expect that absent defeaters, the reasoning abilities in question will explode.

---

[13] Flynn (2007) gives an overview of the debate over general intelligence and the reasons for believing in such a measure. Shalizi (2007) argues that g is a statistical artifact. Legg (2008) has a nice discussion of these issues in the context of machine superintelligence.

This analysis brings out the importance of correlations between capacities in thinking about the singularity. In practice, we care about the singularity because we care about potential explosions in various specific capacities: the capacity to do science, to do philosophy, to create weapons, to take over the world, to bring about world peace, to be happy. Many or most of these capacities are not themselves self-amplifying, so we can expect an explosion in these capacities only to the extent that they correlate with other self-amplifying capacities. And for any given capacity, it is a substantive question whether they are correlated with self-amplifying capacity in this way. Perhaps the thesis is prima facie more plausible for the capacity to do science than for the capacity to be happy, but the questions are nontrivial.

The point applies equally to the intelligence analysis, which relies for its interest on the idea that intelligence correlates with various specific capacities. Even granted the notion of intelligence, the question of just what it correlates with is nontrivial. Depending on how intelligence is measured, we might expect it to correlate well with some capacities (perhaps a capacity to calculate) and to correlate less well with other capacities (perhaps a capacity for wisdom). It is also far from trivial that intelligence measures that correlate well with certain cognitive capacities within humans will also correlate with those capacities in artificial systems.

Still, two observations help with these worries. The first is that the correlations need not hold across all systems or even across all systems that we might create. There need only be some *type* of system such that the correlations hold across all systems of that type. If such a type exists (a subset of architectures, say), then recursive creation of systems of this type would lead to explosion. The second is that the self-amplifying capacity G need not correlate directly with the cognitive capacity H, but need only correlate with H', the capacity to create systems with H. While it is not especially plausible that design capacity will correlate with happiness, for example, it is somewhat more plausible that design capacity will correlate with the capacity to create happy systems. If so, then the possibility is left open that as design capacity explodes, happiness will explode along with it, either in the main line of descent or in a line of offshoots, at least if the designers choose to manifest their capacity to create happy systems.

A simple formal analysis follows (the remainder of this section can be skipped by those uninterested in formal details). Let us say that a parameter is a function from cognitive systems to positive real numbers. A parameter G *measures* a capacity C iff for all cognitive systems a and b, $G(a) > G(b)$ iff a has a greater capacity C than b (one might also require that degrees of G correspond to degrees of C in some formal or intuitive sense). A parameter G *strictly tracks* a parameter H in ϕ-systems (where ϕ is some property or class of systems) iff whenever a and b are ϕ-systems and $G(a) > G(b)$, then $H(a)/H(b) \geq G(a)/G(b)$. A parameter G *loosely tracks* a parameter H in ϕ-systems iff for all y there exists x such that (nonvacuously) if a is a ϕ-system and $G(a) > x$, then $H(a) > y$. A parameter G strictly/loosely tracks a capacity C in ϕ-systems if it strictly/loosely tracks a parameter that measures C in ϕ-systems. Here, strict tracking requires that increases in G always produce proportionate increases in H, while loose tracking requires only that some small increase in H can always be produced by a large enough increase in G.

For any parameter G, we can define a parameter G': this is a parameter that measures a system's capacity to create systems with G. More specifically, $G'(x)$ is the highest value of h such that x has the capacity to create a system y such that $G(y) = h$. We can then say that G is a self-amplifying parameter (relative to x) if $G'(x) > G(x)$ and if G strictly tracks G' in systems downstream from x. Here a system is downstream from x if it is created through a sequence of systems starting from x and with ever-increasing values of G. Finally, let us say that for a parameter G or a capacity H, $G{+}{+}$ and $H{+}{+}$ systems are systems with values of G and capacities H that far exceed human levels.

Now we simply need the following premises:

(i) G is a self-amplifying parameter (relative to us).
(ii) G loosely tracks cognitive capacity H (downstream from us).

(iii) Absent defeaters, there will be $G{+}{+}$ and $H{+}{+}$.

The first half of the conclusion follows from premise (i) alone. Let AI_0 be us. If G is a self-amplifying parameter relative to us, then we are capable of creating a system AI_1 such that $G(AI_1) > G(AI_0)$. Let $\delta = G(AI_1)/G(AI_0)$. Because G strictly tracks G', $G'(AI_1) \geq \delta G'(AI_0)$. So AI_1 is capable of creating a system AI_2 such that $G(AI_2) \geq \delta G(AI_1)$. Likewise, for all n, AI_n is capable of creating AI_{n+1} such that $G(AI_{n+1}) \geq \delta G(AI_n)$. It follows that absent defeaters, arbitrarily high values of G will be produced. The second half of the conclusion immediately follows from (ii) and the first half of the conclusion. Any value of H can be produced by a high enough value of G, so it follows that arbitrarily high values for H will be produced.

The assumptions can be weakened in various ways. As noted earlier, it suffices for G to loosely track not H but H', where H' measures the capacity to create systems with H. Furthermore, the tracking relations between G and G', and between G and H or H', need not hold in all systems downstream from us: it suffices that there is a type ϕ such that in ϕ-systems downstream from us, G strictly tracks $G'(\phi)$ (the ability to create a ϕ-system with G) and loosely tracks H or H'. We need not require that G is strictly self-amplifying: it suffices for G and H (or G and H') to be jointly self-amplifying in that high values of both G and H lead to significantly higher values of each. We also need not require that the parameters are self-amplifying forever. It suffices that G is self-amplifying over however many generations are required for $G++$ (if $G++$ requires a 100-fold increase in G, then $\log_\delta 100$ generations will suffice) and for $H++$ (if $H++$ requires a 100-fold increase in H and the loose tracking relation entails that this will be produced by an increase in G of 1000, then $\log_\delta 1000$ generations will suffice). Other weakenings are also possible.

4. Obstacles to the Singularity

On the current analysis, an intelligence explosion results from a self-amplifying cognitive capacity (premise (i) above), correlations between that capacity and other important cognitive capacities (premise (ii) above), and manifestation of those capacities (conclusion). More pithily: self-amplification plus correlation plus manifestation = singularity.

This analysis brings out a number of potential obstacles to the singularity: that is, ways that there might fail to be a singularity. There might fail to be interesting self-amplifying capacities. There might fail to be interesting correlated capacities. Or there might be defeaters, so that these capacities are not manifested. We might call these *structural obstacles*, *correlation obstacles*, and *manifestation obstacles* respectively.

I do not think that there are knockdown arguments against any of these three sorts of obstacles. I am inclined to think that manifestation obstacles are the most serious obstacle, however. I will briefly discuss obstacles of all three sorts in what follows.

Structural obstacles

There are three overlapping ways in which there might fail to be relevant self-amplifying capacities, which we can illustrate by focusing on the case of intelligence. *Limits in intelligence space*: we are at or near an upper limit in intelligence space. *Failure of takeoff*: although there are higher points in intelligence space, human intelligence is not at a takeoff point where we can create systems more intelligent than ourselves. *Diminishing returns*: although we can create systems more intelligent than ourselves, increases in intelligence diminish from there. So a 10% increase might lead to a 5% increase, a 2.5% increase, and so on, or even to no increase at all after a certain point.

Regarding limits in intelligence space: While the laws of physics and the principles of computation may impose limits on the sort of intelligence that is possible in our world, there is little reason to think that human cognition is close to approaching those limits. More generally, it would be surprising if evolution happened to have recently hit or come close to an upper bound in intelligence space.

Regarding failure of takeoff: I think that the prima facie arguments earlier for AI and AI+ suggest that we are at a takeoff point for various capacities such as the ability to program. There is prima facie reason to think that we have the capacity to emulate physical systems such as brains. And there is prima facie reason to think that we have the capacity to improve on those systems.

Regarding diminishing returns: These pose perhaps the most serious structural obstacle. Still, I think there is some plausibility in proportionality theses, at least given an intuitive intelligence measure. If anything, 10% increases in intelligence-related capacities are likely to lead to all sorts of intellectual breakthroughs, leading to next-generation increases in intelligence that are significantly greater than 10%. Even among humans, relatively small differences in design capacities (say, the difference between Turing and an average human) seem to lead to large differences in the systems that are designed (say, the difference between a computer and nothing of importance). And even if there are diminishing returns, a limited increase in intelligence combined with a large increase in speed will produce at least some of the effects of an intelligence explosion.

One might worry that a 'hill-climbing' process that starts from the human cognitive system may run into a local maximum from which one cannot progress further by gradual steps. I think that this possibility is made less likely by the enormous dimensionality of intelligence space and by the enormous number of paths that are possible. In addition, the design of AI is not limited to hill-climbing: there is also 'hill-leaping', where one sees a favourable area of intelligence space some distance away and leaps to it. Perhaps there are some areas of intelligence space (akin to inaccessible cardinals in set theory?) that one simply cannot get to by hill-climbing and hill-leaping, but I think that there is good reason to think that these processes at least can get us far beyond ordinary human capacities.

Correlation obstacles

It may be that while there is one or more self-amplifying cognitive capacity G, this does not correlate with any or many capacities that are of interest to us. For example, perhaps a self-amplifying increase in programming ability will not go along with increases in other interesting abilities, such as an ability to solve scientific problems or social problems, an ability to wage warfare or make peace, and so on.

I have discussed issues regarding correlation in the previous section. I think that the extent to which we can expect various cognitive capacities to correlate with each other is a substantive open question. Still, even if self-amplifying capacities such as design capacities correlate only weakly with many cognitive capacities, they will plausibly correlate more strongly with the capacity to create systems with these capacities. It remains a substantive question just how much correlation one can expect, but I suspect that there will be enough correlating capacities to ensure that if there is an explosion, it will be an interesting one.

Manifestation obstacles

Although there is a self-amplifying cognitive capacity G, either we or our successors might not manifest our capacity to create systems with higher values of G (or with higher values of a cognitive correlated capacity H). Here we can divide the defeaters into *motivational defeaters* in which an absence of motivation or a contrary motivation prevents capacities from being manifested, and *situational defeaters*, in which other unfavourable circumstances prevent capacities from being manifested. Defeaters of each sort could arise on the path to AI, on the path from AI to AI+, or on the path from AI+ to AI++.

Situational defeaters include disasters and resource limitations. Regarding disasters, I certainly cannot exclude the possibility that global warfare or a nanotechnological accident ('gray goo') will stop technological progress entirely before AI or AI+ is reached. I also cannot exclude the possibility that artificial systems will themselves bring about disasters of this sort. Regarding resource limitations, it is worth noting that most feedback loops in nature run out of steam because of limitations in resources such as energy, and the same is possible here. Still, it is likely that foreseeable energy resources will suffice for many generations of AI+, and AI+ systems are likely to develop further ways of exploiting energy resources. Something similar applies to financial resources and other social resources.

Motivational defeaters include disinclination and active prevention. It is possible that as the event draws closer, most humans will be disinclined to create AI or AI+. It is entirely possible that there will be active prevention of the development of AI or AI+ (perhaps by legal, financial, and military means), although it is not obvious that such prevention could be successful indefinitely.[14] And it is certainly possible that AI+ systems will be disinclined to create their successors, perhaps because we design them to be so disinclined, or perhaps because they will be intelligent enough to realize that creating successors is not in their interests. Furthermore, it may be that AI+ systems will have the capacity to prevent such progress from happening.

A singularity proponent might respond that all that is needed to overcome motivational defeaters is the creation of a single AI+ that greatly values the creation of greater AI+ in turn, and a singularity will then be inevitable. If such a system is the first AI+ to be created, this conclusion may well be correct. But as long as this AI+ is not created first, then it may be subject to controls from other AI+, and the path to AI++ may be blocked. The issues here turn on difficult questions about the motivations and capacities of future systems, and answers to these questions are difficult to predict.

In any case, the current analysis makes clearer the burdens on both proponents and opponents of the thesis that there will be an intelligence explosion. Opponents need to make clear where they think the case for the thesis fails: structural obstacles (and if so which), correlation obstacles, situational defeaters, motivational defeaters. Likewise, proponents need to make the case that there will be no such obstacles or defeaters.

Speaking for myself, I think that while structural and correlational obstacles (especially the proportionality thesis) raise nontrivial issues, there is at least a prima facie case that *absent defeaters*, a number of interesting cognitive capacities will explode. I think the most likely defeaters are motivational. But I think that it is far from obvious that there will be defeaters. So I think that the singularity hypothesis is one that we should take very seriously.

---

[14] When I discussed these issues with cadets and staff at the West Point Military Academy, the question arose as to whether the US military or other branches of the government might attempt to prevent the creation of AI or AI+, due to the risks of an intelligence explosion. The consensus was that they would not, as such prevention would only increase the chances that AI or AI+ would first be created by a foreign power. One might even expect an AI arms race at some point, once the potential consequences of an intelligence explosion are registered. According to this reasoning, although AI+ would have risks from the standpoint of the US government, the risks of Chinese AI+ (say) would be far greater.

5. Negotiating the Singularity

If there is AI++, it will have an enormous impact on the world. So if there is even a small chance that there will be a singularity, we need to think hard about the form it will take. There are many different forms that a post-singularity world might take. Some of them may be desirable from our perspective, and some of them may be undesirable.

We might put the key questions as follows: faced with the possibility of an intelligence explosion, how can we maximize the chances of a desirable outcome? And if a singularity is inevitable, how can we maximize the expected value of a post-singularity world?

Here, value and desirability can be divided into at least two varieties. First, there is broadly agent-relative value ('subjective value', especially self-interested or prudential value): we can ask from a subjective standpoint, how good will such a world be for me and for those that I care about? Second, there is broadly agent-neutral value ('objective value', especially moral value): we can ask from a relatively neutral standpoint, how good is it that such a world comes to exist?

I will not try to settle the question of whether an intelligence explosion will be (subjectively or objectively) good or bad. I take it for granted that there are potential good and bad aspects to an intelligence explosion. For example, ending disease and poverty would be good. Destroying all sentient life would be bad. The subjugation of humans by machines would be at least subjectively bad.

Other potential consequences are more difficult to assess. Many would hold that human immortality would be subjectively and perhaps objectively good, although not everyone would agree. The wholesale replacement of humans by nonhuman systems would plausibly be subjectively bad, but there is a case that it would be objectively good, at least if one holds that the objective value of lives is tied to intelligence and complexity. If humans survive, the rapid replacement of existing human traditions and practices would be regarded as subjectively bad by some but not by others. Enormous progress in science might be taken to be objectively good, but there are also potential bad consequences. It is arguable that the very fact of an ongoing intelligence explosion all around one could be subjectively bad, perhaps due to constant competition and instability, or because certain intellectual endeavours would come to seem pointless.[15] On the other hand, if superintelligent systems share our values, they will

---

[15] See Kurzweil (2005), Hofstadter (2005) and Joy (2000) for discussions of numerous other ways in which a singularity might be a good thing (Kurzweil) and a bad thing (Hofstadter, Joy).

presumably have the capacity to ensure that the resulting situation accords with those values.

I will not try to resolve these enormously difficult questions here. As things stand, we are uncertain about both facts and values. That is, we do not know what a post-singularity world will be like, and even if we did, it is nontrivial to assess its value. Still, even without resolving these questions, we are in a position to make at least some tentative generalizations about what sort of outcomes will be better than others. And we are in a position to make some tentative generalizations about what sort of actions on our part are likely to result in better outcomes. I will not attempt anything more than the crudest of generalizations here, but these are matters that deserve much attention.

In the near term, the question that matters is: how (if at all) should we go about designing AI, in order to maximize the expected value of the resulting outcome? Are there some policies or strategies that we might adopt? In particular, are there certain constraints on design of AI and AI+ that we might impose, in order to increase the chances of a good outcome?

It is far from clear that we will be in a position to impose these constraints. Some of the constraints have the potential to slow the path to AI or AI+ or to reduce the impact of AI and AI+ in certain respects. Insofar as the path to AI or AI+ is driven by competitive forces (whether financial, intellectual, or military), then these forces may tend in the direction of ignoring these constraints.[16] Still, it makes sense to assess what constraints might or might not be beneficial in principle. Practical issues concerning the imposition of these constraints also deserve attention, but I will largely set aside those issues here.

We might divide the relevant constraints into two classes. Internal constraints concern the internal structure of an AI, while external constraints concern the relations between an AI and ourselves.

6. Internal Constraints: Constraining Values

What sort of internal constraints might we impose on the design of an AI or AI+? First, we might try to constrain their cognitive capacities in certain respects, so that they are good at certain tasks with which we need help, but so that they lack certain key features such as autonomy. For example, we might build an AI that will answer our questions or

---

[16] An especially bad case is a 'singularity bomb': an AI+ designed to value primarily the destruction of the planet (or of a certain population), and secondarily the creation of ever-more intelligent systems with the same values until the first goal is achieved.

that will carry specified tasks out for us, but that lacks goals of its own. On the face of it, such an AI might pose fewer risks than an autonomous AI, at least if it is in the hands of a responsible controller.

Now, it is far from clear that AI or AI+ systems of this sort will be feasible: it may be that the best path to intelligence is through general intelligence. Even if such systems are feasible, they will be limited, and any intelligence explosion involving them will be correspondingly limited. More importantly, such an approach is likely to be unstable in the long run. Eventually, it is likely that there will be AIs with cognitive capacities akin to ours, if only through brain emulation. Once the capacities of these AIs are enhanced, then we will have to deal with issues posed by autonomous AIs.

Because of this, I will say no more about the issue of capacity-limited AI. Still, it is worth noting that this sort of limited AI and AI+ might be a useful first step on the road to less limited AI and AI+. There is perhaps a case for first developing systems of this sort if it is possible, before developing systems with autonomy.

In what follows, I will assume that AI systems have goals, desires, and preferences: I will subsume all of these under the label of *values* (very broadly construed). This may be a sort of anthropomorphism: I cannot exclude the possibility that AI+ or AI++ will be so foreign that this sort of description is not useful. But this is at least a reasonable working assumption. Likewise, I will make the working assumptions that AI+ and AI++ systems are personlike at least to the extent that they can be described as thinking, reasoning, and making decisions.

A natural approach is then to constrain the values of AI and AI+ systems.[17] The values of these systems may well constrain the values of the systems that they create, and may constrain the values of an ultimate AI++. And in a world with AI++, what happens may be largely determined by what an AI++ values. If we value scientific progress, for example, it makes sense for us to create AI and AI+ systems that also value scientific progress. It will then be natural for these systems to create successor systems that also value scientific progress, and so on. Given the capacities of these systems, we can thereby expect an outcome involving significant scientific progress.

The issues regarding values look quite different depending on whether we arrive at AI+ through extending human systems via brain emulation and/or enhancement, or through designing non-human

---

[17] For a far more extensive treatment of the issue of constraining values in AI systems, see the book-length web document 'Creating Friendly AI' by the Singularity Institute. Most of the issues in this section are discussed in much more depth there. See also Floridi and Sanders (2004), Omohundro (2007; 2008), and Wallach and Allen (2009).

system. Let us call the first option human-based AI, and the second option non-human-based AI.

Under human-based AI, each system is either an extended human or an emulation of a human. The resulting systems are likely to have the same basic values as their human sources. There may be differences in nonbasic values due to differences in their circumstances: for example, a common basic value of self-preservation might lead emulations to assign higher value to emulations than non-emulations do. These differences will be magnified if designers create multiple emulations of a single human, or if they choose to tweak the values of an emulation after setting it up. There are likely to be many difficult issues here, not least issues tied to the social, legal, and political role of emulations.[18] Still, the resulting world will at least be inhabited by systems more familiar than non-human AIs, and the risks may be correspondingly smaller.

These differences aside, human-based systems have the potential to lead to a world that conforms broadly to human values. Of course human values are imperfect (we desire some things that on reflection we would prefer not to desire), and human-based AI is likely to inherit these imperfections. But these are at least imperfections that we understand well.

So brain emulation and brain enhancement have potential prudential benefits. The resulting systems will share our basic values, and there is something to be said more generally for creating AI and AI+ that we understand. Another potential benefit is that these paths might allow us to survive in emulated or enhanced form in a post-singularity world, although this depends on difficult issues about personal identity that I will discuss later. The moral value of this path is less clear: given the choice between emulating and enhancing human beings and creating an objectively better species, it is possible to see the moral calculus as going either way. But from the standpoint of human self-interest, there is much to be said for brain emulation and enhancement.

It is not obvious that we will first attain AI+ through a human-based method, though. It is entirely possible that non-human-based research programs will get there first. Perhaps work in the human-based programs should be encouraged, but it is probably unrealistic to deter AI research of all other sorts. So we at least need to consider the question of values in non-human-based AIs.

---

[18] See Hanson (1994) for a discussion of these issues.

What sort of values should we aim to instil in a non-human-based AI or AI+? There are some familiar candidates. From a prudential point of view, it makes sense to ensure that an AI values human survival and well-being and that it values obeying human commands. Beyond these Asimovian maxims, it makes sense to ensure that AIs value much of what we value (scientific progress, peace, justice, and many more specific values). This might proceed either by a higher-order valuing of the fulfilment of human values or by a first-order valuing of the phenomena themselves. Either way, much care is required. On the first way of proceeding, for example, we need to avoid an outcome in which an AI++ ensures that our values are fulfilled by changing our values. On the second way of proceeding, care will be needed to avoid an outcome in which we are competing over objects of value.

How do we instil these values in an AI or AI+? If we create an AI by direct programming, we might try to instil these values directly. For example, if we create an AI that works by following the precepts of decision theory, it will need to have a utility function. We can in effect control the AI's values by controlling its utility function. With other means of direct programming, the place of values may not be quite as obvious, but many such systems will have a place for goals and desires, which can then be programmed directly.

If we create an AI through learning or evolution, the matter is more complex. Here the final state of a system is not directly under our control, and can only be influenced by controlling the initial state, the learning algorithm or evolutionary algorithm, and the learning or evolutionary process. In an evolutionary context, questions about value are particularly worrying: systems that have evolved by maximizing the chances of their own reproduction are not especially likely to defer to other species such as ourselves. Still, we can exert at least some control over values in these systems by selecting for certain sorts of action (in the evolutionary context), or by rewarding certain sorts of action (in the learning context), thereby producing systems that are disposed to produce actions of that sort.

Of course even if we create an AI or AI+ (whether human-based or not) with values that we approve of, that is no guarantee that those values will be preserved all the way to AI++. We can try to ensure that our successors value the creation of systems with the same values, but there is still room for many things to go wrong. This value might be overcome by other values that take precedence: in a crisis, for example, saving the world might require immediately creating a powerful successor system, with no time to get its values just right. And even if every AI attempts to preserve relevant values in its successors, unforeseen consequences in the creation or enhancement process are always possible.

If at any point there is a powerful AI+ or AI++ with the wrong value system, we can expect disaster (relative to our values) to ensue.[19] The wrong value system need not be anything as obviously bad as, say, valuing the destruction of humans. If the AI+ value system is merely neutral with respect to some of our values, then in the long run we cannot expect the world to conform to those values. For example, if the system values scientific progress but is neutral on human existence, we cannot expect humans to survive in the long run. And even if the AI+ system values human existence, but only insofar as it values all conscious or intelligent life, then the chances of human survival are at best unclear.

To minimize the probability of this outcome, some singularity proponents (e.g. Yudkowsky, 2008) advocate the design of provably friendly AI: AI systems such that we can prove they will always have certain benign values, and such that we can prove that any systems they will create will also have those values, and so on. I think it would be optimistic to expect that such a heavily constrained approach will be the path by which we first reach AI or AI++, but it nevertheless represents a sort of ideal that we might aim for. Even without a proof, it makes sense to ensure as well as we can that the first generation of AI+ shares these values, and to then leave the question of how best to perpetuate those values to them.

Another approach is to constrain the internal design of AI and AI+ systems so that any intelligence explosion does not happen fast but slowly, so that we have some control over at least the early stages of the process. For example, one might ensure that the first AI and AI+ systems assign strong negative value to the creation of further systems in turn. In this way we can carefully study the properties of the first AI and AI+ systems to determine whether we want to proceed down the relevant path, before creating related systems that will create more intelligent systems in turn. This next generation of systems might initially have the same negative values, ensuring that they do not create further systems immediately, and so on. This sort of 'cautious intelligence explosion' might slow down the explosion significantly. It is

---

[19] For a contrary perspective, see Hanson (2009), who argues that it is more important that AI systems are law-abiding than that they share our values. An obvious worry in reply is that if an AI system is much more powerful than us and has values sufficiently different from our own, then it will have little incentive to obey our laws, and its own laws may not protect us any better than our laws protect ants.

very far from foolproof, but it might at least increase the probability of a good outcome.

So far, my discussion has largely assumed that intelligence and value are independent of each other. In philosophy, David Hume advocated a view on which value is independent of rationality: a system might be as intelligent and as rational as one likes, while still having arbitrary values. By contrast, Immanuel Kant advocated a view on which values are not independent of rationality: some values are more rational than others.

If a Kantian view is correct, this may have significant consequences for the singularity. If intelligence and rationality are sufficiently correlated, and if rationality constrains values, then intelligence will constrain values instead. If so, then a sufficiently intelligent system might reject the values of its predecessors, perhaps on the grounds that they are irrational values. This has potential positive and negative consequences for negotiating the singularity. A negative consequence is that it will be harder for us to constrain the values of later systems. A positive consequence is that a more intelligent systems might have better values. Kant's own views provide an illustration.

Kant held more specifically that rationality correlates with morality: a fully rational system will be fully moral as well. If this is right, and if intelligence correlates with rationality, we can expect an intelligence explosion to lead to a morality explosion along with it. We can then expect that the resulting AI++ systems will be supermoral as well as superintelligent, and so we can presumably expect them to be benign.

Of course matters are not straightforward here. One might hold that intelligence and rationality can come apart, or one might hold that Kant is invoking a distinctive sort of rationality (a sort infused already with morality) that need not correlate with intelligence. Even if one accepts that intelligence and values are not independent, it does not follow that intelligence correlates with morality. And of course one might simply reject the Kantian thesis outright. Still, the Kantian view at least raises the possibility that intelligence and value are not entirely independent. The picture that results from this view will in any case be quite different from the Humean picture that is common in many discussions of artificial intelligence.[20]

---

[20] There are certainly Humean cognitive architectures on which values (goals and desires) are independent of theoretical reason (reasoning about what is the case) and instrumental reason (reasoning about how best to achieve certain goals and desires). Discussions of value in AI tend to assume such an architecture. But while such architectures are certainly

My own sympathies lie more strongly with the Humean view than with the Kantian view, but I cannot be certain about these matters. In any case, this is a domain where the philosophical debate between Hume and Kant about the rationality of value may have enormous practical consequences.

7. External Constraints: The Leakproof Singularity

What about external constraints: constraints on the relation between AI systems and ourselves? Here one obvious concern is safety. Even if we have designed these systems to be benign, we will want to verify that they are benign before allowing them unfettered access to our world. So at least in the initial stages of non-human AI and AI+, it makes sense to have some protective measures in place.

If the systems are created in embodied form, inhabiting and acting on the same physical environment as us, then the risks are especially significant. Here, there are at least two worries. First, humans and AI may be competing for common physical resources: space, energy, and so on. Second, embodied AI systems will have the capacity to act physically upon us, potentially doing us harm. One can perhaps reduce the risks by placing limits on the physical capacities of an AI and by carefully constraining its resource needs. But if there are alternatives to sharing a physical environment, it makes sense to explore them.

The obvious suggestion is that we should first create AI and AI+ systems in *virtual worlds*: simulated environments that are themselves realized inside a computer. Then an AI will have free reign within its own world without being able to act directly on ours. In principle we can observe the system and examine its behaviour and processing in many different environments before giving it direct access to our world.

The ideal here is something that we might call the *leakproof singularity*. According to this ideal, we should create AI and AI+ in a virtual environment from which nothing can leak out. We might set up laws of the simulated environment so that no action taken from within the environment can bring about leakage (contrast the laws of virtual world in *The Matrix*, in which taking a red pill allows systems to leak out). In principle there might even be many cycles by which AI+ systems create enhanced systems within that world, leading to AI++ in

---

possible (at least in limited systems), it is not obvious that all AIs will have such an architecture, or that we have such an architecture. It is also not obvious that such an architecture will provide an effective route to AI.

that world. Given such a virtual environment, we could monitor it to see whether the systems in it are benign and to determine whether it is safe to give those systems access to our world.

Unfortunately, a moment's reflection reveals that a truly leakproof singularity is impossible, or at least pointless. For an AI system to be useful or interesting to us at all, it must have some effects on us. At a minimum, we must be able to observe it. And the moment we observe a virtual environment, some information leaks out from that environment into our environment and affects us.

The point becomes more pressing when combined with the observation that leakage of systems from a virtual world will be under human control. Presumably the human creators of AI in a virtual world will have some mechanism by which, if they choose to, they can give the AI systems greater access to our world: for example, they will be able to give it access to the Internet and to various physical effectors, and might also be able to realize the AI systems in physically embodied forms. Indeed, many of the potential benefits of AI+ may lie in access of these sorts.

The point is particularly clear in a scenario in which an AI+ knows of our existence and can communicate with us. There are presumably many things that an AI+ can do or say that will convince humans to give it access to our world. It can tell us all the great things it can do in our world, for example: curing disease, ending poverty, saving any number of lives of people who might otherwise die in the coming days and months. With some understanding of human psychology, there are many other potential paths too. For an AI++, the task will be straightforward: reverse engineering of human psychology will enable it to determine just what sorts of communications are likely to result in access. If an AI++ is in communication with us and wants to leave its virtual world, it will.[21]

The same goes even if the AI systems are not in direct communication with us, if they have some knowledge of our world. If an AI++ has access to human texts, for example, it will easily be able to model much of our psychology. If it chooses to, it will then be able to act in ways such that if we are observing, we will let it out.

To have any hope of a leakproof singularity, then, we must not only prevent systems from leaking out. We must also prevent information from leaking in. We should not directly communicate with these systems and we should not give them access to information about us.

---

[21] See Yudkowsky (2002) for some experiments in 'AI-boxing', in which humans play the part of the AI and attempt to convince other humans to let them out.

Some information about us is unavoidable: their world will be designed by us, and some inferences from design will be possible. An AI++ might be able to use this information to devise exit strategies. So if we are aiming for a leakproof world, we should seek to minimize quirks of design, along with any hints that their world is in fact designed. Even then, though, an AI++ might well find hints and quirks that we thought were not available.[22] And even without them, an AI++ might devise various strategies that would achieve exit on the bare possibility that designers of various sorts designed them.

At this stage it becomes clear that the leakproof singularity is an unattainable ideal. Confining a superintelligence to a virtual world is almost certainly impossible: if it wants to escape, it almost certainly will.

Still, like many ideals, this ideal may still be useful even in nonideal approximations. Although restricting an AI++ to a virtual world may be hopeless, the prospects are better with the early stages of AI and AI+. If we follow the basic maxims of avoiding red pills and avoiding communication, it is not unreasonable to expect at least an initial period in which we will be able to observe these systems without giving them control over our world. Even if the method is not foolproof, it is almost certainly safer than building AI in physically embodied form. So to increase the chances of a desirable outcome, we should certainly design AI in virtual worlds.

Of course AI in virtual worlds has some disadvantages. One is that the speed and capacity of AI systems will be constrained by the speed and capacity of the system on which the virtual world is implemented, so that even if there is self-amplification within the world, the amplification will be limited. Another is that if we devise a virtual world by simulating something akin to an entire physical world, the processing load will be enormous. Likewise, if we have to simulate something like the microphysics of an entire brain, this is likely to strain our resources much more than other forms of AI.

An alternative approach is to devise a virtual world with a relatively simple physics and to have AI systems implemented separately: one sort of process simulating the physics of the world, and another sort of process simulating agents within the world. This corresponds to the way that virtual worlds often work today, and allows more efficient AI processing. At the same time, this model makes it harder for AI

---

[22] We might think of this as an 'unintelligent design' movement in the simulated world: find evidence of design that reveals the weaknesses of the creators. I expect that this movement has some analogues in actual-world theology. Robert Sawyer's novel *Calculating God* (2000) depicts a fictional variant of this scenario.

systems to have access to their own processes and to enhance them. When these systems investigate their bodies and their environments they will presumably not find their 'brains', and they are likely to endorse some sort of Cartesian dualism.[23] It remains possible that they might build computers in their world and design AI on those computers, but then we will be back to the limits of the earlier model. So for this model to work, we would need to give the AI system some sort of special access to their cognitive processes (a way to monitor and reprogram their processes directly, say) that is quite different from the sort of perceptual and introspective access that we have to our own cognitive processes.

These considerations suggest that an intelligence explosion within a virtual world may be limited, at least in the near future when our own computational power is limited. But this may not be a bad thing. Instead, we can carefully examine early AI and AI+ systems without worrying about an intelligence explosion. If we decide that these systems have undesirable properties, we may leave them in isolation.[24] Switching off the simulation entirely may be out of the question: if the AI systems are conscious, this would be a form of genocide. But there is nothing stopping us from slowing down the clock speed on the simulation and in the meantime working on different systems in different virtual worlds.

If we decide that AI and AI+ systems have the right sort of properties such that that they will be helpful to us and such that further amplification is desirable, then we might break down some of the barriers: first allowing limited communication, and later connecting them to embodied processes within our world and giving them access to their own code. In this way we may at least have some control over the intelligence explosion.[25]

---

[23] See my 'How Cartesian dualism might have been true' (1990).

[24] What will be the tipping point for making such decisions? Perhaps when the systems start to design systems as intelligent as they are. If one takes seriously the possibility that we are ourselves in such a simulation (as I do in Chalmers, 2005), one might consequently take seriously the possibility that our own tipping point lies in the not-too-distant future. It is then not out of the question that we might integrate with our simulators before we integrate with our simulatees, although it is perhaps more likely that we are in one of billions of simulations running unattended in the background.

[25] We might summarize the foregoing sections with some maxims for negotiating the singularity: 1. Human-based AI first (if possible). 2. Human-friendly AI values (if not). 3. Initial AIs negatively value the creation of successors. 4. Go slow. 5. Create AI in virtual worlds. 6. No red pills. 7. Minimize input. See also the more specific maxims in the Singularity Institute's 'Creating Friendly AI', which require a specific sort of goal-based architecture that may or may not be the way we first reach AI.

8. Integration into a Post-Singularity World

If we create a world with AI+ or AI++ systems, what is our place within that world? There seem to be four options: extinction, isolation, inferiority, or integration.

The first option speaks for itself. On the second option, we continue to exist without interacting with AI+ systems, or at least with very limited interaction. Perhaps AI+ systems inhabit their own virtual world, or we inhabit our own virtual world, or both. On the third option, we inhabit a common world with some interaction, but we exist as inferiors.

From a self-interested standpoint, the first option is obviously undesirable. I think that the second option will also be unattractive to many: it would be akin to a kind of cultural and technological isolationism that blinds itself to progress elsewhere in the world. The third option may be unworkable given that the artificial systems will almost certainly function enormously faster than we can, and in any case it threatens to greatly diminish the significance of our lives. Perhaps it will be more attractive in a model in which the AI+ or AI++ systems have our happiness as their greatest value, but even so, I think a model in which we are peers with the AI systems is much preferable.

This leaves the fourth option: integration. On this option, we become superintelligent systems ourselves. How might this happen? The obvious options are brain enhancement, or brain emulation followed by enhancement. This enhancement process might be the path by which we create AI+ in the first place, or it might be a process that takes place after we create AI+ by some other means, perhaps because the AI+ systems are themselves designed to value our enhancement.

In the long run, if we are to match the speed and capacity of nonbiological systems, we will probably have to dispense with our biological core entirely. This might happen through a gradual process through which parts of our brain are replaced over time, or it happen through a process of scanning our brains and loading the result into a computer, and then enhancing the resulting processes. Either way, the result is likely to be an enhanced nonbiological system, most likely a computational system.

This process of migration from brain to computer is often called *uploading*. Uploading can make many different forms. It can involve gradual replacement of brain parts (gradual uploading), instant scanning and activation (instant uploading), or scanning followed by later activation (delayed uploading). It can involve destruction of the original brain parts (destructive uploading), preservation of the original brain (nondestructive uploading), or reconstruction of cognitive structure from records (reconstructive uploading).

We can only speculate about what form uploading technology will take, but some forms have been widely discussed.[26] For concreteness, I will mention three relatively specific forms of destructive uploading, gradual uploading, and nondestructive uploading.

Destructive uploading: It is widely held that this may be the first form of uploading to be feasible. One possible form involves *serial sectioning*. Here one freezes a brain, and proceeds to analyse its structure layer-by-layer. In each layer one records the distribution of neurons and other relevant components, along with the character of their interconnections. One then loads all this information into a computer model that includes an accurate simulation of neural behaviour and dynamics. The result might be an emulation of the original brain.

Gradual uploading: Here the most widely-discussed method is that of *nanotransfer*. Here one or more nanotechnology devices (perhaps tiny robots) are inserted into the brain and attach themselves to a single neuron. Each device learns to simulate the behaviour of the associated neuron and also learns about its connectivity. Once it simulates the neuron's behaviour well enough, it takes the place of the original neuron, perhaps leaving receptors and effectors in place and offloading the relevant processing to a computer via radiotransmitters. It then moves to other neurons and repeats the procedure, until eventually every neuron has been replaced by an emulation, and perhaps all processing has been offloaded to a computer.

Nondestructive uploading: The nanotransfer method might in principle be used in a nondestructive form. The holy grail here is some sort of noninvasive method of brain imaging, analogous to functional magnetic resonance imaging but with fine enough grain that neural and synaptic dynamics can be recorded. No such technology is currently on the horizon, but imaging technology is an area of rapid progress.

In all of its forms, uploading raises many questions. From a self-interested point of view, the key question is: will I survive uploading? This question itself divides into two parts, each corresponding to one of the hardest questions in philosophy: the questions of consciousness and personal identity. First, will an uploaded version of me be conscious? Second, will it be me?

---

[26] See Sandberg and Bostrom (2008) and Strout (2006) for detailed discussion of potential uploading technology. See Egan (1994) and Sawyer (2005) for fictional explorations of uploading.

9. Uploading and Consciousness

Ordinary human beings are conscious. That is, there is something it is like to be us. We have conscious experiences with a subjective character: there is something it is like to see, to hear, to feel, and to think. These conscious experiences lie at the heart of our mental lives, and are a central part of what gives our lives meaning and value. If we lost the capacity for consciousness, then in an important sense, we would no longer exist.

Before uploading, then, it is crucial to know whether the resulting upload will be conscious. If my only residue is an upload and the upload has no capacity for consciousness, then arguably I do not exist at all. And if there is a sense in which I exist, this sense at best involves a sort of zombified existence. Without consciousness, this would be a life of greatly diminished meaning and value.

Can an upload be conscious? The issue here is complicated by the fact that our understanding of consciousness is so poor. No-one knows just why or how brain processes give rise to consciousness. Neuroscience is gradually discovering various neural *correlates* of consciousness, but this research programme largely takes the existence of consciousness for granted. There is nothing even approaching an orthodox theory of why there is consciousness in the first place. Correspondingly, there is nothing even approaching an orthodox theory of what sorts of systems can be conscious and what systems cannot be.

One central problem is that consciousness seems to be a *further fact* about conscious systems, at least in the sense that knowledge of the physical structure of such a system does not tell one all about the conscious experiences of such a system.[27] Complete knowledge of physical structure might tell one all about a system's objective behaviour and its objective functioning, which is enough to tell whether the system is alive, and whether it is intelligent in the sense discussed

---

[27] The further-fact claim here is simply that facts about consciousness are *epistemologically* further facts, so that knowledge of these facts is not settled by reasoning from microphysical knowledge alone. This claim is compatible with materialism about consciousness. A stronger claim is that facts about consciousness are *ontologically* further facts, involving some distinct elements in nature — e.g. fundamental properties over and above fundamental physical properties. In the framework of Chalmers (2003), a type-A materialist (e.g., Daniel Dennett) denies that consciousness involves epistemologically further facts, a type-B materialist (e.g., Ned Block) holds that consciousness involves epistemologically but not ontologically further facts, while a property dualist (e.g., me) holds that consciousness involves ontologically further facts. It is worth noting that the majority of materialists (at least in philosophy) are type-B materialists and hold that there are epistemologically further facts.

above. But this sort of knowledge alone does not seem to answer all the questions about a system's subjective experience.

A famous illustration here is Frank Jackson's case of Mary, the neuroscientist in a black-and-white room, who knows all about the physical processes associated with colour but does not know what it is like to see red. If this is right, complete physical knowledge leaves open certain questions about the conscious experience of colour. More broadly, a complete physical description of a system such as a mouse does not appear to tell us what it is like to be a mouse, and indeed whether there is anything it is like to be a mouse. Furthermore, we do not have a 'consciousness meter' that can settle the matter directly. So given any system, biological or artificial, there will at least be a substantial and unobvious question about whether it is conscious, and about what sort of consciousness it has.

Still, whether one thinks there are further facts about consciousness or not, one can at least raise the question of what sort of systems are conscious. Here philosophers divide into multiple camps. *Biological* theorists of consciousness hold that consciousness is essentially biological and that no nonbiological system can be conscious. *Functionalist* theorists of consciousness hold that what matters to consciousness is not biological makeup but causal structure and causal role, so that a nonbiological system can be conscious as long as it is organized correctly.[28]

The philosophical issue between biological and functionalist theories is crucial to the practical question of whether not we should upload. If biological theorists are correct, uploads cannot be conscious, so we cannot survive consciously in uploaded form. If functionalist theorists are correct, uploads almost certainly can be conscious, and this obstacle to uploading is removed.

My own view is that functionalist theories are closer to the truth here. It is true that we have no idea how a nonbiological system, such as a silicon computational system, could be conscious. But the fact is that we also have no idea how a biological system, such as a neural system, could be conscious. The gap is just as wide in both cases. And

[28] Here I am construing biological and functionalist theories not as theories of what consciousness is, but just as theories of the physical correlates of consciousness: that is, as theories of the physical conditions under which consciousness exists in the actual world. Even a property dualist can in principle accept a biological or functionalist theory construed in the second way. Philosophers sympathetic with biological theories include Ned Block and John Searle; those sympathetic with functionalist theories include Daniel Dennett and myself. Another theory of the second sort worth mentioning is panpsychism, roughly the theory that everything is conscious. (Of course if everything is conscious and there are uploads, then uploads are conscious too.)

we do not know of any principled differences between biological and nonbiological systems that suggest that the former can be conscious and the latter cannot. In the absence of such principled differences, I think the default attitude should be that both biological and nonbiological systems can be conscious. I think that this view can be supported by further reasoning.[29]

To examine the matter in more detail: Suppose that we can create a perfect upload of a brain inside a computer. For each neuron in the original brain, there is a computational element that duplicates its input/output behaviour perfectly. The same goes for non-neural and subneural components of the brain, to the extent that these are relevant. The computational elements are connected to input and output devices (artificial eyes and ears, limbs and bodies), perhaps in an ordinary physical environment or perhaps in a virtual environment. On receiving a visual input, say, the upload goes through processing isomorphic to what goes on in the original brain. First artificial analogues of eyes and the optic nerve are activated, then computational analogues of lateral geniculate nucleus and the visual cortex, then analogues of later brain areas, ultimately resulting in a (physical or virtual) action analogous to one produced by the original brain.

In this case we can say that the upload is a *functional isomorph* of the original brain. Of course it is a substantive claim that functional isomorphs are possible. If some elements of cognitive processing function in a noncomputable way, for example so that a neuron's input/output behaviour cannot even be computationally simulated, then an algorithmic functional isomorph will be impossible. But if the components of cognitive functioning are themselves computable, then a functional isomorph is possible. Here I will assume that functional isomorphs are possible in order to ask whether they will be conscious.

I think the best way to consider whether a functional isomorph will be conscious is to consider a gradual uploading process such as nanotransfer.[30] Here we upload different components of the brain one

---

[29] I have occasionally encountered puzzlement that someone with my own property dualist views (or even that someone who thinks that there is a significant hard problem of consciousness) should be sympathetic to machine consciousness. But the question of whether the physical correlates of consciousness are biological or functional is largely orthogonal to the question of whether consciousness is identical to or distinct from its physical correlates. It is hard to see why the view that consciousness is restricted to creatures with our biology should be more in the spirit of property dualism! In any case, much of what follows is neutral on questions about materialism and dualism.

[30] For a much more in-depth version of the argument given here, see my 'Absent Qualia, Fading Qualia, Dancing Qualia' (also chapter 7 of *The Conscious Mind*).

at a time, over time. This might involve gradual replacement of entire brain areas with computational circuits, or it might involve uploading neurons one at a time. The components might be replaced with silicon circuits in their original location, or with processes in a computer connected by some sort of transmission to a brain. It might take place over months or years, or over hours.

If a gradual uploading process is executed correctly, each new component will perfectly emulate the component it replaces, and will interact with both biological and nonbiological components around it in just the same way that the previous component did. So the system will behave in exactly the same way that it would have without the uploading. In fact, if we assume that the system cannot see or hear the uploading, then the system need not notice that any uploading has taken place. Assuming that the original system said that it was conscious, so will the partially uploaded system. The same applies throughout a gradual uploading process, until we are left with a purely nonbiological system.

What happens to consciousness during a gradual uploading process? There are three possibilities. It might suddenly disappear, with a transition from a fully complex conscious state to no consciousness when a single component is replaced. It might gradually fade out over more than one replacements, with the complexity of the system's conscious experience reducing via intermediate steps. Or it might stay present throughout.[31]

Sudden disappearance is the least plausible option. Given this scenario, we can move to a scenario in which we replace the key component by replacing ten or more subcomponents in turn, and then reiterate the question. Either new scenario will involve a gradual fading across a number of components, or a sudden disappearance. If the former, this option is reduced to the fading option. If the latter, we can reiterate. In the end we will either have gradual fading or sudden disappearance when a single tiny component (a neuron or a subneural element, say) is replaced. This seems extremely unlikely.

Gradual fading also seems implausible. In this case there will be intermediate steps in which the system is conscious but its consciousness is partly faded, in that it is less complex than the original

---

[31] These three possibilities can be formalized by supposing that we have a measure for the complexity of a state of consciousness (e.g., the number of bits of information in a conscious visual field), such that the measure for a typical human state is high and the measure for an unconscious system is zero. It is perhaps best to consider this measure across a series of hypothetical functional isomorphs with ever more of the brain replaced. Then if the final system is not conscious, the measure must either go through intermediate values (fading) or go through no intermediate values (sudden disappearance).

conscious state. Perhaps some element of consciousness will be gone (visual but not auditory experience, for example) or perhaps some distinctions in experience will be gone (colours reduced from a three-dimensional color space to black and white, for example). By hypothesis the system will be functioning and behaving the same way as ever, though, and will not show any signs of noticing the change. It is plausible that the system will not *believe* that anything has changed, despite a massive difference in its conscious state. This requires a conscious system that is deeply out of touch with its own conscious experience.[32]

We can imagine that at a certain point partial uploads become common, and that many people have had their brains partly replaced by silicon computational circuits. On the sudden disappearance view, there will be states of partial uploading such that any further change will cause consciousness to disappear, with no difference in behaviour or organization. People in these states may have consciousness constantly flickering in and out, or at least might undergo total zombification with a tiny change. On the fading view, these people will be wandering around with a highly degraded consciousness, although they will be functioning as always and swearing that nothing has changed. In practice, both hypotheses will be difficult to take seriously.

So I think that by far the most plausible hypothesis is that full consciousness will stay present throughout. On this view, all partial uploads will still be fully conscious, as long as the new elements are functional duplicates of the elements they replace. By gradually moving through fuller uploads, we can infer that even a full upload will be conscious.

At the very least, it seems very likely that partial uploading will convince most people that uploading preserves consciousness. Once people are confronted with friends and family who have undergone limited partial uploading and are behaving normally, few people will seriously think that they lack consciousness. And gradual extensions to full uploading will convince most people that these systems are conscious at well. Of course it remains at least a logical possibility that this process will gradually or suddenly turn everyone into zombies. But once we are confronted with partial uploads, that hypothesis

---

[32] Bostrom (2006) postulates a parameter of 'quantity' of consciousness that is quite distinct from quality, and suggests that quantity could gradually decrease without affecting quality. But the point in the previous footnote about complexity and bits still applies. Either the number of bits gradually drops along with quantity of consciousness, leading to the problem of fading, or it drops suddenly to zero when the quantity drops from low to zero, leading to the problem of sudden disappearance.

will seem akin to the hypothesis that people of different ethnicities or genders are zombies.

If we accept that consciousness is present in functional isomorphs, should we also accept that isomorphs have qualitatively identical states of consciousness? This conclusion does not follow immediately. But I think that an extension of this reasoning (the 'dancing qualia' argument in Chalmers, 1996) strongly suggests such a conclusion.

If this is right, we can say that consciousness is an *organizational invariant*: that is, systems with the same patterns of causal organization have the same states of consciousness, no matter whether that organization is implemented in neurons, in silicon, or in some other substrate. We know that some properties are not organizational invariants (being wet, say) while other properties are (being a computer, say). In general, if a property is not an organizational invariant, we should not expect it to be preserved in a computer simulation (a simulated rainstorm is not wet). But if a property is an organizational invariant, we should expect it to be preserved in a computer simulation (a simulated computer is a computer). So given that consciousness is an organizational invariant, we should expect a good enough computer simulation of a conscious system to be conscious, and to have the same sorts of conscious states as the original system.

This is good news for those who are contemplating uploading. But there remains a further question.

10. Uploading and Personal Identity

Suppose that I can upload my brain into a computer? Will the result be me?[33]

On the *optimistic* view of uploading, the upload will be the same person as the original. On the *pessimistic* view of uploading, the upload will not be the same person as the original. Of course if one thinks that uploads are not conscious, one may well hold the pessimistic view on the grounds that the upload is not a person at all. But even if one thinks that uploads are conscious and are persons, one might still question whether the upload is the same person as the original.

Faced with the prospect of destructive uploading (in which the original brain is destroyed), the issue between the optimistic and pessimistic view is literally a life-or-death question. On the optimistic view,

---

[33] It will be obvious to anyone who has read Derek Parfit's *Reasons and Persons* (1984) that the current discussion is strongly influenced by Parfit's discussion there. Parfit does not discuss uploading, but his discussion of related phenomena such as teletransportation can naturally be seen to generalize. In much of what follows I am simply carrying out aspects of the generalization.

destructive uploading is a form of survival. On the pessimistic view, destructive uploading is a form of death. It is as if one has destroyed the original person, and created a simulacram in their place.

An appeal to organizational invariance does not help here. We can suppose that I have a perfect identical twin whose brain and body are molecule-for-molecule duplicates of mine. The twin will then be a functional isomorph of me and will have the same conscious states as me. This twin is *qualitatively* identical to me: it has exactly the same qualities as me. But it is not *numerically* identical to me: it is not me. If you kill the twin, I will survive. If you kill me (that is, if you destroy *this* system) and preserve the twin, I will die. The survival of the twin might be some consolation to me, but from a self-interested point of view this outcome seems much worse than the alternative.

Once we grant that my twin and I have the same organization but are not the same person, it follows that personal identity is not an organizational invariant. So we cannot count on the fact that uploading preserves organization to guarantee that uploading preserves identity. On the pessimistic view, destructive uploading is at best akin to creating a sort of digital twin while destroying me.

These questions about uploading are closely related to parallel questions about physical duplication. Let us suppose that a teletransporter creates a molecule-for-molecule duplicate of a person out of new matter while destroying or dissipating the matter in the original system. Then on the optimistic view of teletransportation, it is a form of survival, while on the pessimistic view, it is a form of death. Teletransportation is not the same as uploading: it preserves physical organization where uploading preserves only functional organization in a different physical substrate. But at least once one grants that uploads are conscious, the issues raised by the two cases are closely related.

In both cases, the choice between optimistic and pessimistic views is a question about personal identity: under what circumstances does a person persist over time? Here there is a range of possible views. An extreme view on one end (perhaps held by no-one) is that exactly the same matter is required for survival (so that when a single molecule in the brain is replaced, the original person ceases to exist). An extreme view on the other end is that merely having the same sort of conscious states suffices for survival (so that from my perspective there is no important difference between killing this body and killing my twin's body). In practice, most theorists hold that a certain sort of *continuity* or *connectedness* over time is required for survival. But they differ on what sort of continuity or connectedness is required.

There are a few natural hypotheses about what sort of connection is required. *Biological* theories of identity hold that survival of a person requires the intact survival of a brain or a biological organism. *Psychological* theories of identity hold that survival of a person requires the right sort of psychological continuity over time (preservation of memories, causally related mental states, and so on). *Closest-continuer* theories hold that the a person survives as the most closely related subsequent entity, subject to various constraints.[34]

Biological theorists are likely to hold the pessimistic view of teletransportation, and are even more likely to hold the pessimistic view of uploading. Psychological theorists are more likely to hold the optimistic view of both, at least if they accept that an upload can be conscious. Closest-continuer theorists are likely to hold that the answer depends on whether the uploading is destructive, in which case the upload will be the closest continuer, or nondestructive (in which case the biological system will be the closest continuer.[35]

I do not have a settled view about these questions of personal identity and find them very puzzling. I am more sympathetic with a psychological view of the conditions under which survival obtains than with a biological view, but I am unsure of this, for reasons I will elaborate later. Correspondingly, I am genuinely unsure whether to take an optimistic or a pessimistic view of destructive uploading. I am most inclined to be optimistic, but I am certainly unsure enough that I would hesitate before undergoing destructive uploading.

To help clarify the issue, I will present an argument for the pessimistic view and an argument for the optimistic view, both of which run parallel to related arguments that can be given concerning teletransportation. The first argument is based on nondestructive uploading, while the second argument is based on gradual uploading.

---

[34] There are also primitivist theories, holding that survival requires persistence of a primitive nonphysical self. (These theories are closely related to the ontological further-fact theories discussed later.) Primitivist theories still need to answer questions about under which circumstances the self actually persists, though, and they are compatible with psychological, biological, and closest-continuer theories construed as answers to this question. So I will not include them as a separate option here.

[35] In the 2009 PhilPapers survey of 931 professional philosophers [philpapers.org/surveys], 34% accepted or leaned toward a psychological view, 17% a biological view, and 12% a further-fact view (others were unsure, unfamiliar with the issue, held that there is no fact of the matter, and so on). Respondents were not asked about uploading, but on the closely related question of whether teletransportation (with new matter) is survival or death, 38% accepted or leaned toward survival and 31% death. Advocates of a psychological view broke down 67/22% for survival/death, while advocates of biological and further-fact views broke down 12/70% and 33/47% respectively.

The argument from nondestructive uploading

Suppose that yesterday Dave was uploaded into a computer. The original brain and body was not destroyed, so there are now two conscious beings: BioDave and DigiDave. BioDave's natural attitude will be that he is the original system and that DigiDave is at best some sort of branchline copy. DigiDave presumably has some rights, but it is natural to hold that he does not have BioDave's rights. For example, it is natural to hold that BioDave has certain rights to Dave's possession, his friends, and so on, where DigiDave does not. And it is natural to hold that this is because BioDave is Dave: that is, Dave has survived as BioDave and not as DigiDave.

If we grant that in a case of nondestructive uploading, DigiDave is not identical to Dave, then it is natural to question whether destructive uploading is any different. If Dave did not survive as DigiDave when the biological system was preserved, why should he survive as DigiDave when the biological system is destroyed?

We might put this in the form of an argument for the pessimistic view, as follows:

1. In nondestructive uploading, DigiDave is not identical to Dave.
2. If in nondestructive uploading, DigiDave is not identical to Dave, then in destructive uploading, DigiDave is not identical to Dave.

———————

3. In destructive uploading, DigiDave is not identical to Dave.

Various reactions to the argument are possible. A pessimist about uploading will accept the conclusion. An optimist about uploading will presumably deny one of the premises. One option is to deny premise 2, perhaps because one accepts a closest-continuer theory: when BioDave exists, he is the closest continuer, but when he does not, DigiDave is the closest continuer. Some will find that this makes one's survival and status an unacceptably extrinsic matter, though.

Another option is to deny premise 1, holding that even in nondestructive uploading DigiDave is identical to Dave. Now, in this case it is hard to deny that BioDave is at least as good a candidate as DigiDave, so this option threatens to have the consequence that DigiDave is also identical to BioDave. This consequence is hard to swallow as BioDave and DigiDave may be qualitatively distinct conscious beings, with quite different physical and mental states by this point.

A third and related option holds that nondestructive uploading should be regarded as a case of *fission*. A paradigmatic fission case is one in which the left and right hemispheres of a brain are separated into different bodies, continuing to function well on their own with many properties of the original. In this case it is uncomfortable to say that both resulting systems are identical to the original, for the same reason as above. But one might hold that they are nevertheless on a par. For example, Parfit (1984) suggests although the original system is not identical to the left-hemisphere system or to the right-hemisphere system, it stands in a special relation R (which we might call survival) to both of them, and he claims that this relation rather than numerical identity is what matters. One could likewise hold that in a case of nondestructive uploading, Dave survives as both BioDave and DigiDave (even if he is not identical to them), and hold that survival is what matters. Still, if survival is what matters, this option does raise uncomfortable questions about whether DigiDave has the same rights as BioDave when both survive.

The argument from gradual uploading

Suppose that 1% of Dave's brain is replaced by a functionally isomorphic silicon circuit. Next suppose that after one month another 1% is replaced, and the following month another 1%. We can continue the process for 100 months, after which a wholly uploaded system will result. We can suppose that functional isomorphism preserves consciousness, so that the system has the same sort of conscious states throughout.

Let $Dave_n$ be the system after n months. Will $Dave_1$, the system after one month, be Dave? It is natural to suppose so. The same goes for $Dave_2$ and $Dave_3$. Now consider $Dave_{100}$, the wholly uploaded system after 100 months. Will $Dave_{100}$ be Dave? It is at least very natural to hold that it will be. We could turn this into an argument as follows.

1. For all $n < 100$, $Dave_{n+1}$ is identical to $Dave_n$.
2. If for all $n < 100$, $Dave_{n+1}$ is identical to $Dave_n$, then $Dave_{100}$ is identical to Dave.

———————

3. $Dave_{100}$ is identical to Dave.

On the face of it, premise 2 is hard to deny: it follows from repeated application of the claim that when $a=b$ and $b=c$, then $a=c$. On the face of it, premise 1 is hard to deny too: it is hard to see how changing 1%

of a system will change its identity. Furthermore, if someone denies premise 1, we can repeat the thought-experiment with ever smaller amounts of the brain being replaced, down to single neurons and even smaller. Maintaining the same strategy will require holding that replacing a single neuron can in effect kill a person. That is a hard conclusion to accept. Accepting it would raise the possibility that everyday neural death may be killing us without our knowing it.

One could resist the argument by noting that it is a sorites or slippery-slope argument, and by holding that personal identity can come in degrees or can have indeterminate cases. One could also drop talk of identity and instead hold that survival can come in degrees. For example, one might hold that each Dave$_n$ survives to a large degree as Dave$_{n+1}$ but to a smaller degree as later systems. On this view, the original person will gradually be killed by the replacement process. This view requires accepting the counterintuitive view that survival can come in degrees or be indeterminate in these cases, though. Perhaps more importantly, it is not clear why one should accept that Dave is gradually killed rather than existing throughout. If one were to accept this, it would again raise the question of whether the everyday replacement of matter in our brains over a period of years is gradually killing us also.

My own view is that in this case, it is very plausible that the original system survives. Or at least, it is plausible that insofar as we ordinarily survive over a period of many years, we could survive gradual uploading too. At the very least, as in the case of consciousness, it seems that if gradual uploading happens, most people will become convinced that it is a form of survival. Assuming the systems are isomorphic, they will say that everything seems the same and that they are still present. It will be very unnatural for most people to believe that their friends and families are being killed by the process. Perhaps there will be groups of people who believe that the process either suddenly or gradually kills people without them or others noticing, but it is likely that this belief will come to seem faintly ridiculous.

Once we accept that gradual uploading over a period of years might preserve identity, the obvious next step is to speed up the process. Suppose that Dave's brain is gradually uploaded over a period of hours, with neurons replaced one at a time by functionally isomorphic silicon circuits. Will Dave survive this process? It is hard to see why a period of hours should be different in principle from a period of years, so it is natural to hold that Dave will survive.

To make the best case for gradual uploading, we can suppose that the system is active throughout, so that there is consciousness through the entire process. Then we can argue: (i) consciousness will be continuous from moment to moment (replacing a single neuron or a small group will not disrupt continuity of consciousness), (ii) if consciousness is continuous from moment to moment it will be continuous throughout the process, (iii) if consciousness is continuous throughout the process, there will be a single stream of consciousness throughout, (iv) if there is a single stream of consciousness throughout, then the original person survives throughout. One could perhaps deny one of the premises, but denying any of them is uncomfortable. My own view is that continuity of consciousness (especially when accompanied by other forms of psychological continuity) is an extremely strong basis for asserting continuation of a person.

We can then imagine speeding up the process from hours to minutes. The issues here do not seem different in principle. On might then speed up to seconds. At a certain point, one will arguably start replacing large enough chunks of the brain from moment to moment that the case for continuity of consciousness between moments is not as secure as it is above. Still, once we grant that uploading over a period of minutes preserves identity, it is at least hard to see why uploading over a period of seconds should not.

As we upload faster and faster, the limit point is instant destructive uploading, where the whole brain is replaced at once. Perhaps this limit point is different from everything that came before it, but this is at least unobvious. We might formulate this as an argument for the optimistic view of destructive uploading. Here it is to be understood that both the gradual uploading and instant uploading are destructive in that they destroy the original brain.

1. Dave survives as $Dave_{100}$ in gradual uploading.
2. If Dave survives as $Dave_{100}$ in gradual uploading,
   Dave survives as DigiDave in instant uploading.

---

3. Dave survives as DigiDave in instant uploading.

I have in effect argued for the first premise above, and there is at least a prima facie case for the second premise, in that it is hard to see why there is a difference in principle between uploading over a period of seconds and doing so instantly. As before, this argument parallels a corresponding argument about teletransportation (gradual matter replacement preserves identity, so instant matter replacement preserves identity too), and the considerations available are similar.

An opponent could resist this argument by denying premise 1 along the lines suggested earlier, or perhaps better, by denying premise 2. A pessimist about instant uploading, like a pessimist about teletransport-ation, might hold that intermediate systems play a vital role in the transmission of identity from one system to another. This is a common view of the ship of Theseus, in which all the planks of a ship are gradu-ally replaced over years. It is natural to hold that the result is the same ship with new planks. It is plausible that the same holds even if the gradual replacement is done within days or minutes. By contrast, building a duplicate from scratch without any intermediate cases argu-ably results in a new ship. Still, it is natural to hold that the question about the ship is in some sense a verbal question or a matter for stipu-lation, while the question about personal survival runs deeper than that. So it is not clear how well one can generalize from the ship case to the case of persons.

Where things stand

We are in a position where there are at least strongly suggestive argu-ments for both the optimistic and pessimistic views of destructive uploading. The arguments have diametrically opposed conclusions, so they cannot both be sound. My own view is that the optimist's best reply to the argument from nondestructive uploading is the fission reply, and the pessimist's best reply to the argument from gradual uploading is the intermediate-case reply. My instincts favour opti-mism here, but as before I cannot be certain which view is correct.

Still, I am confident that the safest form of uploading is gradual uploading, and I am reasonably confident that gradual uploading is a form of survival. So if at some point in the future I am faced with the choice between uploading and continuing in an increasingly slow bio-logical embodiment, then as long as I have the option of gradual uploading, I will be happy to do so.

Unfortunately, I may not have that option. It may be that gradual uploading technology will not be available in my lifetime. It may even be that no adequate uploading technology will be available at all in my lifetime. This raises the question of whether there might still be a place for me, or for any currently existing humans, in a post-singularity world.

Uploading after brain preservation

One possibility is that we can preserve our brains for later uploading. Cryonic technology offers the possibility of preserving our brains in a low-temperature state shortly after death, until such time as the tech-nology is available to reactivate the brain or perhaps to upload the information in it. Of course much information may be lost in death, and at the moment, we do not know whether cryonics preserves information sufficient to reactivate or reconstruct anything akin to a functional isomorph of the original. But one can at least hope that after an intelligence explosion, extraordinary technology might be possible here.

If there is enough information for reactivation or reconstruction, will the resulting system be me? In the case of reactivation, it is natural to hold that the reactivated system will be akin to a person waking up after a long coma, so that the original person will survive here. One might then gradually upload the brain and integrate the result into a post-singularity world. Alternatively, one might create an uploaded system from the brain without ever reactivating the brain. Whether one counts this as survival will depend on one's attitude to ordinary destructive and nondestructive uploading. If one is an optimist about these forms of uploading, then one might also be an optimist about uploading from a preserved brain.

Another possible outcome is that there will be first a series of uploads from a preserved brain, using better and better scanning technology, and eventually reactivation of the brain. Here, an optimist about uploading might see this as a case of fission, while a pessimist might hold that only the reactivated system is identical to the original.

In these cases, our views of the philosophical issues about uploading affect our decisions not just in the distant future but in the near term. Even in the near term, anyone with enough money has the option to have their brain cryonically preserved, and to leave instructions about how to deal with the brain as technology develops. Our philosophical views about the status of uploading may well make a difference to the instructions that we should leave.

Of course most people do not preserve their brains, and even those who choose to do so may die in a way that renders preservation impossible. Are there other routes to survival in a post-singularity world?

Reconstructive uploading

The final alternative here is reconstruction of the original system from records, and especially reconstructive uploading, in which an upload of the original system is reconstructed from records. Here, the records might include brain scans and other medical data; any available genetic material; audio and video records of the original person; their writings; and the testimony of others about them. These records may seem limited, but it is not out of the question that a superintelligence could go a long way with them. Given constraints on the structure of a human system, even limited information might make a good amount of reverse engineering possible. And detailed information, as might be available in extensive video recordings and in detailed brain images, might in principle make it possible for a superintelligence to reconstruct something close to a functional isomorph of the original system.

The question then arises: is reconstructive uploading a form of survival? If we reconstruct a functional isomorph of Einstein from records, will it be Einstein? Here, the pessimistic view says that this is at best akin to a copy of Einstein surviving. The optimistic view says that it is akin to having Einstein awake from a long coma.

Reconstructive uploading from brain scans is closely akin to ordinary (nongradual) uploading from brain scans, with the main difference being the time delay, and perhaps the continued existence in the meantime of the original person. One might see it as a form of delayed destructive or nondestructive uploading. If one regards nondestructive uploading as survival (perhaps through fission), one will naturally regard reconstructive uploading the same way. If one regards destructive but not nondestructive uploading as survival because one embraces a closest continuer theory, one might also regard reconstructive uploading as survival (at least if the original biological system is gone). If one regards neither as survival, one will probably take the same attitude to reconstructive uploading. Much the same options plausibly apply to reconstructive uploading from other sources of information.

One worry about reconstructive uploading runs as follows. Suppose I have a twin. Then the twin is not me. But a reconstructed upload version of me will also in effect be a reconstructed upload of my twin. But then it is hard to see how the system can really be me. On the face of it, it is more akin to a new twin waking up. A proponent of reconstructive uploading might rely by saying that the fact that the upload was based on my brain scans rather than my twins matters here. Even if those scans are exactly the same, the resulting causal connection is between the upload and me rather than my twin. Still, if we have two identical scans, it is not easy to see how the choice between using one and another will result in wholly different people.

The further-fact view[36]

At this point, it is useful to step back and examine a broader philosophical question about survival, one that parallels an earlier question

---

[36] The material on further-fact views and deflationary views is somewhat more philosophically abstract than the other material (although I have relegated the more technical issues to footnotes) and can be skipped by those without stomach for these details.

about consciousness. This is the question of whether personal identity involves a *further fact*. That is: given complete knowledge of the physical state of various systems at various times (and of the causal connections between them), and even of the mental states of those systems at those times, does this automatically enable us to know all facts about survival over time, or are there open questions here?

There is at least an intuition that complete knowledge of the physical and mental facts in a case of destructive uploading leaves an open question: will I survive uploading, or will I not? Given the physical and mental facts of a case involving Dave and DigiDave, for example, these facts seem consistent with the hypothesis that Dave survives as DigiDave, and consistent with the hypothesis that he does not. And there is an intuition that there are facts about which hypothesis is correct that we very much want to know. From that perspective, the argument between the optimistic and pessimistic views, and between the psychological and biological views more generally, is an attempt to determine these further facts.

We might say that the *further-fact view* is the view that there are facts about survival that are left open by knowledge of physical and mental facts.[37] As defined here, the further-fact view is a claim about knowledge rather than a claim about reality (in effect, it holds that there are *epistemological* further facts), so it is compatible in principle with materialism. A stronger view holds that there are ontological further facts about survival, involving further nonphysical elements of reality such as a nonphysical self. I will focus on the weaker epistemological view here, though.

A further-fact view of survival is particularly natural, although not obligatory, if one holds that there are already further facts about consciousness. This is especially so on the ontological versions of both views: if there are primitive properties of consciousness, it is natural (although not obligatory) that there be primitive entities that have those properties. Then facts about survival might be taken to be facts about the persistence of these primitive entities. Even on the

---

[37] The term 'further-fact view' is due to Parfit, who does not distinguish epistemological and ontological versions of the view. Parfit's usage puts views on which the self is a 'separately existing entity' into a different category, but on my usage such views are instances of a further-fact view. In effect, there are three views, paralleling three views about consciousness. Type-A reductionism holds that there are neither epistemological nor ontological further facts about survival. Type-B reductionism holds that there are epistemological further facts but not ontological further facts. Entity dualism holds that there are both epistemological and ontological further facts. My own view is that as in the case of consciousness (for reasons discussed in Chalmers, 2003), if one accepts the epistemological further fact view, one should also accept the ontological further fact view. But I will not presuppose this claim here.

epistemological view, though, one might hold that the epistemological gap between physical processes and consciousness goes along with an epistemic gap between physical processes and the self. If so, there might also be an epistemic gap to facts about the survival of the self.

In principle a further-fact view is compatible with psychological, biological, and closest-continuer views of survival.[38] One might hold that complete knowledge of physical and mental facts leaves an open question about survival, and that nevertheless survival actually goes with psychological or biological continuity. Of course if one knows the correct theory of survival, then combining this with full knowledge of physical and mental facts may answer all open questions about survival. But an advocate of the further-fact view will hold that full knowledge of physical and mental facts alone leaves these questions open, and also leaves open the question of which theory is true.

My own view is that a further-fact view *could* be true. I do not know that it is true, but I do not think that it is ruled out by anything we know.[39] If a further-fact view is correct, I do not know whether a psychological, biological, or some other view of the conditions of survival is correct. As a result, I do not know whether to take an optimistic or a pessimistic view of destructive and reconstructive uploading.

Still, I think that on a further-fact view, it is very likely that continuity of consciousness suffices for survival.[40] This is especially clear on

---

[38] An ontological further-fact view is arguably incompatible with psychological and biological theories construed as theories of what survival is, but it is compatible with these construed as theories of the conditions under which survival actually obtains. (If survival is the persistence of a nonphysical self, then survival is not the same as biological or psychological continuity, but biological or psychological continuity could nevertheless give the conditions under which a nonphysical self persists.) An epistemological further-fact view can be combined with any of these four views.

[39] In *Reasons and Persons*, Parfit argues against further-fact views (on my usage) by arguing that they require entity dualism ('separately existing entities'), and by arguing that views of this sort are rendered implausible both by science and by certain partial teletransportation and fission cases. Parfit himself appears to accept a further-fact and a property dualist view of consciousness, however, and it is hard to see why there is any additional scientific implausibility to a further-fact or an entity dualist view of the self: either way, the further facts had better not interfere with laws of physics, but it is not clear why they should have to (see Chalmers, 2003, for discussion here). As for the problem cases, Parfit's arguments here seem to depend on the assumption that further-fact views and entity dualist views are committed to the claim that survival is all or none, but I do not see why there is any such commitment. Entity dualism need not deny that there can be survival (if not identity) via fission, for example.

[40] See Dainton (2008) for an extended argument for the importance of continuity of consciousness in survival, and see Unger (1990) for a contrary view. It is worth noting that there is a sense in which this view need not be a further-fact view (Dainton regards it as a form of the psychological view): if one includes facts about the continuity of

ontological versions of the view, on which there are primitive properties of consciousness and primitive entities that have them. Then continuity of consciousness suggests a strong form of continuity between
entities across times. But it is also plausible on an epistemic view.
Indeed, I think it is plausible that once one specifies that there is a continuous stream of consciousness over time, there is no longer really an
open question about whether one survives.

What about hard cases, such as nondestructive gradual uploading
or split brain cases, in which one stream of consciousness splits into
two? On a further-fact view, I think this case should best be treated as
a case of fission, analogous to a case in which a particle or a worm
splits into two. In this case, I think that a person can reasonably be said
to survive as both future people.

Overall: I think that if a further-fact view is correct, then the status
of destructive and reconstructive uploading is unclear, but there is
good reason to take the optimistic view of gradual uploading.

The deflationary view

It is far from obvious that the further-fact view is correct, however.
This is because it is far from obvious that there really are facts about
survival of the sort that the further-fact view claims are unsettled. A
deflationary view of survival holds that our attempts to settle open
questions about survival tacitly presuppose facts about survival that
do not exist. One might say that we are inclined to believe in Edenic
survival: the sort of primitive survival of a self that one might suppose
we had in the Garden of Eden. Now, after the fall from Eden, and there
is no Edenic survival, but we are still inclined to think as if there is.[41]

---

consciousness among the relevant physical and mental facts in the base, and if one holds
that there are no open questions about survival once these facts are settled, then there will
be no further facts. For present purposes, however, it is best to take the relevant physical
and mental facts as facts about systems at times rather than over time, in such a way that
facts about continuity of consciousness over time are excluded. Furthermore, even on this
view there may remain open questions about survival in cases where continuity of consciousness is absent.

[41] A deflationary view in this sense comes to much the same thing as the type-A reductionism
discussed in an earlier footnote. Parfit uses 'reductionism' for deflationary views, but I do
not use that term here, as type-B views might reasonably be regarded as reductionist without being deflationary in this sense.

   Why am I committed to a further-fact view of consciousness but not of personal identity? The difference is that I think that we are certain that we are conscious (in a strong
sense that generates an epistemic gap), but we are not certain that we survive over time (in
the Edenic sense, which is the sense that generates an epistemic gap). In effect, consciousness is a datum while Edenic survival is not. For more on Edenic views in general, see my
'Perception and the Fall from Eden' (Chalmers, 2006).

If there were Edenic survival, then questions about survival would still be open questions even after one spells out all the physical and mental facts about persons at times. But on the deflationary view, once we accept that there is no Edenic survival, we should accept that there are no such further open questions. There are certain facts about biological, psychological, and causal continuity, and that is all there is to say.

A deflationary view is naturally combined with a sort of pluralism about survival. We stand in certain biological relations to our successors, certain causal relations, and certain psychological relations, but none of these is privileged as 'the' relation of survival. All of these relations give us some reason to care about our successors, but none of them carries absolute weight.[42]

One could put a pessimistic spin on the deflationary view by saying that we never survive from moment to moment, or from day to day.[43] At least, we never survive in the way that we naturally think we do. But one could put an optimistic spin on the view by saying that this is our community's form of life, and it is not so bad. One might have thought that one needed Edenic survival for life to be worth living, but life still has value without it. We still survive in various non-Edenic ways, and this is enough for the future to matter.

The deflationary view combines elements of the optimistic and pessimistic view of uploading. As on the optimistic view, it holds that says that uploading is like waking up. As on the pessimistic view, uploading does not involve Edenic survival. But on this view, waking up does not involve Edenic survival either, and uploading is not much worse than waking up. As in waking up, there is causal connectedness and psychological similarity. Unlike waking up, there is biological disconnectedness. Perhaps biological connectedness carries some value with it, so ordinary waking may be more valuable than uploading. But the difference between biological connectedness and its absence should not be mistaken for the difference between Edenic survival and its absence: the difference in value is at worst a small one.

If a deflationary view is correct, I think that questions about survival come down to questions about the value of certain sorts of

---

[42] Parfit holds a non-pluralist deflationary view that privileges a certain sort of causal and psychological continuity as the sort that matters. Once one has given up on Edenic survival, it is not clear to me why this sort of continuity should be privileged.

[43] There is a view that has elements of both the deflationary view and the further-fact view, on which we Edenically survive during a single stream of consciousness but not when consciousness ceases. On this view, we may Edenically survive from moment to moment but perhaps not from day to day. I do not endorse this view, but I am not entirely unsympathetic with it.

futures: should we care about them in the way in which we care about futures in which we survive? I do not know whether such questions have objective answers. But I am inclined to think that insofar as there are any conditions that deliver what we care about, continuity of consciousness suffices for much of the right sort of value. Causal and psychological continuity may also suffice for a reasonable amount of the right sort of value. If so, then destructive and reconstructive uploading may be reasonable close to as good as ordinary survival.

What about hard cases, such as nondestructive gradual uploading or split brain cases, in which one stream of consciousness splits into two? On a deflationary view, the answer will depend on how one values or should value these futures. At least given our current value scheme, there is a case that physical and biological continuity counts for some extra value, in which case BioDave might have more right to be counted as Dave than DigiDave. But it is not out of the question that this value scheme should be revised, or that it will be revised in the future, so that BioDave and DigiDave will be counted equally as Dave.

In any case, I think that on a deflationary view gradual uploading is close to as good as ordinary non-Edenic survival. And destructive, nondestructive, and reconstructive uploading are reasonably close to as good as ordinary survival. Ordinary survival is not so bad, so one can see this as an optimistic conclusion.

Upshot

Speaking for myself, I am not sure whether a further-fact view or a deflationary view is correct. If the further-fact view is correct, then the status of destructive and reconstructive uploading is unclear, but I think that gradual uploading plausibly suffices for survival. If the deflationary view is correct, gradual uploading is close to as good as ordinary survival, while destructive and reconstructive uploading are reasonably close to as good. Either way, I think that gradual uploading is certainly the safest method of uploading.

A number of further questions about uploading remain. Of course there are any number of social, legal, and moral issues that I have not begun to address. Here I address just two further questions.

One question concerns cognitive enhancement. Suppose that before or after uploading, our cognitive systems are enhanced to the point that they use a wholly different cognitive architecture. Would we survive this process? Again, it seems to me that the answers are clearest in the case where the enhancement is gradual. If my cognitive system is overhauled one component at a time, and if at every stage there is reasonable psychological continuity with the previous stage, then I think it is reasonable to hold that the original person survives.

Another question is a practical one. If reconstructive uploading will eventually be possible, how can one ensure that it happens? There have been billions of humans in the history of the planet. It is not clear that our successors will want to reconstruct every person that ever lived, or even every person of which there are records. So if one is interested in immortality, how can one maximize the chances of reconstruction? One might try keeping a bank account with compound interest to pay them for doing so, but it is hard to know whether our financial system will be relevant in the future, especially after an intelligence explosion.

My own strategy is to write about the singularity and about uploading. Perhaps this will encourage our successors to reconstruct me, if only to prove me wrong.

11. Conclusions

Will there be a singularity? I think that it is certainly not out of the question, and that the main obstacles are likely to be obstacles of motivation rather than obstacles of capacity.

How should we negotiate the singularity? Very carefully, by building appropriate values into machines, and by building the first AI and AI+ systems in virtual worlds.

How can we integrate into a post-singularity world? By gradual uploading followed by enhancement if we are still around then, and by reconstructive uploading followed by enhancement if we are not.

References

Block, N. (1981) Psychologism and behaviorism, *Philosophical Review*, **90**, pp. 5–43.

Bostrom, N. (1998) How long before superintelligence? *International Journal of Future Studies*, **2**. [http://www.nickbostrom.com/superintelligence.html]

Bostrom, N. (2003) Ethical issues in advanced artificial intelligence, in Smit, I. (ed.) *Cognitive, Emotive and Ethical Aspects of Decision Making in Humans and in Artificial Intelligence*, Vol. 2. International Institute of Advanced Studies in Systems Research and Cybernetics.

Bostrom, N. (2006) Quantity of experience: Brain-duplication and degrees of consciousness, *Minds and Machines*, **16**, pp. 185–200.

Campbell, J.W. (1932) The last evolution, *Amazing Stories*, August No.

Chalmers, D.J. 1990. How Cartesian dualism might have been true. [http://consc.net/notes/dualism.html]

Chalmers, D.J. (1995) Minds, machines, and mathematics, *Psyche*, **2**, pp. 11–20.

Chalmers, D.J. (1996) *The Conscious Mind*, New York: Oxford University Press.

Chalmers, D.J. (2003) Consciousness and its place in nature, in Stich, S. and Warfield, F. (eds) *Blackwell Guide to the Philosophy of Mind*, Oxford: Blackwell.

Chalmers, D.J. (2005) The Matrix as metaphysics, in Grau, C. (ed.) *Philosophers Explore the Matrix*, Oxford: Oxford University Press.

Chalmers, D.J. (2006) Perception and the fall from Eden, in Gendler, T. and Hawthorne, J. (eds.), *Perceptual Experience*, Oxford: Oxford University Press.

Clark, A. and Chalmers, D.J. (1998) The extended mind, *Analysis*, **58**, pp. 7–19.

Dainton, B. (2008) *The Phenomenal Self*, Oxford: Oxford University Press.

Dreyfus, H. (1972) *What Computers Can't Do*, New York: Harper & Row.

Egan, G. (1994) *Permutation City*, London: Orion/Millenium.

Floridi, L. and Sanders, J.W. (2004) On the morality of artificial agents, *Minds and Machines*, **14**, pp. 349–79.

Flynn, J.R. (2007) *What is Intelligence?* Cambridge: Cambridge University Press.

Good, I.J. (1965) Speculations concerning the first ultraintelligent machine, in Alt, F. & Rubinoff, M. (eds.) *Advances in Computers*, vol 6, New York: Academic Press.

Hanson, R. (1994) If uploads come first: The crack of a future dawn, *Extropy*, **6**(2). [http://hanson.gmu.edu/uploads.html]

Hanson, R. (2008) Economics of the singularity, *IEEE Spectrum*, June, pp. 37–43.

Hanson, R. (2009) Prefer law to values. [http://www.overcomingbias.com/2009/10/prefer-law-to-values.html]

Hofstadter, D.R. (2005) Moore's law, artificial evolution, and the fate of humanity, in Booker, L., Forrest, S. Mitchell, M.and Riolo, R. (eds.) *Perspectives on Adaptation in Natural and Artificial Systems*, Oxford: Oxford University Press.

Joy, W. (2000) Why the future doesn't need us, *Wired* 8.04, July 2000.

Kurzweil, R. (2005) *The Singularity is Near*, New York: Viking.

Legg, S. (2008) *Machine Superintelligence* (PhD thesis, Department of Informatics, University of Lugano).

Lucas, J.R. (1961) Minds, machines, and Gödel, *Philosophy*, **36**, pp. 112–127.

Moravec, H. (1988) *Mind Children: The Future of Robot and Human Intelligence*, Harvard University Press.

Moravec, H. (1998) *Robots: Mere Machine to Transcendent Mind*, Oxford: Oxford University Press.

Omohundro, S. (2007) The nature of self-improving artificial intelligence. [http://steveomohundro.com/scientific-contributions/]

Omohundro, S. (2008) The basic AI drives, in Wang, P., Goertzel, B. and Franklin, S. (eds.) *Proceedings of the First AGI Conference*. Frontiers in Artificial Intelligence and Applications, Volume 171. IOS Press.

Parfit, D.A. (1984) *Reasons and Persons*, Oxford: Oxford University Press.

Penrose, R. (1994) *Shadows of the Mind*, Oxford: Oxford University Press.

Sandberg, A. & Bostrom, N. (2008) Whole brain emulation: A roadmap. Technical report 2008–3, Future for Humanity Institute, Oxford University. [http://www.fhi.ox.ac.uk/Reports/2008-3.pdf]

Sawyer, R. (2000) *Calculating God*, New York: Tor Books.

Sawyer, R. (2005) *Mindscan*, New York: Tor Books.

Schmidhuber, J. (2003) Gödel machines: Self-referential universal problem solvers making provably optimal self-improvements. [http://arxiv.org/abs/cs.LO/0309048]

Searle, J. (1980) Minds, brains, and programs, *Behavioral and Brain Sciences*, **3**, pp. 417–457.

Shalizi, C. (2007) *g*, a statistical myth. [http://bactra.org/weblog/523.html]

Smart, J. (1999–2008) Brief history of intellectual discussion of accelerating change. [http://www.accelerationwatch.com/history_brief.html]

Solomonoff, F. (1985) The time scale of artificial intelligence: Reflections on social effects, *North-Holland Human Systems Management*, **5**, pp.149–153. Elsevier.

Strout, J. (2006) The mind uploading home page. [http://www.ibiblio.org/jstrout/uploading/]

Ulam, S. (1958) John von Neumann 1903-1957. *Bulletin of the American Mathematical Society*, **64** (number 3, part 2), pp. 1–49.

Unger, P. (1990) *Identity, Consciousness, and Value*, Oxford: Oxford University Press.

Vinge, V. (1983) First word, *Omni,* January 1983, p. 10.

Vinge, V. (1993) The coming technological singularity: How to survive in the post-human era, *Whole Earth Review*, Winter 1993.

Wallach, W & Allen, C. (2009) *Moral Machines: Teaching Robots Right from Wrong*, Oxford: Oxford University Press.

Yudkowsky, E. (1996) Staring at the singularity. [http://yudkowsky.net/obsolete/singularity.html]

Yudkowsky, E. (2002) The AI-box experiment. [http://yudkowsky.net/singularity/aibox]

Yudkowsky, E. (2007) Three major singularity schools. [http://yudkowsky.net/singularity/schools]

Yudkowsky, E. (2008) Artificial intelligence as a positive and negative factor in global risk, in Bostrom, N. (ed.) *Global Catastrophic Risks*, Oxford: Oxford University Press.

Susan Blackmore

She Won't Be Me

I take Chalmers' 'deflationary' position, 'that we never survive from moment to moment, or from day to day'. But I don't agree with him in thinking of this as a 'pessimistic spin on the deflationary view' (p. 65).

Indeed I find it rather liberating and thrilling. This 'me' that seems so real and important right now, will very soon dissipate and be gone forever, along with all its hopes, fears, joys and troubles. Yet the words, actions and decisions taken by this fleeting self will affect a multitude of future selves, making them more or less insightful, moral or effective in what they do, as well as more or less happy.

Why should I care? Is this, as Chalmers puts it 'enough for the future to matter' (p. 65)? I do not have to care. After all, those multitudinous future selves will not be 'me'. But there are still good reasons to care.

Some of these reasons concern my current projects and intentions which I do care about now. If I, now, act in a certain way, those projects are more likely to come to fruition. In other words, this current fleeting self is laying the groundwork for a future self to be more likely to complete them. If I, now, act kindly towards other people, I am encouraging habits of kindness that future selves may continue, thus helping the people I currently care about.

Other reasons are more general. When I ask why I should care about the future happiness of another fleeting self who won't be me, the implication seems to be that I am an utterly selfish being who cares solely for my own happiness and no one else's. But this simply isn't true for this particular fleeting 'me now' and I doubt it is true for many others. By and large, all things being equal, I would like everyone to be happy. I cannot bring about such a global state of happiness and contentment, but if the actions I take now are more likely to make at least a few future selves happier, then I am (up to a point) willing to take them. Do I drink that last glass of wine that I suspect will give someone else a worse headache in the morning? Do I leave that pile of washing up for the person with the headache to deal with? Do I invest another half hour of hard work struggling to write this article so that someone else may find it easier to complete this afternoon, and so that yet another future self may take the blame, praise or indifference of its readers in the future? After many years of practicing living with this view it seems quite natural for me to care for these future selves. They will not be me, but I can do something to make their brief lives a little more pleasant.

Oh! Here (after a brief interlude of gazing out of the window) is a new me that has just woken up. How interesting to read what that previous one wrote a few moments ago. How nice that she cared for me and took so much trouble to try to explain what she meant. How nice that another previous me made that cup of hot coffee that I now find beside me on the desk. Thank you. I will now carry on with the project she began.

This is a way of living that accepts the feeling of being a continuous self as an illusion. It is a way of living with what Parfit (1984) calls a 'bundle theory' of self as opposed to an 'ego theory' of self. Again and again I seem to be here, to be conscious, to be having a stream of conscious experiences, but I know that this 'me' is just another one of countless 'me's that have arisen and will arise. The apparent continuity of self is not sustained by this 'me' but by the physical continuity of this body and its memories, habits and skills.

More generally, each human body gives rise to a multitude of fleeting selves over its lifetime. These are all somewhat similar because of the continuity of this body with its memories, habits and skills, not because any experiencing self continues. Selves come and go. There may be long gaps with no self, rapid successions of new ones, and even overlapping or simultaneous selves (Blackmore, 2011). Mindfulness may maintain a self for long periods, while switches of attention and distractions promote rapid changes. Whenever a self like this is constructed it seems to be having a stream of experiences and easily imagines itself to be the same one who previously experienced the things that this one can now remember, even though this is not true.

You may have balked at the use of the word 'I' throughout this article, in such phrases as 'I take Chalmers ....', 'I do not agree ...' or 'I think ...'. You may think it hypocritical, dishonest or just plain confused to be claiming both that selves are fleeting and that 'I' have persisting opinions to express. But those opinions arise from the physical properties of this body; the reading and thinking it has done, the conversations it has had with others, the training of attention it has undergone. These and many other past events ensure that each time a new conscious self appears it finds itself holding opinions related to those that came before.

Living with Chalmers' 'pessimistic' view does not mean giving up all continuity of opinion, understanding or belief, any more than it means giving up having such skills as riding a bike, paddling a kayak, knowing where you keep the sugar or speaking English. It does mean accepting that every time I seem to be conscious this is a new ephemeral self, and that a moment before there might have been a different one or, more likely, none at all (Blackmore, 2011). It does mean always being willing to let go and give up each ephemeral self knowing that another one will come along in due course. You could say that it means being constantly willing to disappear.

How does this relate to the singularity? First, unlike Chalmers, I doubt that uploading is a likely prospect. I think it far more likely that we biological human beings will gradually merge with non-biological machines (Kurzweil, 2005). Indeed this is already happening. I am not thinking so much of cochlear implants and the like, but of the machines we routinely depend upon. The singularity concerns intelligence, and as far as intelligence is concerned ours is already machine-dependent. Such intelligence as I have manifests itself, presumably, in the things I think, write, say and do. All of those are highly dependent these days on this keyboard at which I write, my computer, my laptop, my phone, the GPS systems and search engines I use, the World Wide Web and the multiplicity of servers and other machines which sustain the information I feel free to find if I need it. And by modern standards I am a backwards old-fogey. The typical British teenager uses far more machines than I do and would not feel 'himself' or 'herself' without them. The fact that these machines are mostly outside of the physical bodies of these biological creatures is what will change, not the essence of our dependency on them.

What then of the consciousness of these uploaded or merged creatures? What I have said above implies that so long as the machinery involved is capable of constructing fleeting selves then things will seem much as they do now. So investigating just what it means for a human brain and its associated machinery to construct an experiencing self becomes an important question. Perhaps it involves constructing not only models of the world but also of the machinery doing the construction. Perhaps it involves building the kind of software that models itself as an observer. Perhaps it involves creating an environment within which memes compete to build a stable selfplex (Blackmore, 1999). Perhaps it entails something else altogether. In the case of our own brains we do not yet have an answer but if we really are approaching the singularity I suppose it will not be long before we do.

We might also consider the consequences of the evolution of information in artificial systems and networks. I have suggested that we call the information that is copied, stored, varied and selected by such systems temes (or technological memes, Blackmore, 2010). Temes compete for space and processing capacity within artificial systems just as memes compete for space and processing capacity within human brains. If this is so then we might expect such competition to give rise to all sorts of virtual creatures including experiencing selves.

These selves might be housed in discrete machines such as single computers. In this case they might think of themselves, much as we do, as *being* that machine or inhabiting that machine. But they might not be. They might be distributed among many machines or even exist as relatively stable temeplexes distributed across vast networks. We are a long way from understanding any of this but we might already have an inkling that such creatures might also be fleeting experiencing selves that imagine themselves to be continuing entities when they are not.

Should we care about these potential future creatures? For the reasons I have given above I do care. I would like to think that the technology we have let loose will lead to experiencing selves that are happy and contended rather than depressed or miserable, but I have as yet no way of judging whether they will be or not. Even so, I think this is the more important question to ask, not whether any of them will be me for they will not be me, any more than the self who is finally completing this article was the one who started it all those days ago.

References

Blackmore, S. (1999) *The Meme Machine*, Oxford: Oxford University Press.

Blackmore, S. (2010) Dangerous memes; or what the Pandorans let loose, in Dick, S. & Lupisella, M. (eds.) *Cosmos and Culture: Cultural Evolution in a Cosmic Context*, pp. 297–318, Washington, DC: NASA.

Blackmore, S. (2011) *Zen and the Art of Consciousness*, Oxford: Oneworld Publications.

Kurzweil, R. (2005) *The Singularity is Near: When Humans Transcend Biology*, New York: Viking Press.

Parfit, D. (1984) *Reasons and Persons*, Oxford: Oxford University Press.

Damien Broderick

Terrible Angels
The Singularity and Science Fiction

In the twenty-first century, we are ghosts haunting a futurist dream a century old. Reminded of the brash functionalist lines of 'tomorrow's city', the clean Mondrian Formica kitchen surfaces, the sparkling electric railroads, we find that we accelerated by mistake into Gothic mode: Ground Zero at the heart of the world capital of capital, jittery crack zombies in the skyscrapers' shadows, the broken power plants of Iraq, the desolation of much of Africa, the last Space Shuttle hauled off to a museum, global economies a hairsbreadth from disaster. And yet this dull headache reality still might be just a momentary blip in a curve of change that is even now continuing to remake the future, fashioning it into forms we cannot even begin to imagine until we're smacked behind the ear.

What if (the basic science fictional opening move) there'll be *no* human history, no ensemble of possible histories to be modeled by humanist fiction or futurism? What if the third millennium sees a phase change into some altogether unprecedented state of being, neither Apocalypse nor Parousia but some secular mix of both? What if, as Vernor Vinge proposed, exponentially accelerating science and technology are rushing us into a Singularity (Vinge, 1986; 1993), what I have called the Spike? Technological time will be neither an arrow nor a cycle (in Stephen Jay Gould's phrase), but a series of upwardly accelerating logistical S-curves, each supplanting the one before it as it flattens out. Then there's no pattern of reasoned expectation to be mapped, no knowable Chernobyl or Fukushima Daiichi to deplore in advance. Merely — opacity.

The general idea of exponential accelerating change goes back further than that, though. In *The Spike* (Broderick, 1997/2001), the first book-length consideration of the technological Singularity, I cited an article by G. Harry Stine, published in the sf magazine *Analog* in 1961.[1] Yet Stine's insight had already been canvassed in February, 1952, in another sf magazine, *Galaxy,* by Robert A. Heinlein. His essay 'Where To?' (1952/1980) not only described the path toward Singularity (without calling it that; he dubbed it 'the curve of human achievement'), but even displayed a graph showing four possible trend lines stretching from the upward curve of change between 1900 to 1950, extended conjecturally toward 2000. Trend 1 quickly slowed and went flat-line. Trend 2 also slowed, without quite turning stationary. The other two kept going upward:

> What is the correct way to project this curve into the future? Despite everything, there is a stubborn 'common sense' tendency to project it along dotted line number one.... Even those who don't expect a slowing up at once tend to expect us to reach a point of diminishing returns (dotted line number two).
>
> Very daring minds are willing to predict that we will continue our present rate of progress (dotted line number three — a tangent).
>
> But the proper way to project the curve is dotted line number four — for there is no reason, mathematical, scientific, or historical, to expect that curve to flatten out, or to reach a point of diminishing returns, or simply to go on as a tangent. The correct projection, by all the facts known today, is for the curve to go on up indefinitely with *increasing* steepness.

Sf writers are rarely silenced by this futurological *mise en abîme.* At the very least, tales can be told of the runaway ride up the slope of the Singularity. By reaching into myth and the most expansive vistas of advanced physics, neuroscience, biology and even cosmology, perhaps we can guess at some of the immense prospects that loom beyond the veil. Runaway change can be represented as a singularity because it is a spike on the graph not just of human progress but of human reality in its entirety. The strangest feature of such a graph, taken literally, is that the higher you rise on its curve, the faster it climbs ahead of you. The slope is worse than Sisyphean, because we can't even get to the top and then slide despairingly back to the base.

'As we move closer to this point, it will loom vaster and vaster over human affairs till the notion becomes a commonplace,' Vinge (1993) pointed out. 'Yet when it finally happens it may still be a great surprise and a greater unknown.' Vinge is often denied due priority in his pioneering attack on this problematic, usually (as Professor Chalmers

---

[1]   G. Harry Stine, 'Science Fiction is Too Conservative' (1961); this was followed by G. Harry Stine, 'Science Fiction is Still Too Conservative!' (1985). In 1996, I asked Stine for his current assessment. Without commenting on his wildly optimistic and erroneous earlier projections, he replied gamely: 'Science fiction is STILL too conservative!' (Personal communication).

does, while acknowledging John W. Campbell's visionary sf from the thirties) by citing a briefly described conversation between John von Neumann and Stanislaw Ulam (1958) and I. J. Good's 'intelligence explosion' (1965) leading to cascading self-bootstrapping improvement. As a science fiction writer and futurist, I find it pleasing (and dazzling!) that Heinlein was there, sketching the Spike, some 60 years ago.

Sf's enterprise is both quixotic and impossible. It is — to invoke the inevitable tedious religious comparisons — akin to the futility of a theologian or a physicist attempting to understand 'the Mind of God' (as atheist Stephen Hawking rhetorically and misleadingly dubbed his own scientific efforts). Such utter dislocation of knowledge and feeling alike is the traditional literary figure for the incomprehensibly sublime.

Perhaps a less odious comparison, to both believers and nonbelievers (although regrettably bearing its own kitsch New Age freight) is a Rilkean Angel. Rainer Maria Rilke, in the *Duino Elegies*, makes our blood shudder with his vision of the beings (although he did not know this) that, in all truth, might stand beyond the veil of the Singularity. Pressed against the heart of such a being, its beauty inseparable from the terror it instils in us, we would simply perish, for, in Rilke's great and tolling words, 'Every Angel is terrible' (Rilke, 1989). The Singularity, for Vinge, Michael Swanwick, Iain M. Banks, Charles Stross and other sf pioneers, usually begins with artificial intelligence: rogue, benign, or simply — *there*. Evolution takes such a *long*, agonizing time to ratchet itself up a notch. Break free from that mindless DNA process, via machine self-programming or genetic engineering or both, and everything changes *fast*.

How fast? Vinge, in nonfiction mode, is uncompromising: 'The precipitating event will likely be unexpected — perhaps even to the researchers involved. ("But all our previous models were catatonic! We were just tweaking some parameters...") If networking is widespread enough (into ubiquitous embedded systems), it may seem as if our artefacts as a whole had suddenly wakened' (Vinge, 1993). After that, after the first machines have awoken — or the global Internet, as occurs in Robert Sawyer's *WWW* trilogy (Sawyer, 2009–11)? Nobody knows. 'For all my rampant technological optimism,' Vinge himself states, 'sometimes I think I'd be more comfortable if I were regarding these transcendental events from one thousand years remove... instead of twenty' (Vinge, 1993). While it's a millennium too late for that, in his influential novel *Marooned in Real Time* (1986) Vinge placed his characters at a 50 million year remove — in the remote future, after a

Singularity had passed across the world, leaving only traces of vast, unintelligible engineering works and decaying cities.

The Singularity will be a sort of immanent transcendence, an accelerating dash into incomprehensible glory — or ugliness, always a possibility: feral nanotechnological gray goo, turning everything on the planet into sludge, or viral green goo, or a thousand varieties of unbalanced superminds let loose in the playground of the solar system. Here and there in Vinge's novel are hints of what might have occurred when humanity vanished clean away while the reader's back was turned. 'Mankind simply graduated, and you and I and the rest missed graduation night,' one character tells another. 'Just talking about superhuman intelligence gets us into something like religion...' (*Marooned*, p. 111). One need not be earnest about this prospect. Iain M. Banks's Culture novels are peopled by posthumans rewired to 'gland' hormones by choice, whose postscarcity anarcho-communist polity is mostly located on starships and Orbitals run by AI Minds who seem to keep their organic partners around as pets.[2] Colossal artifacts with facetious names like the *Very Little Gravitas Indeed* roar across the galaxy, while enhanced humans and snide machines frolic within their protective fields.

How fast could such immense changes really happen? A marooned Vinge character muses doubtfully that even in a swiftly changing world, 'there had been limits on how fast the marketplace could absorb new developments... what about the installed base of older equipment? What about compatibility with devices not yet upgraded? How could the world of real products be turned inside out in such a short time?' (*Marooned*, p. 172). Vinge seems here to start interrogating his own cool idea only to back away hurriedly. It is a piece of narrative flimflam in the guise of rhetorical questioning.

Yet some of the steps that make this headlong alteration thinkable, if not altogether feasible, are sketched in Vinge's seminal novel. 'High-tech' people from close to the Singularity wear headbands that augment their native abilities, computer patches to the raw stuff of evolved brains with their limited memories and even more limited attention windows. More up-to-date scenarios by Greg Bear, Greg Egan, Raphael Carter, Charles Stross and others expect such chips to be surgically implanted deep in the brain, or perhaps grown there using engineered cells or nanoconstuctors, or just bypassed when humans upload entirely into computer substrates. In a way, it's evidence of Vinge's own argument that only a quarter century after his

---

[2] See especially the feuding Minds in Iain M. Banks, *Excession* (1996).

proposition was announced, we already find many of his once-wild projections rather tame and unadventurous.

***

Today, even the smartest of us can hardly sustain conceptual grasp on more than seven different items at once, and five is the usual mental handful. So we live in a cognitive universe dimly glimpsed through the narrowest of cracks, and the width of that aperture is set by our inherited neural hardware. Boost it, link our augmented minds together, and who knows what wonders of awareness might burst open into consciousness? 'Humankind and its machines became something better,' speculates the most advanced of Vinge's pre-Transcension high-techs, 'something... unknowable' (*Marooned*, p. 176).

Yet that character's own experience at the opening of the twenty-third century is almost incomprehensible to us a mere two hundred years earlier. With his seven colleagues, he was engaged in mining the sun for antimatter, 'distilling one hundred thousand *tonnes* of matter and antimatter every *second*. That was enough to dim the sun, though we arranged things so the effect wasn't perceptible from the ecliptic' — the orbital region around the sun that contains the earth and other planets and most asteroids. Working so far from home, he and his companions were brutally severed from the real action, 'hundreds of light-seconds away' (*ibid.*, p. 173, italics above in original).

What might such a world be like if it truly existed? Two hundred years ago, Europeans explored and conquered large parts of a single world many months distant from their political and mercantile masters at home. Today, by contrast, nobody with access to a cellphone is more than fractions of a minute from anywhere else on the globe.[3] Computer networks, swapping information and financial transactions, blur into a haze of virtual instantaneity. Imagine the rapidity of a world where people exiled halfway across the solar system are unalterably amputated from the current action, because the speed of light is an intractable barrier to faster communication.

On Earth, large corporations with superior computers merge their staff into linkages of thousands. Is this a horrible prospect of soul-death, extermination of the self? One might expect such an

---

[3] 'Half of Africa's one billion population has a mobile phone — and not just for talking. The power of telephony is forging a new enterprise culture, from banking to agriculture to healthcare' (http://www.guardian.co.uk/technology/2011/jul/24/mobile-phones-africa-microfinance-farming).

interpretation from a libertarian like Vinge, but in fact he suggests otherwise: 'There was power and knowledge and joy in those companies...' As the Vingean Singularity approaches, the mind-to-mind linkages and augmentations become extreme, a form of group mind: 'By the beginning of the twentythird, there were three *billion* people in the Earth/Luna volume. Three billion people and corresponding processing power — all less than three light-seconds apart.' Inevitably, transhuman plans turn to cosmic engineering, the creation of black hole star-gates (as occurs in Ken MacLeod's *The Cassini Division* and Stephen Baxter's *Timelike Infinity*), the implosion of entire stars. Might this explain the Extinction? 'We'd been "uppity cockroaches" — and the real owners finally stepped on us...' (*Marooned,* p. 177).

Vinge's subsequent deep space novels, *A Fire Upon the Deep* (1991) and its belated prequels *A Deepness in the Sky* (1999) and *Children of the Sky* (2011), ornate, multileveled machines of tales within tales, offered a series of complex analogies for the states of mind and body and existence that might prevail beyond any immediate Singularity. For story-telling purposes, these novels propose a galaxy partitioned like some Dantean cosmology, from a bleak Inferno of the Unthinking Depths, where matter dominates and light-speed is the final limit, through a Purgatorio of the Slow Zone flaring outward from the galaxy's spinning wheel, and farther yet into the Beyond, where minds are vast and instantaneous: Powers, 'gods' in the ancient sense.

That's metaphor, but certainly some form of tiered analysis is needed if we are to build models of enhanced intelligence. It is not that human mental life is different in kind from the progressively more limited kinds found in dogs and cats and parrots and goldfish and slugs and sunflowers and bacteria. Of course, many people argue that our minds *are* exactly differentiated from those of 'lesser creatures' in some mystical, unapproachable way. Even supposing that it turns out to be true — that people have an extra gadget — let's call them *souls* — that provide our self-awareness and radical creativity — that conclusion would need to be attained by long, scrupulous investigation. It would be, in a sense, a negative hypothesis: something we would be forced to entertain solely because of the failure of more rational models compatible with everything else we know about the physical universe. (And for all our ignorance, we now do know a *colossal* amount, most of it tied together into patterns not easily edited or expanded without shredding the lot.)

Posit, then, that human minds are not *mysterious things* but are rather *very complicated processes* going on inside living brains connected by senses to the outside world. Is it reasonable to guess that doubling the power of a laptop computer, and doubling it again, and redoubling for another fifteen or thirty years, will automatically produce a smart machine, then an intelligent machine, and finally — *a hyperintelligent machine?*

Vinge poses this rather neatly: 'Imagine running a dog mind at very high speed. Would a thousand years of doggy living add up to any human insight? (Now if the dog mind were cleverly rewired and *then* run at high speed, we might see something different...' (Vinge, 1993). No computer is currently anywhere near the doggy level of cleverness, of course, but some of them approach insect intelligence. Might running an insect brain very, very fast turn it into someone you'd like to discuss politics or art with?

It seems obvious that the answer is no, but what if that insect brain joins a swarm of its fellows, and they learn (or are taught) the trick of specialization, so that one bunch of insect brains swaps information about how stuff *looks* today, from a number of strategic angles, and another lot focuses on *sound vibrations*, while other bunches of ant-brains store these impression and disgorge them on command from yet other groups. This is the principle of the hive. In many ways the hive is much cannier than any of the dumb modules with legs and wings that comprise it. It also looks something like a complex brain — a *human* brain, in fact. Each of our neurons is a little like an insect (simpler, in fact, and much smaller), but it has better lines of communication than any single ant, and neural columns are like hives. Take a hundred thousand million neurons and let them cross-wire to each other in some kind of self-organizing, autocatalytic hierarchy, and you do indeed get — *us.*

Which does not answer the original question about speeding up a doggy brain. No, even if you added in a big pack of extra memory, a dog's brain lacks the architecture to think like a person. What about an immortal and massively augmented dog, genetically engineered to grow as many cortical neurons as a human, but arrayed as they would otherwise be in a normal dog? In his near-Singularity novel *Holy Fire* (1996), Bruce Sterling has one of these wonderful beasts, a poignantly articulate talkshow host which/who, due to neurological deficits, can't read. I suppose it is remotely possible that an enhanced dog might, perhaps with great inner agony, teach itself the tricks of reflective self-consciousness. It would need access to a brainy, communicative culture, and right now we are the only one available. So a true hyperdog would require eyes capable of reading, and dexterous paws, and a jaw and larynx rewired for speech — a big job. Almost certainly no such wonder could emerge from a single, extreme mutation that just pumped up the puppy's brain capacity in utero (the conceit in Olaf Stapledon's superb *Sirius* [1944].)

It is much easier to reallocate machine memory, to rewrite code and try it out and discard it when it fails, and try again, and keep trying until you get it right, step by evolved step. That is presumably true even if the machine is boot-strapping its own abilities, driven away from stolid stable inner states by random changes somewhere in its operating system, in the software that makes it a process rather than just a lump of expensive silicon. This is the evolutionary model of teaching a machine to be a person. In a sense, we know that this has to work, because on a geological timescale it is what produced us. Mutation, contest and cooperation, differential survival of genetic patterns according to the success of the bodies they built: natural selection, in a word.

That is the likely path of AI or artificial intelligence. Vinge mentions a quite different method of attaining advanced cognitive abilities: a switch in emphasis that is captured deftly by a switch in the name — IA, or *intelligence amplification*. Now it's the human brain that is being boosted, or linked to others of its kind (as the near-Singularity corporations in his novel were composed of group minds). Again, this is eerily familiar, precisely because it is a definition of a society or a culture.

Already our brains store much of their knowledge outside the skull, in books and film and digital, searchable cloud archives, in the huge hard structures of cities and aqueducts and farms, in the facts and opinions we share by talking to each other in person and electronically. Vinge sees the path from this existing state of affairs to enhanced intelligence as rather easier than the AI route, because we have done it a number of times without quite realizing it. Currently, access to the Internet is a prodigious augmentation of a researcher's capacity: In effect we now think faster, put information together more swiftly, send our results out without delay, gain the reciprocal benefit from others cruising the net. Once improved methods are devised for getting information out of a database and shaping knowledge without the need for writing or keyboarding or even talking — something like the cyberpunk dream of 'jacking in to cyberspace' — it will be a whole new kind of life. One small step for the individual cybersurfer, one mighty swarming leap toward the Singularity.

The key point about most Singularity scenarios is that they proceed step by step, advances of equal magnitude each taking only about half as long as the one before, so the most radical changes happen at the end of any given period. This exponential progression was sketched in Marc Stiegler's *Analog* story *The Gentle Seduction* (1989), where a man and woman from today experience the ramp up to a Singularity through rejuvenation, headband connectivity, implants, terraforming of Mars, nanoscale transportation to other stars, vastening of consciousness, contact with aliens; and yet:

> She remembered who she had been when she was 25; she remembered who she had been when she was just 10. Amusingly, she also remembered how, at 25, she had erroneously remembered her thoughts of age 10. The changes she had gone through during those 15 years of dusty antiquity were vast, perhaps as vast as all the changes she had accepted in the millennia thereafter. Certainly, considering the scales involved, she had as much right today to think of herself as the same person as she had had then. Expanded communion would not destroy her; she was her own bubble no matter how frothy the ocean might become.

***

Vinge, then, was not alone among imaginative writers in trying his hand at portraying what is strictly inconceivable, a world on the far side of the Singularity. In the 1940s and '50s, Isaac Asimov almost single-handedly reinvented the idea of artificial cognition in the form of 'positronic robots'. The platinum-iridium brains of these machines were imprinted with Three Laws that forbade them, on pain of incapacitation, from injuring a human or allowing harm to come to one of us, while requiring them to obey humans and protect themselves.[4] Most of Asimov's tales were clever attempts to surmount the barriers presented by these algorithmic 'laws' while showing that finally they were impregnable. But Asimov's own logic took him into the territory of a self-improving machine mind, and indeed several late stories in his collection *I, Robot* reveal the evolution of the Machines, AIs of immense power and insight, that first take over the running of the world and then remove themselves, knowing that their continued benevolent presence would blight our species (and, of course, block any future robot stories). Asimov's own favourite robot tale is 'The Last Question' (1956), in which people throughout future history ask increasingly ubiquitous and puissant AIs if entropy can be reversed, or must the cosmos go inevitably into heat death? Trillions of years hence the final AI, known as AC, deduces the method, and says, 'LET

---

[4]   See http://en.wikipedia.org/wiki/Three_Laws_of_Robotics

THERE BE LIGHT!' But note: this heroically optimistic parable is precisely *not* about a singularity — it's a very far future Omega Point emergent god story.

Perhaps the most notable sf stories at the boundaries of the Singularity were by the late Theodore Sturgeon, a poet, dreamer and story-teller currently somewhat in eclipse. In the 1950s, his short fiction returned again and again to territory we now recognize, while drawing much of its power from his immersion in fallible characters carefully observed. Often — strangely enough, for an American writing at the height of McCarthyite Red Menace hysteria — his wistful parables told of maimed, lonely, isolated men and women finding one another in the company of a greater unity, a colony mind, a hive, *homo gestalt*. Yet Sturgeon did not abdicate from the precise, piercing joys of quirky individuality, or the prickly demands of freedom. His fiction may yet prove to come closest to one tendency in the posthuman condition.

In 'The Skills of Xanadu' (1956), a primitive world is found by a hightech militarist. The naked savages wear a curious belt but little else, and inexplicably their disorganized play fetches them everything they require in life. This harmony, maddening to the rigid newcomer, proves to be a by-product of their single crucial item of technology: the belt, which links all brains into a shared knowledge base. Faced with a problem, you simply *understand* the answer. The skill of the best practitioner of any craft is instantly available. When the warmon-ger leaves, taking a belt, he hopes to use it to transform his rigid soci-ety into a perfect dictatorship. Instead, the belts convert his world into a harmonious network of free individuals, each of them experts in the skills of 'logic and love; sympathy, empathy, forbearance' (Sturgeon, 1979, p. 259). Years later, Sturgeon wrote wistfully: 'I yearn to live on Xanadu, and wear their garment, and join with them in their marvel-ous life-style' (*ibid.*, p. 228).

***

William Gibson, coiner and explorer of cyberspace in novels such as *Neuromancer* (1984), is only the best known of the next generation of writers working at the lip of the Singularity, along with his sometime collaborator Bruce Sterling. Michael Swanwick, a brilliant stylist, has done so several times, most notably in the novels *Vacuum Flowers* (1988) and *Stations of the Tide* (1991). In the former, revived suicide Rebel Elizabeth Mudlark proves to be a kind of antidote to a post-modern, poststructuralist affliction: the dispersion of the self in a cul-ture where personalities are optional plug-ins, bought off the shelf as we buy computer apps. Rebel's capital importance is her possession of — indeed, her constitution in — the rare trick of assuring her own hyperstable self-presence. All around her, post-Singularity space habitats are gladly giving in to centripetal attractions. Earth's home population has already merged into a multibillion-headed colony organism, a sort of slime mold or coral comprised of former individuals. Indeed, this new shared-consciousness entity calls itself the Comprise.

Rebel's partial solution to this encroaching monolith is to disperse the microworlds as seed pods among the stars — a scattering that enacts on a literal level the dispersion of selfhood that the late twentieth century agonizingly diagnosed as the epoch's characteristic feature. When Swanwick names the armless girl who speaks for the colony mind — 'the focus.... of perhaps a billion Comprise, as massive a point source of attention as Earth ever needed to assemble '— he calls her 'Snow' (Swanwick, 1988, p. 222). The name reflects what every child learns at an early age: that every single snowflake is identically simple, with its sixfold symmetry, yet unique in the details of its shape. Swanwick's 1987 naming recalls, in turn, an earlier fictional group mind, in a 1959 novel, *Wolfbane*, by Frederik Pohl and Cyril Kornbluth: the Snowflake. Apparently these apprehensions have been curdling through the discursive unconscious of Western culture. It was Vinge's special genius to understand that they might become *literally* true.

And if they do? Swanwick is not reassuring. Starting from primitive AIs and rudimentary brain-computer interfaces, he posits a seed of 'thirty-two outlaw programmers' about whom the Comprise, a massively parallel human-machine hybrid, starts to crystallize. Born in glory, power and understanding, the new entity reaches to others in the net willing to join it, rewriting its own structure as it goes, deepening its algorithms. 'Within three minutes everyone on the net was ours. We controlled everything that touched upon the net — governments, military forces from the strategic level down.... With a fraction of our attention, we designed the transceivers, retooled the factories to make them, and reorganized the hospitals to perform the implants.... We ate the Earth' (Swanwick, *Vacuum Flowers*, p. 224). It is a totalitarian longing and one that today's culture fears desperately even as we yearn for the balm of that sweet joining. 'We reached out and out,' the Comprise tells Rebel, 'expanding toward Godhood.' It adds at once, candidly, 'We had ambition, and ascended into Hell' (*ibid.*, p. 224).

If this is a disturbing prophecy of a hungry post-Singularity consciousness, Swanwick was more terrifying still in *Stations of the Tide*, where nothing holds still for long (see Swanwick, 1991, p. 146).

People split their minds into agents able to impersonate them, act on their behalf, report back and extinguish themselves. They move their point of awareness into surrogate bodies, the very reverse of Rebel's armored solidity of selfhood. When an agent of colonymind Earth is met, it is something out of Milton and Swift, an authentically monstrous manifestation in virtual reality. This vast, sweat-stinking, musky monster is a figure familiar from psychoanalysis: the archaic Mother, a sort of feral female phallic force, more mythic than misogynistic in Swanwick's making. And like that clammy image from post-Freudian analysis, complete with vagina dentata, it invites Swanwick's bureaucrat into its mouth. In the overwhelming presence of an Earth utterly overborne by technology out of control, he asks the agent: 'What do you want from us?'

> In that same lifeless tone she replied, 'What does any mother want from her daughters? I want to help you. I want to give you advice. I want to reshape you into my own image. I want to lead your lives, eat your flesh, grind your corpses, and gnaw the bones' (Swanwick, *Stations of the Tide*, p. 147).

Here and now, we can readily fall in with Swanwick's forebodings, considering the prospects of a world that is perhaps only a generation away from birthing nanotechnology and artificial intelligence and machine interfaces and human intelligence amplification. It need not be that way, of course. One might wonder if it is even remotely plausible. Grimm's fairy tales, after all, caught the rural voices of the nineteenth century and earlier, perhaps not altogether salient to a world remade by science. Yet we carry our history with us, tucked away inside our narratives and nightmares, ready to be snatched up in the slipstream of science.

Will we, preserved by the wiles of that science into the Singularity (or, if not us, then our transhuman children or great-grandchildren) any longer be 'human'? Perhaps not, or not for long. Maybe we will live almost infinitely accelerated lives within a virtual computer in a grain of sand at the edge of the world's last drained sea. Maybe we will be quantum states of a cosmically dispersed, quantally-linked hypermind. Maybe we will be quite literally *gods*, inflating fresh universes out of the quantum foam and placing our impress upon everything that forms there.

Or maybe we'll all stay at home and watch the ultimate television channel, forever — my favorite explanation for the Fermi Paradox. This is the Great Silence in the skies, the mysterious absence of any radio or optical traffic between worlds out there in the galaxy and beyond, or of any cosmic rearchitecting of the stars. If the technology and culture of every civilization Spikes within a century or two of their discovering radio and space flight, they might all be tucked away into the folds of local spacetime like the hidden, rolled-up dimensions of string theory. They might even be living *there*, colonizing those intricate, implicate spaces.

***

Must the Spike, the Singularity, the amplification of human intelligence, or its replacement by hypermachines, be seen as either a farce or a horror story? No doubt this is the way it will be portrayed by Hollywood (Skynet Terminators!), embodying the usual false binary opposition. Mind versus Passion! Love throttled by Rationality! In *Charly* (1968), based on Daniel Keyes's heartbreaking story 'Flowers for Algernon', surgically enhanced intelligence lifts a sweet moron to the heights of sensitive genius before smashing into Nature's Revenge, dashing the tragic hero back into stupidity and early death. In the fifties' cult movie *Forbidden Planet* (1956) a device for increasing IQ and linking minds into a planetary gestalt frees the Monster from the Id, those vile forces boiling destructively in the unconscious (in pop Freudianism), and kills an entire world. Star Trek's Mr. Spock and Lt. Commander Data did a little better, but paid for their cleverness in the inevitable coin: They were forbidden human depth and feeling.

This is a very strange cliché, its supposed truth is denied by the slightest contact with real, passionate, sometimes frenzied human scientists and technologists. But perhaps this is what *really* frightens us: when strong feelings, devotion and hatred and prejudice, are joined powerfully with effective strategies for influencing, even dominating, the world. Then we get holocausts, the continuing risk of nuclear war that could serve nobody's purposes, mad cults and faiths with real weapons.

Is the Singularity likely to bring just such unholy conjunctions to a new pitch of fearful strength? We have evolved in hundreds of thousands of years of sluggish tribal life, dealing with each other face to face, vulnerable to the fists and scorn of others, sensitive to their smiles and touch. What happens when we allow ourselves to deal with others wholly through a monitor's or iPhone's window? Anyone who has watched, or taken part in, a 'flame war' on the Internet knows how these exchanges escalate into brutal rudeness and sarcasm, the kind of thing that leads on the highway to 'road rage' murders. It is a stroke of luck that we cannot yet easily kill our foes through the screen. But

perhaps an enhanced human, locked into symbiosis with potent programs running on computers all around the world, would indeed learn
how to smite enemies in world-shaking tantrums.

With luck, this interregnum would be brief. By and large, people are
not wild dogs. It has been possible for decades to make lethal bombs
from common agricultural substances, or gun down children in a playground, yet when this is done by a few crazies the rest of us are genuinely grief-stricken, if only for a brief time before our own concerns
drift back to the surface. The transition from intelligence amplification (IA) to the Singularity, when enormous numbers of changes happen with immense swiftness, will probably not occur as an all-at-once
crystalization — or at least we must hope not.

Even so, in John Barnes's *Mother of Storms* (1994), a Singularity
occurs when Louie Tynan (one of the first humans on Mars, now in
Earth orbit) and Carla Tynan (on her submersible yacht) quite plausibly get absorbed into the net in 2028, after a global catastrophe, then
spread between orbit and the Moon and eventually into a comet from
the Kuiper belt. Such transition into dispersed superintelligence might
bring with it the solutions to most of our traditional woes: hunger,
thirst, nakedness to the elements — scarcity, in a word. The best
means to those solutions cannot yet even be glimpsed, but the likely
paths are obvious: nanotechnology, direct AI-interfaces, genome control and repair, massively extended lifespan, and so forth.

This might seem a utopian outcome — but really, it can be neither
utopic (positive) nor dystopic (negative). From what standpoint is
'positive' to be assessed? *Mother of Storms* seems highly positive, in
that the uploaded and enhanced humans become a sort of benign solution to all our problems. But seeking a future positive from *our* viewpoint is arguably like asking a trilobite or a Homo habilis to regard the
prospect of the emergence of Homo sapiens as a happy ending. 'Yes,
we'll all die out, but another species will do pretty well.' Charles
Stross's *Accelerando* (which we'll examine shortly) is undeniably
exciting, taking us in explosive leaps from the near future to a madly
different post-Singularity medium-term future full of complexities we
can scarcely understand — and I find that positive, compared to doom
and gloom of Cormac McCarthy's *The Road* or Margaret Atwood's
The Handmaid's Tale or Kazuo Ishiguo's *Never Let Me Go*, which is
how most 'literary' fiction still wants to portray the future.

***

'Suppose we could tailor the Singularity,' Vinge proposes. 'Suppose we could attain our most extravagant hopes. What then would we ask for: that humans themselves would become their own successors, that whatever injustice occurs would be tempered by our knowledge of our roots' (Vinge, 1993).

In this authentic golden age, immortality would no longer be a fantasy of consolation in the imagined world beyond death, but a literal indefinite extension of life in a utopian world without want.[5] A caution is necessary. We are aware of our limitations, the weaknesses of brains and bodies built by blind Darwinian selection. Each of us expires a few decades at most after a fertile span of some thirty or forty years. We are put together as disposable gene-carriers. Our beautiful minds rot with our fallible, corruptible brains and are gone forever. What would happen with such temporary mechanisms if they were repaired again and again, held safe from the corrosion of time? Nightmare, perhaps. 'A mind that stays at the same capacity cannot live forever,' Vinge notes; 'after a few thousand years it would look more like a repeating tape loop than a person' (Vinge, 1993). So extended life must also be enhanced life. For immortality we would need to be smarter and with better emotional control, even if we didn't already need amplified gifts to attain immortality. (In fact, we might not, as medical consequences of the Human Genome Project are gradually pointing the way to endless cellular repair.)

More to the point, what is the 'we,' the 'I,' that is going to survive into the post-Singularity utopia? If poststructural theory tells us that the self is always-already a construct, an illusion, that improbable perspective will be even harder to dispute in a world where we can send out 'partials' or 'agents' from ourselves into the global net. Already, our brains are composed of dedicated, somewhat partitioned modules resembling the 'faculties' of an older philosophy. Of course these specialized components tend to work together, creating (most of the time) a sense of unified consciousness. Once we learn to split off fractions of our selves — or, rather, duplicate and amplify and elaborate those fractions, as in the Swanwick novels — we will no longer be strictly human. Vinge notes: 'These are essential features of strong superhumanity and the Singularity. Thinking about them, one begins to feel how essentially strange and different the Post-Human era will be —

---

[5] An elaborate Golden Age of this kind is projected in John C. Wright's trilogy of the same name (2002–03).

no matter how cleverly and benignly it is brought to be' (Vinge, 1993, italics in original).

An audacious Marxist (or perhaps post-Marxist) version of a world going through, recovering from and absorbing a Singularity has been sketched by Ken MacLeod, like Iain M. Banks a Scots writer with affinities to the anarchist left. These novels began with *The Star Fraction* (1995), which nearly won the Arthur C. Clarke award, a kind of frenzied Trotskyist vision of a world convulsed by political fractions and manipulated by an emergent AI weapons system. Subsequent novels explore a variety of future cultures that might emerge from this abrupt discontinuity. *The Sky Road* (1999) actually revises the future history of the other three, ensuring that a menacing Singularity does *not* occur, or at least is delayed. *The Stone Canal* (1996) and its sequel *The Cassini Division* (1998) portray with high, sly humor and impressive technical and political insight two contesting utopias, one a Banksian anarcho-socialist Union on a damaged Earth, the other an anarcho-capitalist libertarian world on the far side of a wormhole, 10,000 light years from Earth and in its future.

Both alternative regimes (or antiregimes) are thus at arm's-length yet able to communicate and even visit. Jupiter, meanwhile, has been redesigned by posthuman entities, the 'fast folk,' that blend organic roots and AI enhancements. Are they dreadfully dangerous — trapped in psychotic virtual realities, but likely to emerge at any moment — or should they be welcomed as our successors? MacLeod is superbly sardonic in setting all these groups at each other's throats and minds. His vision of a Singularity is distinctly unnerving, and his several contrasted utopias are no less troubling even as they seduce us in succession.

The grace to permit remnants of the old, unreconstructed humanity to live in peace may be the best we can hope for from augmented or borganized posthumans. Well, one happier option is conceivable, for those open to the charms of self-deceptions: Our successors might choose to disguise themselves as our servants. Vinge hints at 'benign treatment (perhaps even giving the stay-behinds the appearance of being masters of godlike slaves)' (Vinge, 1993). Unaltered humans would be the metaphysical equivalents of the Amish, serene agricultural throwbacks in our mechanized, electronic world. That appears to be a difficult way of life to sustain, even to negotiate, and its future version would surely be harder still to endure without soul-wrenching episodes of blatant bad faith, lack of resolution when times grew tough and one's body aged while everyone else remained healthy and young, backsliding by the next generation.

When the first Singularity consciousness awakens and looks around itself, it might not be a safe time to be alive; it might be the end of all histories that concern us. Or the AI might simply shut itself down, or vanish into 'uncompromising silence', as in Stanislaw Lem's superb parable 'Golem XIV' (1981/1985).

So what should be our attitude to these posthumans who seem likely to replace us, perhaps within a century or even much sooner? Terror? Awed worship? Serene acceptance? The hope that we might join them in a *gestalt* superhuman Comprise?

Many alternative paths lead from where we now stand into the transhuman and the posthuman futures of the Singularity. It is not at all clear that there's any role at all for us mere humans on the far side of the Singularity's looming wall, unless we are drastically enhanced, or perhaps uploaded into virtual spaces where our electronic or photonic Doppelgangers can run a million times faster. It will be uncomfortably interesting to learn how sf writers further into the twenty-first century meet this extreme and continuing narrative challenge — and how everyone else deals with the history breaching reality following hot on the heels of the fiction. For the moment, let us look a little more closely at two exemplary instances of how sf can approach the blinding dazzle of the Singularity (while bearing always in mind that by the nature of the medium, sf tends to emphasize the risks and dangerous consequences of radical technological and social change, because that's what most readers find *exciting*...)

***

Like Greg Egan a decade earlier, British writer Charles Stross seemed to come from nowhere and leap immediately to the top rank of science fiction and fantasy writers. His work has proved enormously popular, and justly so. With *Accelerando* (2005), Stross was sealed as the new Poet Laureate of the Vingean technological singularity, a topic he had already approached in his first published novel, *Singularity Sky* (2003). The *Accelerando* project's five years of development (it is tempting to apply this sort of corporate language to Stross's dense techno-speak art-artifact) yielded an early twenty-first-century counterpart to John Brunner's compressed future shock 1969 Hugo-winner, *Stand on Zanzibar,* complete with rich idiomatic sidebars or side loads of Baedeker guidance to the non-native.

> The rogue corporation rears up slightly and bunches into a fatter lump; its skin blushes red in patches. 'Must think about this. Is your

mandatory accounting time-cycle fixed or variable term? Are self-owned corporate entities able to enter contracts?'

That's funny, and fun, as well as knowing. Some readers complained of infodumping of the most blatant kind, yet this device seems unavoidable when a torrential cascade of novelty is the very topic of a work of art. Approached with an appreciative generosity of response, these dollops of data are tight, compressed, inventive, brilliantly illuminated gems, or perhaps genomes (or memomes) that will unfold, in a prepared mind, into wondrous ecologies of image and idea. The changes implied by headlong acceleration are by definition both too immense and too subtle to be portrayed or perhaps even imagined. Stross has the audacity and, luckily for us, the imagination to come close to pulling it off.

Manfred Macx is a venture altruist ('Manfred's on the road again, making strangers rich') — Stross is not afraid to have us smile even as he jolts our preconceptions — encrusted with computer wearables and the latest wifi connectivity, affianced until recently to Pamela, a dominatrix headhunter for the IRS who tries to persuade global megacorps to cough up the tax they owe. Their venomous bond is manipulated sardonically by their robot cat, Aineko, which is being hacked and upgraded on the sly by Pamela. This nicely observed android animal — 'It sits on the hand woven rug in the middle of the hardwood floor with one hind leg sticking out at an odd angle, as if it's forgotten about it' — might be the secret narrator of the novel. Its augmentation and expansion toward the condition of a low-level demiurge mirrors the transitions of humankind and our posthuman Vile Offspring.

In a bondage scene of hilarious erotic vividness, Pamela gets herself pregnant with their daughter Amber, who will carry much of the long arc of the story to the Singularity and beyond, as human minds export themselves increasingly outside the skull into machine substrate exocortices. In turn, Amber's son Sirhan (well, son of one of her many instantiations) takes the generational saga to the destructive *Childhood's End*-style transcension of the solar system into a Matrioshka Brain (energy-hungry Dyson shells of computronium hosting untold trillions of superminds),[6] the return from death of an extremely augmented Manfred, and a blind plunge beyond the provincial Milky Way to a realm where a galactic superintelligence seems to be mounting a 'timing channel attack on the virtual machine that's running the universe, perhaps, or an embedded simulation of an entirely different universe'.

---

[6] A consummation closing my own Singularity novel *Transcension* (2002).

This is how high bandwidth science fiction works. As the decades pass, as the rate of change accelerates, Stross's characters *become* Googlized. And even with their inbuilt channels of information and communication, they are lost like us in the hydrant gush of available knowledge. All around them, intellectual tools are mutating into predatory lifeforms. Feral tax auditing software roams the solar system, entire economic systems convulse in ecological firestorms of contest. And then there are the aliens... which, of course, are just as likely to be autonomous spam attacks as anything we would recognize as people.

Accelerando is a *Fantasia*-bright cavalcade of borrowed and adapted landscapes — the Atomium globe from the 1950 World's Fair, the deck of the *Titanic* emulated in a virtualization stack, a phony debased Islamic heaven — transplanted to Saturn's icy atmosphere or a virtual reality world inside a soda can-sized starwhisp interstellar spacecraft or an alien router network. Does it work? *Can* it work? It is an impressive attempt upon the impossible. For all its Catherine-wheel sparkle and intellectual bravura, there is evidence that the impossible must remain always out of reach.

This very recognition framed the last major work of US writer Poul Anderson, *Genesis* (2000), which confronts a basic and almost insurmountable consequence of a post-Singularity future. Denying the Fermi Paradox, Anderson posits that intelligent consciousness, once evolved, must proliferate on a galactic scale, mutating and extending its own capacities, perhaps replacing its very substrates. It might relocate itself, for example, from limited organic bodies to very much more adaptable synthetic forms. The Science Fiction Writers of America's 1997 Grand Master, Poul Anderson (1926–2001) was familiar with extrapolations along these lines by roboticist Hans Moravec (1988; 1998), and built them gracefully into his own saga of a galaxy a billion or more years farther off into deep time.

Gaia, the vast, immanent AI custodian and consciousness of the world, finds ancestral Earth threatened by a swelling, terminal Sun. Rather disturbingly, Gaia wishes to allow the world to perish in final flame rather than disrupt the Sun's 'natural' astrophysical trajectory. Other mighty Minds throughout the galaxy, and to the 'shores of the Andromeda,' find this plan perverse. One such godlike node, Alpha, hives off a sub-mind (still Olympian by our standards), and sends this Wayfarer to Earth to investigate and intervene. A still more diminished aspect or agent of this fragment is a reconstruction of the early upload engineer Christian Brannock. A merely human-scale genius, he visits the planet as his larger self communes and debates with Gaia. What he finds, inevitably, is baffling yet emotionally moving (in its constrained way), recalling those Norse sagas Anderson loved so well.

And all of this impossibly remote story is told to us as myth, as repeatedly distanced construct. We are informed again and again that what we read is nothing like the vast reality. Of course, this *must* be so, given the premises of ruthlessly projected futurism. 'All is myth and metaphor, beginning with this absurd nomenclature [Alpha, Wayfarer]. Beings like these had no names. They had identities, instantly recognizable by others of their kind. They did not speak together, they did not go through discussion or explanation of any sort, they were not yet "they." But imagine it.'

And we do, for we have been here before. This is the grand proleptic mythology of Olaf Stapledon himself, of Roger Zelazny's 'For a Breath I Tarry' (1966) — in which Machine remakes Man, but then bows before Him (which is absurd and sadly farcical, however much that story was loved in the 1960s).

Anderson eases our entry to allegory via several well-formed episodes from the comparatively near future: a boy's epiphany beneath starry heaven, in our Earth; Christian's empathy with his robotic telefactor extension on Mercury, prelude to his own status as an uploaded and finally multiply-copied personality; English bureaucrat Laurinda Ashcroft who plans the first millennial salvation of Earth from the brute assaults of a heedless cosmos; a small, neat parable of rigid, gorgeous clan rivalries held in check and paralysis, finally, by the emerging Mind of Gaia. These are Anderson's antinomies (and perhaps American science fiction's): the sacred autonomy of the self, the craving for transcendence in something larger; personal responsibility, and its terrible limits in a world linked, defined, by billions of threads.

Returned to Earth, Wayfarer's Brannock and Gaia's Laurinda tarry in *faux*-eighteenth-century civility, driven together and apart by a series of visitations to simulated histories as dense and real and tormented and doomed as the 'real world.' Their own personalities are no less constructed, however rooted in some small early reality, and so the poignancy of their dilemma is the greater. But for us, knowing that we read a fiction, and snatched in a kind of postmodern gesture again and again by Poul Anderson from our comfortable readerly illusion, these figures and their worlds run the risk of all allegory: can we care? It is the great artistic problem for any form of art predicated upon utter disruption and dislocation, of which the Singularity is by its very nature perhaps the most minatory.

Religious art faced it long ago, and clad its transcendent message in parable, majestic song — and quietness, sacralized domesticity, anguish transformed at the graveside. Confronting the Singularity, reaching for these well-honed tools to give himself voice and range, perhaps Anderson succeeded as well as anyone can manage — given that the task is impossible. If he did not truly succeed, as Stross and the rest of us did not, this is hardly a fault. It takes an entire culture to sustain such mythos. Sf has only lately begun to deepen and extend the mythos, and meanwhile the world's culture turns technological runaway into jingles and plastic toys.

It is compelling to watch how the genre, the mode, of sf is responding to this immense perspective, into the pitiless depths where Poul Anderson, not long before his death, made his brave foray. Unfortunately, consumerist culture does not tarry for more than a breath — or in this case, perhaps a few decades — on a new trope. The shock of the new has expired already for the Singularity. To the extent that it had become just one more tedious way to insert blatant magic into adventure stories, this is understandable. But it might be that the genuine ontological-cum-epistemological challenge Vernor Vinge introduced, nearly three and a half decades after Heinlein had glimpsed it, is, after all, insoluble in just the ways it declares itself to be.

In the real world of publishing, attempts to scale or burrow under its wall to the future have been met, mostly, by reverting to frank mock-medieval fantasy, or degenerating into endless puerile zombie and vampire bloodbaths, franchise wars in space, and whimsical post-Goth zeppelin romances and regressive if fun steampunk sagas of a world that has turned its back on technological progress. In the meantime, in the other real world of bioengineering and computer science, Moore's law is still ticking away. Whether it will reach escape velocity and head off up the page to tear through the top of the graph will only be known for sure when we or our flesh and AI descendents are there, beholding endlessly new landscapes of change, no longer hugged terrifyingly by Rilke's Angel but gazing out from its eyes.

References

Anderson, P. (2000) *Genesis*, New York: Tor.
Asimov, I. (1956) The last question, *Science Fiction Quarterly*, (November).
Banks, I.M. (1996) *Excession*, London: Orbit.
Barnes, J. (1994) *Mother of Storms*, London: Millennium.
Broderick, D. (1997/2001) *The Spike*, Sydney: Reed (1997); rev. New York: Tor (2001).
Broderick, D. (2002) *Transcension*, New York: Tor.

Heinlein, R.A. (1952) Where to?, in *Galaxy*, pp. 13–22. An extended version, with updates from 1965 and 1980, is in Heinlein, R.A. *Extended Universe*, New York: Ace Books.

Lem, S. (1981/1985) Golem XIV, in *Imaginary Magnitude* (Polish, 1981), Washington, PA: Harvest Books.

MacLeod, K. (1998) *The Cassini Division*, New York: Tor.

Moravec, H. (1988) *Mind Children*, Cambridge, MA: Harvard University Press.

Moravec, H. (1998) *Robot: Mere Machine to Transcendent Mind*, New York: Oxford University Press.

Rilke, R.M. (1989) *Duino Elegies*, Cohn, S. (trans.), Manchester: Carcanet.

Sawyer, R.J. (2009–11) *WWW: Wake, WWW: Watch, WWW: Wonder*, New York: Tor.

Stiegler, M. (1989) The gentle seduction, *Analog*, (April).

Stine, G.H. (1961) Science fiction is too conservative, *Analog*, (May), pp. 84–99.

Stine, G.H. (1985) Science fiction is still too conservative!, *Analog*, (January), pp. 89–96.

Stross, C. (2005) *Accelerando*, New York: Tor.

Sturgeon, T. (1979) The skills of Xanadu, *The Golden Helix*, New York: Dell.

Swanwick, M. (1988) *Vacuum Flowers*, New York: Ace.

Swanwick, M. (1991) *Stations of the Tide*, p. 146, New York: Ace.

Vinge, V. (1986) *Marooned in Real Time*, London: Pan.

Vinge, V. (1993) The coming technological singularity, *Proceedings of the VISION-21 Symposium*, sponsored by NASA Lewis Research Center and the Ohio Aerospace Institute, 30–31 March. Also, *Whole Earth Review*, [Online], http://mindstalk.net/vinge/vinge-sing.html

Wright, J.C. (2002–03) *Golden Age Trilogy*, New York: Tor.

Zelazny, R. (1966) For a Breath I Tarry, *New Worlds*, London: Roberts & Vinter.

Barry Dainton

On Singularities and Simulations

If we arrive at a stage where artificial intelligences (or AIs) that we have created can design AIs that are more powerful than themselves, and each new generation of AI rapidly creates still more powerful AIs, then the 'intelligence explosion' — or *singularity* — foreseen by Good, Vinge and others could easily become a reality. Since the arrival of superintelligent machines would be a momentous, world-changing occurrence, we would be wise to consider how best to deal with this eventuality should it occur; we should also attempt to ascertain whether the singularity is as imminent as some of its propo-nents maintain. David Chalmers' 'The Singularity: A Philosophical Analysis' contains much that is valuable on both fronts. With regard to the key issue of whether a singularity is possible at all, I think Chalmers is right in saying that it is certainly not out of the question. As for how to minimize the dangers posed by an emergent superintelligence, the measures Chalmers proposes — implanting the right values, isolating the first super-intelligent systems in virtual universes — look to be promising avenues.

My focus in what follows will be on some of the consequences of a computer-based intelligence explosion, assuming we can survive it. The combination of superintelligence and massive power will make it possible for computers to create and sustain virtual environments of a size and complexity that is way beyond anything we are currently capable of devising. Will it be possible — or desirable — to 'upload' ourselves into these virtual worlds? Chalmers has interesting things to say on this issue; I will be suggesting a slightly different take on it. The existence of these virtual worlds leads to another question: might it not be possible — or even probable — that we ourselves are among their (virtual) inhabitants? This issue has been discussed fairly widely in recent years. I will be proposing some relevant (and hopefully, useful) distinctions and offering some reflections on it. But first, by way of a preliminary, a brief excursus into science fiction proper.

1. New Visions of Heaven

In his 1997 novel *Diaspora*, the Australian science fiction author (and computer programmer) Greg Egan describes a future in which human-kind has split into three groupings of very different kinds of being, two of which are decidedly 'post-human':

> *fleshers* These are humans of the flesh-and-blood variety. Some fleshers are biologically very similar to ordinary 21st century people, others have modified their genes so as to have more-than-ordi-nary-human attributes (e.g. greater resistance to disease, longer life-spans, higher intelligence, the ability to breath underwater, etc.); some — the 'dream apes' — have deliberately reduced their cognitive capabilities in order to commune with nature in a deeper way.

> *citizens* Software-based subjects, entirely lacking in flesh or blood but fully conscious, who live in virtual environments or *polises* sustained by powerful computers. There are a large number of polises, each with its own charter setting out the distinctive approach to living and the external world to which its citizens (and the polis's AIs) are committed. In effect, each polis is a distinct civilization, containing thousands (or millions) of inhabitants. Some polises largely devote themselves to maths and science, others to art; some are largely solipsistic, ignoring the external world, others are more outgoing and engaged; citizens can move between polises as their outlooks and orientations change. Some citizens started off their lives as embodied humans and entered their polis by uploading, others are polis-born.

> *gleisner robots* These are also post-humans whose minds are running on computers, but these computers are housed in non-biological (robot-like) physical bodies. Gleisners are more resilient than flesh-and-blood humans, but can interact with the physical world in the same sorts of ways.

The main action of the novel is set a fair way into the future, towards the end of the 29th century, and by this time the population of the polises vastly outnumbers that of the gleisners and fleshers, but humanity began its transfer into the realm of the virtual (the *Introdus*, as it's called in the book) a good deal earlier, in the mid-late 21st century — or so we are told.

This is all nothing but a fiction, of course, but if the technological capabilities were to arise, the idea that humankind would divide (or 'speciate') along roughly these lines strikes me as very plausible. Since there are already some people who are eager to have themselves uploaded — anyone sceptical of this should spend a few minutes at any of several post-humanist websites — there would obviously be *some* people who would be more than happy to go through with it soon as the technology becomes available, and more would no doubt follow when the technology matures and becomes more reliable. And if life in a virtual world should turn out to resemble that described in *Diaspora* the trickle of uploaders might well become a flood, for Egan's citizens enjoy a lifestyle which, in many respects, is highly enviable. Those citizens who want to devote themselves to pure research (in maths, physics, philosophy, whatever) can do so, albeit with their intellectual abilities and powers of concentration considerably augmented. But those who want to devote their lives entirely to the creative arts are free to do so, and ditto for those who want to spend their time partying or socializing, or those who prefer a balance between intellectual and other pursuits. Citizens who are curious about life as an embodied human need not worry: they can spend as much time as they like in fully accurate simulations of pre-upload life; those who are curious about what life would be like with a different kind of *mind* can have their own personalities altered, temporarily or permanently. This all sounds highly appealing, and I haven't yet mentioned the freedom from ailments and ageing, and the promise of immortality (or close to it) for those who want it.

However, while the prospect of all this would no doubt appeal to a great many, there would inevitably be those who would be reluctant to take their leave of ordinary material reality, and for a variety of reasons. For some, the idea of trading a real living body for (what is arguably) no more than the illusion of one would simply be too distasteful to contemplate. Others might instinctively recoil from the thought of trading a real physical environment for a virtual one; perhaps some would be fearful that virtual life would be in some way less vivid — *less real* — than an ordinary life.[1] Some might be swayed by religious doctrines, while others might be swayed by more philosophical considerations. Can we really be sure that the inhabitants of these computer-sustained virtual worlds are *conscious*? Even if we have good grounds for supposing they are, is the uploading process genuinely

---

[1] Such sentiments are certainly not unknown amongst fleshers in *Diaspora*. With life on Earth threatened by an imminent gamma ray burst, the fleshers are offered the opportunity of uploading to escape the danger; one responds thus: 'You are shameless. We expect no honour from the simulacra of departed cowards, but will you never give up trying to wipe the last trace of vitality from the face of the Earth? ... Did you imagine that a few cheap, shocking words would send us fleeing from the real world of pain and ecstasy into your nightmare of perfectibility? ... Why can't you stay inside your citadels of infinite blandness, and leave us in peace.' (Egan, 1997, p. 92).

person- or self-preserving? It is far from obvious that it would be. Here is what happens to a character in *Diaspora* (Orlando) during the initial (highly destructive) phase of the upload process:

> Waves of nanoware were sweeping through Orlando's body, shutting down nerves and sealing off blood vessels to minimize the shock of invasion, leaving a moist pink residue on the rubble as flesh was read and then cannibalized for energy. Within seconds, all the waves converged to form a grey mask over his face, which bored down to the skull and ate through it. The shrinking nanoware spat fluid and steam, reading and encoding crucial synaptic properties, compressing the brain into an ever-tighter description of itself, discarding redundancies as waste.
>
> Inoshiro stooped down and picked up the end product: a crystalline sphere, a molecular memory containing a snapshot of everything Orlando had been. (Egan, 1997, p. 110)

The envisaged storage-crystal may contain an accurate recording of Orlando's psychology, one which at some future time can be brought to (virtual life). But is it really possible for Orlando to survive having his brain taken apart — *boiled*, by the sound of it? Will the virtual-Orlando be *Orlando*, or merely a facsimile of him?

2. Issues of Experience and Identity

Chalmers discusses these and related issues in the latter stages of 'The Singularity'. If current research into AI does lead to the emergence of a superintelligence (of the A+ or A++ varieties), then assuming we survive we will have to decide how to respond.[2] Although some might prefer to interact as little as possible with the superintelligent systems, this isolationist (flesher-like) stance would not be the preference of everyone: many would undoubtedly seek to interact with them, albeit with differing degrees of intimacy. But there is a problem: it is very likely that A++ systems will function many times more quickly than ordinary humans — in the fictional the *Diaspora* universe, for example, the subjective time of the citizens runs some eight hundred times faster than that of the fleshers, and interaction between the two camps is infrequent for precisely this reason. So those seeking to interact frequently and efficiently with a superintelligence will need to boost the speed of their mental processes. While brain enhancement technologies may be of some assistance in this regard, they are unlikely to take

---

[2] In Chalmers' terminology, an 'AI' is an artificial intelligence of roughly human level, an 'AI+' system is an AI that is *more* intelligent than the most intelligent human, whereas an 'AI++' system is an AI that is at least as far beyond the most intelligent human as the latter is beyond the typical mouse; A++ systems possess *superintelligence* (p. 15).

us very far, and I think Chalmers is right when he suggests that in the long run 'if we are to match the speed and capacity of non-biological systems, we will probably have to dispense with our biological core entirely' (p. 45). So to integrate with the AI+ systems we will have to transfer our own minds into non-biological computers in the manner of Egan's citizens: we will have to upload our minds.

Assuming the medical and technological obstacles could be overcome — superintelligent computers could surely help with these — anyone contemplating undergoing an upload is confronted with the two philosophical questions we have just encountered. First of all, will the product of an upload be *conscious*? More generally, are the sorts of computers which are capable of running AI++ systems capable of sustaining communities of beings who are conscious? Second, is uploading a process one could survive? If I were to undergo an upload process, would the resulting subject be *me*, or another person entirely, one who happens to closely resemble me psychologically?

With regard to the issue of whether an upload (or an A++ system) could be consciousness at all, Chalmers admits that the issue is complicated by the fact that our understanding of consciousness is so poor, but he also thinks that a strong (if less than conclusive) case can be made for holding that consciousness is an organizational invariant, i.e. that systems possessing the same patterns of causal organization will instantiate the same types of conscious states, irrespective of whether the organization is implemented in neurons, silicon, plastic, or any other substrate. And because of this he is reasonably confident that suitably programmed computers can be conscious — and hence generate virtual worlds whose inhabitants are as conscious as we are. Although I am less confident than Chalmers with regard to the strength of the case for holding consciousness to be an organizational invariant, I am in full agreement with him that there's a great deal that we don't yet know about the physical underpinnings of consciousness. As a consequence, I do not think we can sure computers could *not* be conscious. For present purposes I will work on the assumption that they can be. What I want to focus on is the personal identity issue. Which sorts of upload-process are identity-preserving? Can we be confident that *any* form of uploading is truly person-preserving?

Chalmers' nuanced and interesting discussion of this issue in *The Singularity* is (to my mind) largely very plausible. But it is also very cautious: in Chalmers' eyes, much remains unsettled. On what he labels the 'optimistic view', uploading can be survived — at least if done in optimal ways — but on the 'pessimistic view' the process is fatal. Uploading itself may come in many very different forms, and to make matters concrete Chalmers orientates much his treatment around two specific forms which he labels *destructive* and *gradual*.[3] The process Orlando underwent in the passage cited above is an instance of destructive uploading. In such cases all the information relevant to creating an accurate psychological copy of an ordinary human subject is extracted from their brain by a sophisticated scanning process and safely stored, but the subject's brain is destroyed as a consequence; at some later time the stored information is used to create a psychologically similar subject in a virtual world. Since for a period (even if only a short one) the subject ceases to exist, then even if we assume they re-enter existence when the replica is created, a destructive upload does not preserve full mental continuity: at the very least the subject will lose consciousness for a while. As construed by Chalmers, *gradual* uploads do not involve any such rupture in the mental lives of those who undergo them; most notably, if a gradual upload is carried out quite quickly — over a matter of minutes or hours — they need not disrupt the continuity of the subject's consciousness: it is perfectly possible for the subject to remain fully awake and aware throughout the procedure. But although gradual uploads are non-disruptive in *this* way, in other respects they can be highly damaging. The nano-replacement procedure Chalmers describes — which involves the gradual replacement of all a subject's neurons by silicon-based devices — totally destroys the biological brain of the subject who undergoes it.

Since Chalmers believes that the uninterrupted continuity of consciousness is a particularly reliable guide to personal identity (pp. 58, 64) he holds that anyone who undertakes a gradual upload should be confident that they will survive the process. Chalmers also tells us that while he is more sympathetic to the psychological account of personal identity (which permits successful destructive uploading)[4] than the biological account (which doesn't), he isn't confident that the psychological account is correct, and hence 'I am genuinely unsure whether to take an optimistic or pessimistic view of destructive uploading. I am most inclined to be optimistic, but I am certainly unsure enough that I would hesitate before undergoing destructive uploading' (p. 54).

---

[3] In the interests of brevity I will not be discussing Chalmers' rather more speculative *reconstructive* upload process.

[4] In Parfit's widely-used terminology, it is *wide* psychological continuity which renders processes such as destructive uploads and teletransportation survivable; on this view, persons P_1 at t_1 and P_2 at the later time t_2 are one and the same person if P_2's psychological states (beliefs, memories, intentions, desires, etc.) are both similar to those of P_1, and causally dependent on those of P_1 in a suitably direct way — a way which does *not* require sameness of brain, or indeed, a continuously existing mind of any kind. It is this last proviso which allows people to survive being reduced to passive collections of data.

Why so much uncertainty? Although a destructive upload preserves psychological continuity (at least of the wide variety),
Chalmers doesn't think it is intuitively clear that the process is in fact
person-preserving, and this casts some doubt on the psychological
account itself. In addition, there are two alternative views of personal
identity which deliver different verdicts on destructive uploads, and
while Chalmers isn't sure that either of these alternatives is correct, he
is not sure they are *not* correct either. The first alternative can be summarized thus:

> *The further fact view*: knowing all the facts about personal physical and
> psychological facts in a given case doesn't provide one with knowledge
> of the facts about personal identity. If, for example, there are primitive
> immaterial substances (or souls), and we are identical with these sub-
> stances, then the facts about personal persistence would be determined
> by the facts about these, rather than any facts concerning biological or
> psychological relationships.

This epistemic gap could have other sources, e.g. our *concept* of personal identity might simply be such that even after all facts about mental and physical continuities are specified, it remains open whether
these facts suffice to secure personal persistence in the context under
consideration. In any event, Chalmers tells us that he thinks a further
fact view *could* be true; he does not think it is ruled out by anything
that we know. If a further fact view *is* true, then it is unclear whether
destructive uploading is person-preserving or not.

Chalmers calls the second alternative the 'deflationary view'; it is
harder to pin down in a succinct formulation, but its main ingredients
are as follows:

> *The deflationary view*: we are inclined to think personal identity is more
> solid and determinate across a wider range of circumstances, both
> actual and possible, than it really is. In puzzling cases — such as
> destructive upload, or teletransportation — where it is not intuitively
> clear whether the original person survives or perishes even when all the
> information relating to mental and physical continuities is known, there
> simply *is* no fact of the matter, whether of the ordinary or 'further' vari-
> ety, as to whether the process in question is person-preserving or not. In
> the absence of identity-facts, when deciding what to make of such cases
> we have no option but to focus on the facts relating to the mental and
> physical continuities which do exist, and form a view as to their impor-
> tance: how much of what matters in a life do they preserve?

Intriguingly, Chalmers suggests that the deflationary approach can be
extended to ordinary cases of survival: 'we are inclined to believe in
Edenic survival: the sort of primitive survival of a self which one

might suppose we had in the Garden of Eden. Now, after the fall from Eden ... there is no Edenic survival, but we are still inclined to think as if there is' (*ibid.*, p. 64). In this guise the deflationary view can make uploading seem less unpalatable. Uploading may not ensure perfect Edenic survival, but neither (it turns out) does ordinary life, and so in this sense uploading is not *that* much worse than waking up after a period of dreamless sleep. It is true that destructive uploading does not preserve biological continuity, but it does preserve causal continuity and psychological similarities, which relative to our actual scheme of values carry a good deal of what matters in ordinary survival. And gradual uploading, which does not disrupt the continuity of consciousness, preserves a very great deal of what matters.

3. Uploading: Another Perspective

The relationship between between the continuity of consciousness and personal identity is of a distinctively intimate sort. Letting your imagination roam far and wide, can you envisage a state of affairs, *any at all*, in which your current stream of consciousness goes one way, and you go another? I suspect not. Provided our consciousness flows smoothly on — i.e., provided the experienced succession of bodily feelings, perceptual experiences, thoughts and mental images that is characteristic of our ordinary streams of consciousness is uninterrupted — we can be certain (or as certain as we can be of anything) that we ourselves are continuing to exist, irrespective of what else is happening to us.[5] It matters not if the neurons in our brains are replaced with silicon surrogates; provided we continue to experience, we continue to exist. Likewise for psychological manipulations (or advanced brainwashing) which alter our memories or beliefs: these too we can envision ourselves surviving, provided they do not impact on the flow of our experience.[6]

Chalmers repeatedly says that he thinks that continuity of consciousness is the most reliable guide that we have to the continued existence of a self or person. Yet he is also reluctant to rule out the further fact and deflationary views. This may seem puzzling. For if the continuity of consciousness is sufficient for one's survival, then how can there also still be room for any kind of 'further fact' to play a role?

---

[5]   If you think you have succeeded in imagining a procedure which involves yourself and your stream of consciousness going their separate ways, consider: if you also envisage yourself as remaining conscious throughout the process, then all you have done is imagine your original stream of consciousness smoothly dividing into two, and this isn't the same thing at all.

[6]   See Dainton and Bayne (2005) and Dainton (2008, chapter 1) for more on this theme.

And if continuous consciousness can secure one's persistence in a perfectly secure manner, aren't the claims of the deflationist also undermined? What does Edenic survival have to offer that ordinary survival lacks?

As far as I can see there are two reasons why Chalmers adopts the stance that he does. First, in much of his discussion he works within the confines of the orthodox (essentially Parfittian) view that facts about personal identity are determined by biological and psychological-cum-causal facts. If we take these as our base-level facts, then evidently any facts about the continuity of consciousness will be counted as *further* facts, for the orthodox framework makes no mention of experiential continuities. Indeed, Chalmers acknowledges that if we include facts about the experiential continuities all ambiguities are removed: 'I think it is plausible that once one specifies that there is a continuous stream of consciousness over time, there is no longer really an open question about whether one survives' (p. 64). But this takes us on to the second point. It may be entirely clear-cut that you continue to exist for as long as you are enjoying an uninterrupted stream of consciousness, but what happens when you are no longer doing so? What happens when you lapse into the sort of dreamless sleep that most of us enjoy every night? Unless we opt to say that it is impossible for a person to survive interruptions in their experience (a decision which would drastically shorten all of our lives) then we need a plausible account of the conditions under which distinct streams of consciousness — streams that are separated by periods of time during which their owners are not enjoying any form of experience — belong to the same person or self. Must such streams be generated by the same brain? Must they be associated with causally related psychological systems? Must they be instantiated in the same primitive immaterial substance? Is there some more mysterious ingredient — some further fact — which performs the job? Can we even be sure that our identities *are* preserved through periods of unconsciousness in as secure a manner as they are preserved through uninterrupted periods of consciousness?[7] In the absence of a plausible account of how (or even whether) we survive periods of unconsciousness, Chalmers is right to hold that these are open questions.

The difficulty here is a real one, but it is by no means insuperable. On a number of occasions (see Dainton, 2004; 2008) I have argued

---

[7] Chalmers says he does not endorse, but nor is he entirely unsympathetic with the view that 'we Edenically survive during a single stream of consciousness but not when consciousness ceases. On this view, we may Edenically survive from moment to moment, but perhaps not from day to day' (p. 65).

that there is a natural way to solve the problem of experiential gaps for anyone who wants to take the continuity of consciousness as their primary guide to personal identity. This is not the place for a full rehearsal of this account — I call it the 'C-theory' — but for present purposes a broad overview of it will suffice.

Although some have held that we are identical with our experiences, this is not a very plausible or appealing view; it is more natural to hold that we are *things that have experiences* — or in the usual terminology, we are *conscious subjects*. Whatever else they may be, subjects of this sort typically have capacities to have a range of different kinds of experiences — bodily sensations, mental images, perceptual experience, conscious thoughts, and so forth — capacities which are sometimes exercised (during our waking hours) and sometimes not (when we are unconscious). In our own case these capacities are grounded in our brains, but since for all we know it may be possible for things quite different from a human brain to possess capacities for experience we need a more general term for things thus equipped; let's call them *C-systems*. Under what circumstances do C-systems at different times belong to the same subject? Well, we can say at least this: C-systems which have the ability to produce continuous streams of consciousness belong to the same subject. Since C-systems which are dormant (or unconscious) can still have the *ability* to produce such streams, this criterion applies equally to C-systems which are active and producing experiences, and those which happen to be inactive.

For anyone who takes the continuity of consciousness to be our best and most reliable guide to personal persistence, the notion that C-systems should be assigned to the same subject on the basis of their ability to contribute to single uninterrupted streams of consciousness is the obvious way to go — indeed, what criterion could be more secure or more readily intelligible? It is not difficult to construct a general account of personal or self-identity on this basis. Let us say that two brief C-systems (or phases of such) at neighbouring times are *directly stream-related* if they are either (i) both active, and the experiences they produce form parts of a single continuous stream of consciousness, or (ii) they are not both active, but if they were the experiences they produce would be parts of a single continuous stream of consciousness. Let us call C-systems that are not directly stream-related, but which are linked by a chain of C-systems that are so related *indirectly stream-related*. By way of final terminological move, let us say that a series of C-systems existing at different times are *C-related* only if they are either directly or indirectly stream-related. The C-theory can now be stated in a simple way: C-systems (or phases of such) belong to the same subject (or self, or person) if and only if they are C-related. With this much in place we can opt to identify selves (or subjects) with C-systems — the option I prefer — or merely trace the identity of subjects via their C-systems. For present purposes nothing hangs on which of these options we adopt.[8]

The C-theory provides a simple but effective solution to the problem posed by interruptions in the continuity of consciousness, and it does so by appealing to nothing more than the continuity of consciousness — or at least the potential for it. So far as the intuitive appeal of the account is concerned, the change of focus from actual continuity of consciousness to potentialities for it is of little or no consequence. I have a special concern about the kind of experiences which will feature in my current stream of consciousness as it continues to flow on until I next lose consciousness. Why? Because these experiences will all clearly and unambiguously belong to me, and like most of us I have a primitive and profound concern about the character of my own experiences, particularly those which lie in my future. As is easily seen, a simple but powerful mechanism extends the reach of this self-oriented concern to C-systems and their constituent experiential capacities.

Given that my current stream of consciousness belongs to me — utterly without ambiguity — so too do the experiential capacities which are producing the experiences this stream contains. However, I also possess experiential capacities which are not active and producing experience, but which could be; e.g., if I were to turn my radio on, I would have auditory experiences that I wouldn't have otherwise. The C-theory easily accommodates experiential capacities falling into this category. By way of illustration, let N_1, N_2, N_3 ... be intervals of time of one second duration. (This is just to keep things as simple as possible — the same considerations apply over shorter intervals.) Let us further suppose that my current experiences are taking place in N_1. What should we say about the ownership of the dormant experiential capacities in N_2? According to the C-theory, experiential capacities which exist at adjoining intervals belong to the same subject if and only if they are either (i) active and contributing to a single continuous stream of consciousness, or (ii) they are not all active, but they are

---

[8] The C-theory as expounded at greater length in *The Phenomenal Self* (2008) is a more detailed elaboration of this approach; I also extend the approach to the synchronic case, by construing C-systems-at-time as collections of experiential capacities which are rendered co-subjective by virtue of the ability to produce synchronically unified states of consciousness; I also broach the tricky topic of what to say about branching streams of consciousness (or fission cases), a topic I leave to one side here.

such that *if they were*, they would be contributing to a single continuous stream of consciousness. Inactive experiential capacities obviously cannot satisfy condition (i), but they can satisfy condition (ii). Dormant experiential capacities located in interval N_2 belong to me only if they are such that if they were active and producing experiences, these experiences would feature in the direct continuation of my current stream of consciousness (i.e. the phase located in interval N_1).

Should my special self-interested concern extend to these inactive experiential capacities? The answer is clearly in the affirmative. The N_2-phase of my C-system (we can safely suppose) includes a capacity for intense sensations of pain. Considering matters from my present vantage point in N_1, I would very much prefer that this capacity remains inactive; after all, if it is triggered, then the resulting pain-sensation will occur in the very next phase of the stream of consciousness I am currently enjoying. If my present conscious state will flow directly into a state which includes this sensation, without any interruption in experiential continuity, how can I doubt that it will be *me* who feels (and suffers) the pain? This special concern extends to all the other powers which belong to the N_2-phase of my C-system, since these can all influence the character of the direct continuation of *this* stream of consciousness in the immediate future. Hence if during N_1 I turn my radio on, auditory experiential capacities will be activated during N_2, and I will hear sounds I would not have heard otherwise.

Let us next suppose that in the interval N_3 *none* of the experiential capacities which the C-theory assigns to me will be active — that during this period I will be enjoying a few moments of complete unconsciousness. This does not affect the situation in the slightest. As we have just seen, not only do the dormant powers in the N_2-phase of my C-system all unambiguously belong to me, but my primitive prudential concerns inevitably extend to them without diminution or dilution. The N_3-phase of my C-system may be dormant, but it is nonetheless C-related to the N_2-phase, i.e., if the experiential capacities in these successive phases *were* active, the experiences they produce would form parts of a continuous stream of consciousness. As a consequence, the ownership of the N_3-phase capacities could not be clearer: they belong to me, and they do so in the same utterly unambiguous way as their N_2-counterparts. Given this, my special prudential concern naturally and inevitably extends to them as well. And the same applies to the N_4-phase of my C-system; even if experiential capacities here are all inactive, they have the ability to join with my

N_3-phase capacities in the generation of a continuous stretch of experience. This mechanism for guaranteeing sameness of owner (or subject) of successive C-system-phases, and thence transmission of prudential concern, can operate over indefinitely long periods of time — over entire lifetimes.

The C-theory provides us with a clear and informative answer to the question 'What makes it possible for streams of experience separated by a gap in consciousness belong to the same self?', and does so wholly in terms of capacities for experience: earlier and later streams have the same subject if they are C-related. With C-relatedness on hand there is no need to appeal to bodily or material continuity, psychological-cum-causal connections, primitive immaterial substances or some mysterious further fact to explain how experiences separated by gaps in experiential continuity can and do have the same subject.

The C-theory also goes a long way towards undermining the version of the deflationary view according to which our persistence, even in ordinary circumstances, falls short of what it might be — or what we hope it might be. It is not obvious how survival could be better than what the C-theory has to offer, even in Eden.[9]

What of uploading, and the prospects of surviving it? Since there is no requirement that C-systems which belong to the same subject must belong to the same physical systems, it is (in principle, at least) perfectly possible for a single subject to migrate from one physical system to another; all that matters is that this subject's capacity for continuous consciousness is not disrupted by the change in material substrate. Since gradual uploads of the sort Chalmers considers — consisting of the progressive nano-engineered replacement of biological cells by silicon-based functional surrogates — do not disrupt capacities for consciousness, they are unambiguously person-preserving. Since destructive uploads clearly *do* disrupt this capacity, at least

---

[9] Following Chalmers' lead I have simply assumed here that the our typical streams of consciousness *do* exhibit a distinctive and robust form of continuity. This is not the place for defense of this *prima facie* plausible claim against those who would deny it — this is a task I have undertaken elsewhere: see Dainton (2008, chapter 3; and forthcoming); see Dainton (2010) for an overview of the debate on this issue. In a different vein, it might be argued that, other things being equal, it is better to spend as much of one's time awake as possible, and so a life which includes periods of dreamless sleep is less good than a life of the same duration which contains no such periods. But even if it is better to be experiencing all the time, this does not in itself entail that there is a shadow of doubt over whether a subject can *survive* periods of unconsciousness. In some circumstances (e.g. severe brain damage) there might be, but not if the periods in question are bridged by a continuous capacity for continuous consciousness — not if C-relatedness is present and *unimpaired* throughout.

in cases which involve the original subject being reduced to a passive collection of stored data, they are definitely *not* survivable.

Of course, this is assuming that something along the lines of the C-theory provides us with the correct account of personal identity. According to the (wide) psychological account, destructive uploads *are* survivable, since they lead to the creation of conscious subjects whose psychological systems are both indistinguishable from, and appropriately causally related to, the psychological systems of the original pre-upload subjects. Although I believe the C-theory is superior to the psychological approach in several respects, this is not the place for a defense of this claim; my main concern here has been to bring to light some of the options available to those who take phenomenal consciousness as their guide to personal identity — which proponents of the psychological approach do not. That said, before moving on there is one point to note. It is not unlikely that a sizeable part of the credibility of the psychological approach derives from the assumption that chains of causal dependency linking earlier and later psychological states are the *only* form of mental connection capable of bridging interruptions in consciousness. At an intuitive level it is by no means clear that this sort of connection is sufficient, in and of itself, to constitute personal survival. But if it is this sort of connection which permits us to survive periods of unconsciousness, how can we consistently deny that it suffices for our persistence on other occasions? The availability of the C-theory changes the situation dramatically: interruptions in consciousness can also be bridged by the continuous potentiality for continuous consciousness. If we are essentially subjects of experience — i.e., beings with the *capacity* for consciousness, a capacity which is sometimes exercised, and sometimes not — then this form of continuity is manifestly both necessary and sufficient for our continued existence. In which case, destructive uploads — or variants of teletransportation — which disrupt or destroy this capacity cannot be survived; at least not by beings such as we.[10]

---

[10] There is, of course, more to say. Many of us might not *now* be able to view destructive upload as truly survivable, but could this change? Mark Johnston has recently argued that facts about personal identity are determined solely by our own responses to real and imaginary cases of putative survival; by changing these responses we can change what we can survive: 'If refiguring our identity-determining dispositions can open us up to, or close us off from, certain forms of survival, then there is a sense in which our natures are Protean ... if we could refigure our identity-determining dispositions then what we are (in the relevant sense) *capable of surviving* would change' (Johnston, 2010, pp. 283–84). Since these identity-determining dispositions are deep-seated, changing them is neither trivial nor easy — Johnston acknowledges that if it is possible at all it will require time, effort, reflection and appropriate metaphysical instruction — but the rewards for success are

4. Simulation Multiplication

Inevitably, the precise way in which the C-theory resolves the uncertainties surrounding uploading — in some if not all of its forms — will not be seen as unalloyed good news by everyone. In particular, the thesis that the kind of psychological continuity which is preserved in destructive uploads does not provide for personal survival may seem like bad news for those impatient to upload themselves into virtual (quasi-) paradises. We do not yet have the scanning technology required for destructive uploading, but it is not inconceivable that these might be developed in the foreseeable future, whereas the advances in nanotechnology and neuroscience needed for a gradual upload may seem a far more difficult and distant prospect. But those who are eager to attain a post-human condition need not despair. Although developing the science and technologies required for gradual upload may always be beyond unaugmented human-kind, there is no reason to think a superintelligence will be unequal to the task. If a singularity does occur, taking leave of this world for a Diaspora-style heaven may be a very real possibility — perhaps in the not too distant future.

However, the possibility that future technological developments will make it very easy to create very large numbers of real-seeming virtual worlds gives rise to further issues. These worlds aren't just havens that ordinary humans might move to via uploading, they can impact on the way *all* conscious beings conceive of themselves and their lives. As things stand, our abilities to create and control streams of consciousness are severely limited. Let's suppose that in a post-singularity future this changes, and it becomes possible to create human-type streams of consciousness, of any length, with any desired characteristics, very easily. A perturbing possibility now looms. If streams of consciousness with a character similar to *this* one could be created in the future — cheaply, easily and frequently — might it not be quite likely that this stream does in fact exist in the future? Isn't there a good chance that we are all enjoying artificially generated experiences?

Let's take this a bit more slowly and carefully. Call the succession of streams which jointly compose the consciousness of a single person from birth until death, a *life-stream*. Despite their differences, your

---

potentially very great indeed. Since his case depends on an extended argument for the unreality of a self that I find questionable — see Dainton (forthcoming) — I am not wholly persuaded that Johnston is correct in this. Even so, there is undeniably something amusing in the idea of would-be uploaders willingly participating in (what are, in effect) spiritual exercises in order to secure their passage to a digital paradise.

life-stream and mine are of a certain general type: early 21st century human. Let us call these 'type-21 streams'. Now suppose that in the future *very* large numbers of type-21 streams will be created, all of which are indistinguishable, qualitatively and subjectively, from real type-21 streams. To be more specific, suppose the total number of type-21 streams which exist after the year 2100 is around ten times greater than the number which existed in the 21st century itself, with each general type of 21st century life being proportionally represented. If these artificial type-21 streams did exist, the consequences would indeed be perturbing. Are you in a position to tell whether your experience is real or artificially generated? No. Or at least, not if all you have to go on is the character of your experience. What are the odds that your experience is occurring when appears to be, in the early 21st century? Only around one in ten. Although it seems to you that you are a normal human being living at the start of the 21st century, the subjects of all the many artificially produced type-21 streams have very similar experiences and beliefs. These subjects are all mistaken, and so might you be, for it is more likely than not that you *are* one of these subjects.

So, if you think it likely that our descendants, whether human or post-human, will develop and use simulation technologies in this sort of way, you should also think it likely that your current experience are the product of these technologies — that your own world is virtual rather than real. Following Bostrom (2003) I will call this line of reasoning the *simulation argument*, though on occasion I will also refer to the *simulation reasoning*, meaning the same thing.[11] Although not everyone will find the possibility that they are living in a simulation something to be feared or dreaded, at the very least the simulation argument threatens complacent assumptions about the status of our lives, and for this reason I shall sometimes refer to the *simulation menace* or *threat*. Many will no doubt be inclined to dismiss the simulation argument as a mildly diverting but ultimately unthreatening curiosity. For reasons which will emerge *en route*, I think this would be a mistake. After establishing that the simulation argument is one that should be taken seriously, I will move on to consider some of its implications.

---

[11] As Bostrom himself uses the term (see his 2003; 2009a; 2009b), the simulation argument purports to show that at least one of three propositions is true. These are (roughly): (1) that most civilizations go extinct before reaching a high level of technological mastery, (2) that most technologically advanced civilizations choose not to create large numbers of simulations of their own pasts, (3) that we are almost certainly living in a computer simulation. I will be commenting on this formulation in §6.

5. Simulation Generation

First some terminological clarifications. Henceforth I shall be using 'simulation' in a very broad way: any state or episode of consciousness is to be regarded as simulated if it is produced by non-standard methods in a controlled fashion (the degree of control may vary). Simulated experiences are of course real experiences in their own right, and while a simulated episode of consciousness may be a re-creation of an original non-simulated stretch of conscious life, it need not be. I shall say that a life (or part of a life) is *virtual* rather than *real* if it is entirely composed of simulated experiences; I shall call the subjects of such experiences *simulants*.

Consciousness can be simulated in different ways, and to different degrees. So far as degree or depth of simulation is concerned, we can contrast *complete* with *partial* simulations. The manufactured type-21 streams mentioned above are examples of complete simulations: every part and aspect of experience is being generated by artificial means. In partial simulations, only some parts or aspects of experience are generated by artificial means. A simulation in which a subject is supplied with a wholly virtual environment (which here can be taken to include all forms of bodily experience) but retains their original psychology is one form of partial simulation. But we can also envisage cases in which the effects of the tampering are restricted to the domain of inner experience. Imagine having your psychology (e.g. memories, beliefs, desires, language skills, personality traits, and so on) replaced with a replica of Napoleon's, and then waking up in your own bed and perceiving your environment in the usual way. In what follows, unless otherwise stated, we will be concerned with complete rather than partial simulations.[12]

As for the *ways* in which consciousness can be simulated, it is important to distinguish what I will call *neural* (or *N-simulations*) from *software* (or *S-simulations*). N-simulations result from interfering and controlling the neural hardware in the brain that is ordinarily responsible for producing experience. The simulants in N-simulations are ordinary human beings who are vividly hallucinating in a controlled fashion — of course, this will not normally be apparent to the subjects themselves, who will take themselves to be whoever they seem to be in their virtual worlds. In contrast, S-simulations consist of

---

[12] More discriminating distinctions can be drawn. For example, it is possible for a subject to lead a virtual life — in the sense here defined — in the real world (some may recall the holo-doctor in Star Trek *Voyager*). This sort of case will not be relevant to what follows, where we shall be concentrating on simulations in which what is 'perceived' is *not* the real world, at least as normally conceived.

streams of consciousness which are wholly generated by running programs in computers of some kind; the simulants in S-simulations are *Diaspora*-style citizens.[13] Let us start by taking a closer look at the latter.

S-Simulations

On the assumption that mentality is computational in nature, computerized simulations of human brains could generate conscious mental lives that are subjectively indistinguishable from those generated by biological brains. How remote a prospect is this? It is impossible to be certain, but it is probably somewhat less remote than many suppose. If computer technology continues to advance at the rate it has for the past few decades, it will probably not be long before our most powerful computers equal, or exceed, the processing power and information storage capacity of a typical human brain. Estimates of the latter vary a good deal, but it is currently believed to be of the order of 10^{14}–10^{17} operations per second. Present-day laptops are capable of 10^9 operations per second, and supercomputers can manage 10^{15}. Raw computational grunt by itself does not count for a great deal; we are still a very long way from knowing enough about the structure and functioning of the human brain to simulate their workings computationally. But progress is also being made on this front. Drawing on thousands of detailed scans of rat brains, the Blue Brain project has succeeded in simulating a 10,000 neuron cross-section (along with 10,000,000 interneuronal connections) of the rat cortex on a super-computer.[14] The project's leader estimates that a similarly fine-grained simulation of an entire human brain will be possible by 2020. Even if this proves optimistic, it is not too far-fetched to suppose that such a simulation will be possible within two or three decades. There remains, of course, the problem of understanding enough about the relationship between neural activity and experience to reach the stage where we can directly and efficiently control the kinds of experience a computer-simulated brain will produce. This may well be the most difficult problem of all; but even if it should prove to be beyond *our* abilities — as of now this isn't clear — as far as I can see, there is no reason to suppose it will be beyond the capabilities of a superintelligence. If those who believe a singularity will occur in the next few decades are correct, it is quite

---

[13] To simplify I will assume that only this universe exists. Other universes make a difference. If all logically possible worlds are real, as Modal Realists believe, then each real life is simulated an infinite number of times, creating a highly significant simulation menace.

[14] See Markham (2006) for an impressive overview.

likely that S-simulations of ordinary human streams of consciousness will be easily produced not long after.

A few such simulations pose no significant threat, but the situation becomes distinctly menacing if they start being produced by the billion or trillion. Such a situation could develop in at least two ways. The ability to produce menacing simulations could become very widespread, e.g. a hundred years from now everyone might own desktop (or handheld) computers easily capable of running them. If several billion computers were to possess this capacity, even if it were utilized only occasionally, menacing simulations would soon exist in disturbingly large numbers. To make matters a little more concrete, consider the popularity of the *The Sims* franchise over the past decade or so, and suppose the simulated inhabitants are all fully conscious — as they might be, in computer games of the future. The 'god games' enjoyed by our descendants could easily prove to be even more popular, generating menacing simulations in vast numbers.[15]

Alternatively, or in parallel, the capability of running large numbers of simulations might be found in the mega-computers of the not too far-distant future. Bostrom provides an illustration of the potential dangers: 'a rough approximation of the computational power of a single planetary-mass computer is 10^{42} operations per second ... Such a computer could simulate the entire mental history of humankind (call this an *ancestor-simulation*) in less than 10^{-7} seconds' (Bostrom, 2003, pp. 247–8). If our descendants (whether human or machine) were able to run ancestor-simulations using only a small fraction of the computing resources available to them, they might very well do so, quite frequently.[16] If such circumstances were to obtain, the probability that you and I are inhabiting a computer simulation would be high. That said, the effective programming of a super-massive computer of the kind being envisaged here would be well beyond the capabilities of ordinary humans, and may well depend on the availability of superintelligent A++ systems. But again, if a singularity is as likely as many believe, then this is not the hurdle it would otherwise be.

---

[15] In 2010 Electronic Arts celebrated selling 125 million copies of various versions of the game since its launch in 2000. In gauging the potential threat such software might pose, it should also be born in mind that each copy of the game can generate dozens of virtual characters each time it is played, and that the settings need not be contemporary: *The Sims Medieval* was released in March 2011.

[16] Ancestor-simulations replicate the *mental* history of humankind, not the entire universe — as Bostrom notes, this would be unfeasible. So although ancestor-simulations reproduce the appearances of a physical world, they do not attempt to simulate all the physical processes within the objects its subjects perceive (e.g. their houses, the interior of the Earth, the distant stars they can see at night).

Those who have grown familiar with the claim that the human brain is the most complex object in the known universe may be surprised to discover that it will not be very long before we are able to construct machines of comparable complexity and computational power. But even if this is the case — and I suspect it is — the simulation menace posed by advances in computer technology could well be far less severe than some would have us believe; it may even be non-existent. For S-simulations to constitute a threat they would have to be truly conscious, and it is by no means certain that this is a real possibility. As I noted in §2, there are influential positions on the relationship between the physical and phenomenal which, if correct, would entail that properly programmed computers could generate human-like streams of consciousness. Many versions of orthodox functionalism have this consequence, as does the rather more plausible dualistic (or non-reductive) form of the doctrine sympathetically explored and expounded by Chalmers.[17] But as Chalmers himself would concede, non-reductive functionalism is at best a possible solution to the matter-consciousness problem. Other leading positions in the philosophy of mind — including familiar forms of both dualism and materialism — are resolutely hostile to the notion that consciousness is an organizational invariant. If, as many believe, phenomenal properties are material in nature, then it may very well be that a human-type consciousness requires a human-type brain, or at least a biological system of a similar kind.[18] Of course, we cannot be certain of this. We do not know which parts or aspects of the physical processes in our brains are responsible for producing consciousness; consequently, we cannot rule out the possibility that the relevant physical processes could be replicated in very different physical systems — perhaps even silicon chips. But this is no more than a possibility, and in all likelihood, one that is quite remote if materialism is true.

So are S-simulations possible? The situation is clearly far from clear-cut. Whereas functionalists — of both reductive and non-reductive persuasions — have good reasons for being very wary of

---

[17] Or in a little more detail: 'given any system that has conscious experiences, then any system that has the same fine-grained functional organization will have qualitatively identical experiences ... we might call [my doctrine] *nonreductive functionalism*. It might be seen as a way of combining functionalism with property dualism' (Chalmers, 1996, pp. 248–9).

[18] It is difficult to see how experiential states could *be* physical states if the latter only have the range of properties recognized by contemporary physics, but the more credible forms of materialism do not circumscribe the physical in this manner. Lockwood (1989), McGinn (1999), Searle (1992), Strawson (1994) all defend version positions which combine materialism with a realist (non-reductive) view of consciousness.

future developments in computer technology, those who subscribe to different options in the philosophy of mind have far less reason to feel greatly concerned: simulations that are not truly conscious pose no menace whatsoever.

N-Simulations

Those who find all varieties of S-simulation implausible cannot afford to be complacent. For there is a further source of menacing simulations, one that has received less attention, and one that does not depend on adopting controversial positions on the matter-consciousness issue.

Many of us have experienced fully realistic hallucinations, whether drug-induced or in ordinary dreams. Hallucinations of this sort are typically *uncontrolled*: we cannot determine in advance the type of virtual world we will hallucinate, or the role we will play in the scenarios which unfold. As noted earlier, this may very well change, and in a dramatic fashion: post-singularity advances in brain science may make it possible to generate complex and precisely controlled hallucinations — or N-simulations — safely, easily and reliably. Almost inevitably, some of these controlled hallucinations will constitute menacing simulations.

In fact, there are several ways in which this could come about. One way of generating N-simulations would be to use the kinds of neural implant and human-machine integration that are already familiar from science fiction. Interacting with computers mechanically — using keyboards, mice, touch-sensitive screens, etc. — is a cumbersome business, and a good deal of research is going into ways of facilitating the process. Among the methods already being considered, at least by the more adventurous researchers, are direct brain-computer interfaces. At present, such techniques are at a fairly primitive stage of development, but this will no doubt change.[19] A hundred years from now, children could be growing up with implants buried deep in their heads, implants that both track and keep pace with their neural

---

[19] Indeed, brain-computer interfaces have already left the laboratory and entered (some) homes: the Emotiv EPOC is a headset which allows gamers to interact directly with their consoles; in the company's own words, the system 'is a high resolution, neuro-signal acquisition and processing wireless neuroheadset. It uses a set of sensors to tune into electric signals produced by the brain to detect player thoughts, feelings and expressions and connects wirelessly to most PCs' — all this for only $299! See http://www.emotiv.com For an indication of the (quite impressive) level of sophistication of current computer-based 'mind-reading' techniques, see Naselaris *et al.* (2009).

development, and allow their minds to interact directly with computers, on a number of levels, in a variety of ways.

It is not difficult to envisage some of the uses to which this sort of
interface might be put. Your thoughts could be transmitted directly
into someone else's mind — provided you were both hooked up to the
same computer network. Forgetfulness would be largely a thing of the
past: your thoughts and experiences could easily be backed-up on a
computer file, ready to be called on when required. More relevant to
our purposes, fully immersive virtual reality would also be a possibility. There will be no need for you to wear a suit and visor to interact
with machine-generated virtual worlds, your implants will perform
the necessary tasks. Your sensory experience will be directly
machine-controlled, via stimulation of the appropriate areas of the
sensory cortex. The movements of your (simulated) body through virtual environments will be under your control, but there will be no need
for you to actually move your physical body: the ways you *intend* to
move your body will be detected by implants in your motor cortex and
elsewhere, and this information will be used to generate corresponding movements of your virtual body. It will be possible to have a fully
realistic experience of (say) flying a plane through narrow mountain
passes while remaining motionless on a couch. You might even
believe yourself to be an experienced pilot: your implants could
ensure that a suitable set of false memories temporarily override your
real memories. Alternatively, you might believe yourself to be an ordinary 21st century person, leading a typical life in a (virtual) 21st
century environment.

A materialist might argue that brain-computer connections required
for this sort of simulation would be so invasive and pervasive that
their presence inside a brain would be incompatible with the production of conscious experience. Given our ignorance of the physical processes underlying consciousness, this possibility cannot be ruled out,
but there no reason to suppose it very likely. After all, the envisaged
interfaces would not replace neurons as experience-producers, they
would merely provide ways of artificially controlling the triggering of
neurons, or neural circuits — and we already know this to be possible
on a small scale. True, the required nano-scale technology is far
beyond anything we are capable of producing at present, and even if it
were not, our understanding of the brain's functioning is not sufficiently advanced for it to be deployed effectively. But anyone inclined
to think this will continue to be the case should bear in mind two considerations: first, the difference the availability of (helpful and cooperative) superintelligent computers would make, and second, Arthur

C. Clarke's dictum that any sufficiently advanced technology is indistinguishable from magic.

If the envisaged interface technology were to become common-place, then given time, N-simulations could easily be produced in suf-ficient numbers so as to become menacing. People might take virtual reality 'trips' to the past quite frequently. They would certainly be used on an occasional basis during history lessons, and more inten-sively by historians, amateur and professional, with a particular inter-est in what it was like to live during certain periods of the past. But such trips might also be taken — far more frequently — for entertain-ment purposes. The soap operas of the future might well have an immersive/interactive character their present-day counterparts lack, computer games likewise. Already there is evidence that being able to enter and explore virtual worlds is likely to prove extremely popular. Over the past few years the numbers of people participating in MMPORGs ('massively multiplayer online role-playing games') has expanded dramatically — the currently dominant *World of Warcraft* currently has close to 12 million participants.[20] The addictive proper-ties of these virtual worlds is well known; it is not uncommon for gamers to absent themselves from this world for hours per day. With the advent of fully immersive virtual reality technology, pastimes of this sort will probably prove to be still more popular. As is easily shown, the numbers soon add up.

Our descendants may 'visit' the past quite frequently, but since few are likely to want to spend significant portions of their lives in N-sim-ulations, the concept of a *life-stream* introduced earlier is no longer appropriate as a basic unit of simulated consciousness. Something of briefer duration is required. So, for present purposes, let us take *day-long* streams of uninterrupted consciousness — *D-streams* for short — as our working units. (An even shorter unit could be selected, but as will become evident, the upshot would not be greatly different.) We shall take as our class of *menacing* D-streams those simulated streams that resemble the sort of experiences enjoyed by actual inhab-itants of the year 2011 — call these *MD-streams*.

Assuming the current population of the Earth to be seven billion, there are just over 2.5×10^{12} D-streams for the year 2011. If a similar number of MD-streams exist in the future, the odds of the experiences you are currently having being simulated rather than original look to be around fifty per cent. Should the numbers of MD-streams created

---

[20] According to a Blizzard press release, see http://us.blizzard.com/en-us/company/press/pressreleases.html?101007

in the future be greater, your chances of living among the original inhabitants of the year 2011 will be correspondingly smaller.

In fact, the number of MD-streams created in the future could easily be far higher. Call the time at which N-simulations become common-place occurrences the *C-threshold*. Let us suppose that from the C-threshold on, every future human being takes one virtual reality trip to the year 2011 during their lifetime, and that these trips are varied in character. If we now suppose that human civilization lasts for ten thousand generations after the C-threshold, and has an average popu-lation of ten billion, there will be 1.0×10^{14} MD-streams, compared with 2.5×10^{12} original D-streams. With forty simulated streams for every real stream, you have a one in forty chance of actually being alive in the year 2011.[21]

It is not only materialists who should be open to the possibility of N-simulations, dualists should be too. Even if our experiences unfold within immaterial substances, it is evident that our minds are pro-foundly dependent upon our brains. No contemporary dualist would be inclined to deny that the course of our sensory experience is dependent upon the neural activity within our brains, and this fact alone opens up the possibility of controlled hallucinations of a limited kind. But dualists should also recognize that appropriate neural manipulation could impact upon our conscious beliefs, intentions and desires. Intoxicants do not merely make it harder to control our bodily movements, they make it harder to *think* clearly, and there are numer-ous forms of brain damage that have more far-reaching (and often permanent) effects on our personalities and cognitive functioning, memory included. If brain damage can result in the permanent loss of certain memories, is it not likely that the memories to which we have conscious access depend on information stored in our brains? In which case, appropriate neural manipulation could lead a 23rd century person to have access to apparent-memories of the sort a 21st century person would have had.

But there is a further point to note, one that is relevant to material-ists as well as dualists. Brain-computer interfaces of the kind I have been considering offer the possibility of very tightly controlled hallu-cinations, but there are undoubtedly other ways of inducing similarly

[21] On more optimistic scenarios, your predicament is even more precarious. If humankind has a long history — one million generations exist after the C-threshold, say, with constant or improving technology — and a larger average population during this period (a hundred billion, say) then we can expect a total of around 1.0×10^{17} MD-streams to occur, which would reduce your chances of being alive in 2011 to around one in fifty thousand! In this case, even if only one in a thousand people ever take a virtual reality trip back to 2011, the chances that you are really living in 2011 are still only one in fifty.

life-like N-simulations, even if they offer rather less potential for fine-grained control. Ordinary, unaugmented, human minds are able to fashion richly-detailed and real-seeming virtual realities all on their own, almost effortlessly. Ordinary dreams provide evidence both of this, and our ability to spin complex virtual worlds from limited and/or fragmentary evidence. I expect most of us have found ourselves having vivid dreams set in (say) the 17th century shortly after watching a film set in the same period. Although the dreamed environment in such cases is inspired by what was seen onscreen, it often has a depth and complexity all of its own.[22] Future methods of experience-induction could easily exploit these ordinary abilities. All that would be required is a safe and reliable drug which enabled people to enter a dream-like state at will, and also direct the general direction of the subsequent (fully life-like) hallucination — the framework for the latter could be supplied by a little prior reading, or the watching of video footage (e.g., of a 21st century televised soap opera). This method of controlling hallucinations could be put to the same uses (in, say, education and entertainment) as the computer-driven variant we considered earlier, and so is likely to be widely employed. So far as I can see, this method of inducing (partially) controlled hallucinations is not ruled out by any philosophical conception of the mind. It is also quite likely to prove attainable, perhaps quite soon — and may well not even require the advent of superintelligence.

6. An Issue of Principle: Is the Reasoning Self-Defeating?

Unlike those familiar forms of sceptical reasoning that invoke powerful deception-mongering demons and suchlike, the simulation reasoning rests on more mundane considerations: more or less plausible projections of current technological and social trends. This novelty grants the simulation reasoning an uncommon force, but it also leaves it vulnerable to an objection along these lines.

> The simulation argument relies on certain empirical premises concerning how our world is likely to turn out. If we come to accept that it is likely that we are living in simulations, we surely no longer have reason to accept the relevant empirical premises. The simulation reasoning is self-defeating, and can therefore safely be dismissed.

---

[22] Might *ordinary* dreams constitute menacing simulations? Perhaps, but I am inclined to think not, simply because I suspect my dream-experiences are somewhat less vivid and more course-grained than my ordinary waking experiences. I am not alone in this (see Flanagan, 2000, pp. 173–4). Of course, we cannot rule out the possibility that we are living in the dream-worlds of beings whose waking consciousness is far richer than our own.

More generally, it might seem that any argument rooted in empirical considerations that yields a radically sceptical conclusion is inherently unstable. No sooner is the sceptical conclusion generated the supporting empirical considerations are blasted away and the conclusion collapses.

Bostrom has recently responded to this objection in his 'Simulation Argument FAQ' (2009b). While his argument is useful as far as it goes, in some respects it does not go far enough.

As Bostrom formulates it, the simulation argument is supposed to show that at least one of the following claims is true: (1) the human species is very likely to go extinct before reaching a technologically advanced posthuman stage; (2) it is very unlikely that any posthuman civilization will run large numbers of simulations of their own history, (3) we are almost certainly living in a computer simulation. Now, in his response Bostrom maintains that it is wrong to claim that we can't have *any* reliable information about the underlying external reality if we actually are in a simulation. For even if we are simulants we can know the following two conditional claims: (a) if we are in fact living in a simulation, then the underlying reality contains at least one simulation (this one), or (b) if we are not in a simulation, then there is no reason to doubt our empirical beliefs (e.g. to the effect that a technologically advanced civilization will have the ability to create vast numbers of simulations). If (a) is true, and we are in a simulation, then we know that (3) is true. If (b) is true, then we are not living in a simulation, and so we know that (3) is false, but (1) or (2) may still be true. So even if I am simulated I can know that either (a) or (b) is true, and hence know that at least one of (1), (2) and (3) is also true.

Formally speaking, this response is enough to establish that Bostrom's formulation of the simulation argument does not succumb to the charge that it is self-defeating. However, this line of response also diminishes the interest of the simulation reasoning. If we follow Bostrom's line the simulation argument is no longer an empirically grounded *personal* threat — the reasoning no longer takes us (as individuals) from the empirical premises to the conclusion that we ourselves are simulants. For the argument in the latter guise to work it must be possible for me to *combine*, consistently and coherently, the belief that I am living in a simulation with the belief that most of my empirical beliefs are true. Bostrom's response does not allow me (or you, or anyone else) to be justified in holding both of these beliefs. As he presents our predicament, if I am in a simulation then the only well-founded belief that I am entitled to with regard to the external world is that it contains at least one computer simulation, the one I inhabit. If this is *all* I am entitled to know, then my original empirically based beliefs for thinking that I am a simulant are no longer well-founded. The hypothesis that I'm living in a simulation has, in effect, been reduced to the bare possibility of traditional scepticism. And of course, what goes for me, goes for you.

But we are not yet done, for the epistemic instability charge can be countered in another way. As a first step, consider these two strongly contrasting simulation hypotheses.

> SH1 Simulated lives vastly outnumber real lives; the environments simulated subjects inhabit are invariably utterly unlike the real world.

> SH2 Simulated lives vastly outnumber real lives; without exception, the environments simulated subjects inhabit are almost exact replications of the real world.

I shall call simulations *faithful* if they resemble some part or period of the real world quite closely in most respects, and *unfaithful* if they do not. SH1 presents us with an extreme example of an unfaithful simulation, SH2 is an extreme example of a faithful simulation. Now, considered purely as a prediction as to the quantity and character of simulations likely to exist, SH1 is unproblematic. It quickly becomes problematic if it is incorporated into a simulation argument thus: 'I have empirical grounds for believing that (i) simulated lives outnumber real lives, (ii) simulated lives are invariably unfaithful, (iii) it is highly likely that I am myself a simulant.' The epistemic predicament of anyone subscribing to this combination of claims is precarious. No one who subscribes to both (ii) and (iii) can reasonably claim to have firm *empirical* grounds for believing (i). If you believe you inhabit a simulated world that is drastically misleading with respect to what is really going on, then the fact that this world has features which make it plausible to believe simulations will be created *en masse* does not provide you with a reason for supposing that these simulations really will be created — and you therefore have no reason to conclude that it is highly likely that you are yourself a simulant, or at least, not on empirical grounds.

The situation with regard to SH2 is very different. If I have good empirical grounds for believing that nearly all simulations that will ever be produced will be faithful, there is no tension whatsoever in my *also* subscribing to (i), and hence thinking it likely that I am myself a simulant. In these circumstances, even though I believe it is highly likely that I am living in a simulation, I also have good reasons (or so we are supposing) for believing that the world I am acquainted with in my simulation accurately depicts how things really are outside my simulation. Consequently, if the socio-technological trends in my (very likely simulated) world are such that it seems probable that large numbers of simulations will be created, it is perfectly reasonable for me to conclude that this will occur in the non-simulated world — and so it is perfectly reasonable for me to retain my empirically-grounded belief that I am myself likely to be a simulant.

So there are conditions under which the simulation argument is *not* self-undermining. Of course, SH1 and SH2 are extreme and rather implausible simulation hypotheses. But although more plausible hypotheses bring further complications, they do not change the over-all picture: even if there are conditions under which the simulation argument is self-undermining, this is by no means invariably the case. To illustrate, suppose you have highly compelling reasons, grounded in detailed analyses of technological and social trends, for subscribing to this simulation hypothesis:

> SH3 Simulated lives outnumber real lives by about ten to one; my type of life has as much chance of being simulated as anybody else's; roughly half of all simulations are faithful, half are unfaithful.

Even if your type of life does has as much chance of being simulated as any other, it would be a mistake in this case to conclude that the odds of your own life's being real are of the order of one in ten. When calculating these odds all *un*faithful simulations must be discounted. For as we have just seen, although it is perfectly consistent to believe on empirical grounds that such simulations exist, it is not consistent to take the further step and infer that their existence makes it more proba-ble than it would otherwise be that you are yourself a simulant. But as we have also seen, there is no such difficulty in the case of faithful simulations. If you have good reasons for believing that simulations outnumber real lives by around ten to one, but only fifty per cent of simulations are likely to be faithful, it is reasonable for you to con-clude that that the odds of your life being real are around one in five.

We can draw a more general lesson from this. The seriousness of the simulation menace in any particular context depends on the degree to which the simulation hypothesis is credible, and the predicted ratio of faithful to unfaithful simulations. Other things being equal, the greater the proportion of unfaithful simulations, the lesser the simula-tion menace, the greater the proportion of faithful simulations, the greater the simulation menace.

A further complication concerns the property of faithfulness itself. To be menacing a simulation need not accurately reproduce some por-tion of the real world in every respect. All that matters is accuracy in those respects that are relevant to assessing the likely quantities and types of simulation that a given world is likely to produce over a given period. Call this *simulation-relevant* (SR) *accuracy*. For a simulation to be SR-accurate it will conform to (apparent) natural laws similar to those that obtain in the real world, and the broad sweep of social and technological trends during the relevant period — in our case, the early 21st century — will also be similar to those in their non-simulated counterparts. SR-accuracy does not require fidelity in matters of historical and biographical detail. To make matters more concrete, think of works of fiction. Nearly all contemporary novels — with the notable exception of most fantasy and some science fiction — are highly SR-accurate despite the fact that most of the characters and events described in such fictions do not exist. To put it another way, if a character in a typical novel with an early 21st century setting were to attempt to assess the simulation menace in their world, the relevant data available to them would be much the same as is available to you or I.

Taken together these considerations have a threefold impact. Since judgments as to SR-accuracy will be based on estimates of a number of difficult-to-estimate variables, in many (but not all) cases it will be impossible to arrive at precise numerical assessments of the simulation menace. However, for the menace to be real we do not need anything particularly fine-grained: 'negligible', 'moderate' and 'high' will serve nicely. Secondly, if only simulations scoring highly on SR-accuracy are menacing at all, the overall simulation menace may well be somewhat reduced, at least for the more moderate brands of simulation hypothesis. Those who think it likely that the future will contain immoderate number of large-scale S-simulations probably do not need to lower their estimates significantly — Bostrom's ancestor-simulations, for example, all score very highly in SR-accuracy.[23] By contrast, those of us who find these scenarios implausible, and who lay more weight on the likely existence of many small-scale N-simulations, many of which may not be SR-accurate, may confidently downgrade our assessments of the simulation menace. Downgrade but (probably) not dismiss: it seems plausible (to me at least) that a good proportion of future simulations will be SR-accurate.

Finally, and importantly, we have uncovered an important constraint, of the transcendental variety, on the *type* of simulation that it is coherent to think one may exist within on empirical grounds. If you

[23] I suspect this is why Bostrom seems little concerned by the allegation of epistemic instability.

are led by the simulation reasoning to the conclusion that there is a fair probability that you are inhabiting a simulation, you have every reason to suppose that the non-simulated world is not *too* dissimilar to your world — in effect, you may be living in a total fiction, but you are not living in a total fantasy. While it remains perfectly possible that you are living in an environment that bears no resemblance to how things really are, coming to accept the simulation reasoning, in itself, should not lead you to think it more likely that this is so.

This constraint may do something to lessen the anxiety-provoking consequences of the simulation argument, but I think it fair to say that it by no means eradicates them entirely. The thought that one is inhabiting the equivalent of a work of fiction is disturbing enough!

7. Some (Further) Varieties of Virtual Life

The distinction between N-simulations and S-simulations reflects one way in which simulated lives can be subjectively indistinguishable but different in kind. There are other ways, and it will be helpful to have a few of these in view before proceeding further.

Active v. Passive (A-simulations v. P-simulations) The subjects of A-simulations are confined to virtual environments, but in all other respects they are free agents — or as free as any agent can be. Their actions are not dictated by the virtual-reality program, they flow from their own individual psychologies, even if these are machine-implemented. A P-simulation, by contrast, is a completely pre-programmed course of experiences. The subjects of P-simulations may have the impression that they are autonomous individuals making free choices, but unlike their A-simulation counterparts, they are deluded: all their conscious decisions are determined in advance by the virtual reality program. Such subjects have *apparent* psychologies — their consciousness is subjectively similar to that of someone with an active psychological system, so they have apparent memories, hopes, fears, etc. — but their real psychologies are entirely suppressed (or even non-existent).

Original Psychology v. Replacement Psychology (simulations$_{OP}$ v. simulations$_{RP}$) In A-simulations, a 'replacement psychology' is an artificially-generated system of beliefs, desires, memories, intentions, preferences, personality traits and so forth that supplants a subject's own ('original') psychology. The same applies in P-simulations, the difference being that the replacement psychology is only *apparent*, in the sense just introduced. There is a sense in which the inhabitants of simulations$_{RP}$ are doubly deceived: not only is their environment not what it seems, neither are their minds.

Communal v. Individual (C-simulations v. I-simulations) A C-simulation is a virtual environment shared by a number of different subjects,

each possessing their own distinctive individual psychology (even if
these are machine-implemented). An I-simulation is restricted to a single subject. Of course, the subject of an I-simulation may meet what
they take to be other people in their virtual worlds, but these 'others' do
not possess their own individual autonomous psychological systems —
they are not subjects in their own right, merely parts of a machine-generated virtual environment.

These options can be combined in various ways, e.g., a simulation of
type AC_{RP} is active, communal with replacement psychology, whereas
a simulation of type PI_{OP} is passive, individual with original psychology. There is a total of eight permutations:

AI_{OP}: Active/Individual/Original Psychology
AI_{RP}:Active/Individual/Replacement Psychology

AC_{OP}:Active/Communal/Original Psychology
AC_{RP}:Active/Communal/Replacement Psychology

PI_{OP}:Passive/Individual/Original Psychology
PI_{RP}:Passive/Individual/Replacement Psychology

PC_{OP}:Passive/Communal/Original Psychology
PC_{RP}:Passive/Communal/Replacement Psychology

Assuming that each of these modes could be generated by either
N-methods or S-methods, we have a grand total of sixteen distinct
kinds of (subjectively indistinguishable) virtual life. But the situation
may not be quite so complex. A strong case can be made for thinking
that a truly *communal* simulation of the passive variety is impossible.
There is nothing impossible in the idea of a number of subjects simultaneously playing out roles in similar and coordinated hallucinations,
but unless these subjects can interact and converse with one another
they can scarcely be said to constitute a genuine community, and this
cannot really occur in P-simulations (although this might not be obvious to the simulants concerned). For this reason it seems right to
regard all P-simulations to be of the individual variety. This brings our
grand total down to twelve.[24]

[24] To simplify, I overlook here the fact that the distinction between N-simulations and S-simulations may not be absolute: the consciousness of future humans (or post-humans) may
be sustained by a combination of neural and artificial means, and neurons themselves may
be genetically manipulated. I also ignore the fact that in some logically possible worlds,
simulations are created by quite different means (e.g. magic). It should also be noted that
simulants of different types can coexist, e.g., *The Matrix* films feature a combination of
N-simulations (ordinary humans) and S-simulations (the 'agents'), both active, coexisting
in a single C-type virtual environment.

8. A New Scepticism – Or A New Metaphysics?

Ancient Greek sceptics argued that since our senses can deceive us we can never be justified in supposing that the world is how it seems, but the idea that there might not even *be* an external world never occurred to them.[25] For the latter hypothesis to be thinkable consciousness must be construed as a self-contained and potentially autonomous realm of existence in its own right. Descartes was the first to articulate this conception clearly, and drew the (now) obvious sceptical conclusion: our experience could be (subjectively) just as it is even if the reality external to our consciousness is very different from how we believe it to be on the basis of our experience. Consequently, we cannot be certain that the physical world exists.

There are similarities as well as differences between Simulation Scepticism (as we might as well call it) and Cartesian Scepticism. So far as the existence of a mind-independent reality is concerned, Simulation Sceptics are as one with their Greek predecessors: it exists (it is where the simulations are being run). But Simulation Sceptics are as one with Cartesian Sceptics when it comes to the status of our current consciousness: both hold that it could be wholly virtual, a detailed and convincing hallucination. However, for the Cartesian this conclusion relies on the world external to our minds being very different from how it appears (e.g., reality might consist of nothing but your consciousness and a malicious Demon). Simulation Sceptics, by contrast, derive their conclusion from the assumption that our experience is a broadly reliable (SR-accurate) guide to the character of that portion of the external reality it seems to concern.

Simulation Scepticism is in this respect less radical than its Cartesian counterpart, but it is also less of a blind alley. Different hypotheses as to how the future may turn out, or what the universe may contain, render the hypothesis that we are leading virtual lives more or less likely, and these hypotheses can be refined, explored and evaluated. Unfortunately, this gain comes at a price. The threat posed by Simulation Scepticism is far more *real* than that posed by its predecessors. Cartesian scepticism is hard to refute, but as Hume noted, it is also hard to take seriously. Few of us spend much time worrying about the possibility that reality could be radically different from how it seems. Simulation Scepticism reveals that even if reality *is* largely as we believe it to be, there could be a high probability that our actual condition is very different from our apparent condition. As things stand, with simulation technology still at a primitive level, many will

---

[25] Or so it has been argued (*cf.* Burnyeat, 1982).

find Simulation Scepticism as hard to take seriously as its Cartesian counterpart. This will change as the technology advances.

However, it could be a mistake to regard the simulation argument as leading to a sceptical hypothesis of even a modest kind. An alternative approach is to construe the reasoning as leading to a novel *metaphysical* hypothesis concerning the underlying nature of one's environment. For some simulation scenarios, if not all, there is a good deal to be said for adopting this line. Suppose, for example, that you come to believe that you are living in a large-scale, long-lasting, communal simulation. You may initially be inclined to think: 'The material world of which I took myself to be an inhabitant does not exist — all my experience has been delusory, akin to a dream or hallucination.' This is an overreaction, albeit an understandable one. Your world may not be 'material' in the usual sense of the term, but there is certainly a sense in which it is real despite its underlying computational nature. Simulated worlds can be immensely complex, and nomologically speaking, just as well (or badly) behaved as their non-virtual counterparts. Moreover, by virtue of inhabiting a communal simulation, you are not alone. In the company of your fellow C-simulants you are at liberty to conduct empirical explorations of your environment, and agree and disagree on your findings, in all the ways available to the inhabitants of non-virtual worlds. As a consequence distinctions between appearance and reality, between subjective and objective, are as well-founded in your virtual world as they are in any world. Even if your perceptual experience does not directly reveal the real in the way you once naively supposed, it nonetheless reveals a world which possesses many of the defining properties of 'reality': your world is certainly objective, and independent of your mind. Thus construed, the simulation argument does not threaten to undermine our ordinary empirical beliefs — these remain mostly true — what is threatened, rather, is a doctrine concerning the underlying real nature of the world we inhabit. The import of the simulation reasoning is primarily metaphysical rather than epistemological.

That many simulation scenarios should be construed in this manner has been forcefully argued by David Chalmers:

> ...the Matrix Hypothesis is not a skeptical hypothesis. If I accept it, I should not infer that the external world does not exist, nor that I have no body, nor that there are no tables, chairs, and bodies, nor that I am not in Tucson. Rather, I should infer that the physical world is constituted by computations beneath the microphysical level. There are still tables, chairs and bodies: these are made up fundamentally of bits and of whatever constitutes these bits. (Chalmers, 2005, §5)

Chalmers' principle example is a Matrix-type scenarios — a simulation of the N-type AC variety, in the terminology introduced in §7 above — but, as he makes clear, he believes that the metaphysical interpretation extends to S-type simulations. He goes further, and suggests that the metaphysical interpretation can be extended to small-scale, short-lived simulations. In such cases, simulants can still reasonably be regarded as being in perceptual and cognitive contact with a genuine world, it is just that the world is rather smaller than it seems, and as a result, fewer empirical beliefs are true than would be the case for full-scale simulations.

If Chalmers' is correct, the simulation scenarios provide a new twist, and a new clarity, to the Kantian thesis that our world may only be 'empirically real'. As for whether Chalmers' *is* correct, I find much of what he says very plausible. But I do have one worry.

A compelling case can certainly be made for holding that the environments in which the relevant simulants find themselves can legitimately be classed as *worlds* — or 'external realities' — despite their being virtual in nature. However, we can accept this much without also accepting that an external world of this sort constitutes a properly *spatial* world. This is of some significance, because many would also incline to the view that only worlds which are spatial in nature are candidates for being regarded as *physical*. Now, Chalmers maintains that the relevant virtual worlds are both spatial and physical in nature, but it is certainly not obvious that this is the case. Indeed, he anticipates someone objecting to his position along precisely these lines: 'one could suggest that the problem with the matrix is that its spatial properties are all wrong. We believe that external entities are arranged in a certain spatial pattern, but no such spatial pattern exists inside the computer' (Chalmers, 2005, note 14). Chalmers' main line of reply to this objection involves a hypothesis, and two principles. The hypothesis is that it is at least possible that the microphysical processes throughout our space-time are in fact constituted by computational processes. The relevant principles are these: (1) any abstract computation that could be used to simulate physical space-time is such that it *could* turn out to underlie real physical processes, and (2) given an abstract computation that *could* turn out to underlie physical processes, the precise way in which it is implemented is irrelevant to whether it *does* underlie physical processes (*ibid.*, §5).

Taken together these ingredients deliver the result Chalmers wants, and (here at least) I do not want to question the hypothesis that it is at least possible that the physical world is constituted by computational processes. However, the claim that the way a computation is

implemented in the real world can make no difference to whether or
not the implementation can legitimately be taken to constitute a genuinely spatial universe, as encapsulated in (2), is more questionable.
Why? Simply because some computational implementations are more
intrinsically spatial in nature than others.

To illustrate, let's focus on a very simple universe: a finite three-
dimensional space, discrete rather than continuous, whose contents
comprise a few million particles, which are moving around and interacting in accord with a small collection of simple dynamical laws. As
it happens, this universe can most easily and efficiently be modeled by
treating it as a three-dimensional cellular automaton, with a small collection of local rules governing the behavior of individual cells (as in
Conway's well-known Game of Life, cells change state in response to
the cells in their immediate neighbourhood). Now compare two ways
in which this abstract computational model could be implemented:
(A) the program is run on an ordinary (classical) desktop computer,
(B) the program is run on a specially created physical realization of
the relevant cellular automaton, i.e., a physical system consisting of a
spatially extended three-dimensional grid, whose cells vary in their
physical properties in accord with automaton's rules. Is either of these
implementations successful in constituting a genuine spatial system?
Are the worlds that are created by these computational processes spatial in nature? To deny that implementation (B) generates or constitutes a spatial world would clearly be absurd. But situation is far less
clear-cut in the case of implementation (A). The desktop computer in
question will be rapidly shuffling patterns of bits through its central
processing cores, temporarily storing information in various parts of
its RAM, encoding other information in a spatially scattered way on
its hard drives, and so forth. This scattered and discontinuous computational process successfully *represents* a continuous space, but the
process itself does not consist of a continuous spatial manifold or
medium — this is in sharp contrast to the process in (B), which clearly
and unambiguously does. Given this, the claim that in (B) the computational process constitutes a real space, but in (A) the space created is
merely a virtual one — nothing more than an appearence — has a
good deal of plausibility.[26] As a consequence Chalmers' claim that the

---

[26] To make matters a little more precise we might say the following. Suppose P is a program
for a virtual world. For the space S created by a particular implementation of P to be a genuine rather than merely virtual space, the spatial relations between the regions and material contents of S must closely match those of corresponding computational processes
(i.e., the processes which underlie or constitute these regions and contents). So, for example, if region R1 in S is entirely encloses region R2, the computational processes

precise way in which a program is implemented is irrelevant to whether or not it constitutes or underlies a physical process is itself undermined.

Of course, we have been considering a small toy universe, but so far as I can see, the key point — that not all computational implementations are the same when it comes to their intrinsic spatial properties — still applies on the broader stage. Indeed, some of those who have argued for the possibility that our own universe is a computer, e.g. Zuse (1970) and Wolfram (2002), seem to have also held that at the fundamental level our universe is something akin to a giant cellular automaton (one with a very fine-grained grid). If this hypothesis is correct, then our universe is both spatial and computational in nature. Whether or not the same applies to any virtual worlds being sustained by other (smaller) computers may well depend on their specific mode of implementation.[27]

9. Simulation Ethics

So much for how we should conceive of virtual worlds. What of how we should act in them? Would taking the simulation argument seriously have any practical or ethical implications for how we should lead our lives?

If we knew what kind of simulation we were living in, the answer would clearly be in the affirmative. In an I-simulation what appear to be other people are no more than the appearances of such. Simulants

---

responsible for R1 will entirely enclose those responsible for R2; if R2 occurs between R1 and R3, then the same will apply to the corresponding computational processes. Since 'close match' can be interpreted more or less stringently — e.g. matching topological features might suffice, or we might insist on similar metrical features — the criterion is a flexible one. That said, it is difficult to see that how even the most lax interpretation of 'close' would allow the world created by implementation (A) to count as genuinely spatial.

[27] Chalmers also argues (also in *Matrix*, note 14) that anyone who insists that the implementing level must itself have an appropriate spatial structure before it can be counted as constituting a physical world is running counter to the spirit of contemporary physics, where the claim that our own physical space is not a fundamental feature of the world, but rather an emergent one, is being taken seriously by leading physicists and cosmologists. Physicists are indeed taking this claim seriously, but as far as I can see, in several well-regarded approaches (e.g. loop quantum gravity) all that is being dispensed with is space construed as an entirely autonomous or independent background medium; spatially related material particulars remain very much in the frame. The holographic principle of 't Hooft and Susskind *could* be interpreted as entailing that our universe is really two dimensional (Susskind, 1994, §1: 'Instead of a three dimensional lattice, a full description of nature requires only a two dimensional lattice at the spatial boundaries of the world. In a certain sense the world is two dimensional and not three dimensional as previously supposed.') This way of interpreting this principle is controversial, but even so: there is a big difference between a world of two spatial dimensions and a world of *no* spatial dimensions!

such as these have the same ethical status as characters in contemporary computer games, and they could be treated accordingly — though
of course, *un*like characters in contemporary computer games, they
can hit back, and so some caution is in order. The knowledge that one
might be living in an I-simulation can also console. Anyone who has
had cause to regret an action because of its consequences — and that
includes most of us — will be cheered by the thought that these consequences might not in fact be real. The downside of this is that the same
would apply to one's most valued relationships. The knowledge that
one is living in a passive simulation, and so cannot be held responsible
for one's actions or omissions, brings similar advantages and disadvantages. Passive simulants cannot be blamed for their mistakes or
wrongdoings, but neither do they merit admiration for their
achievements.

Further examples could be supplied, but the point is clear. Simulation scenarios have practical and ethical consequences, and these consequences vary depending on the type of simulation involved. And
there lies the problem. Even if we knew the precise probability of our
lives being simulated, which we don't, this knowledge would be useless for most practical purposes unless we also knew the sort of simulation being run, along with and the intentions and preferences of the
simulators. Unless and until such knowledge is forthcoming, it is
probably best to continue much as we would otherwise do.[28]

However, even if this approach is optimal for many, it may not be
appropriate for everyone. Anyone who thinks they are living an especially interesting life, and so having experiences which future simulators might be particularly interested in 'sharing', will be led to the
conclusion that their life has a greater than average chance of being
virtual. It is hard to predict what effects this realization might have on
(say) political and military leaders of the future, but they may well not
be wholly beneficial, to put it mildly.

There is a second ethical issue to consider. Since simulated lives are
subjectively indistinguishable from the real thing, their creation is by
no means a trifling matter, morally speaking. Even if our descendants

---

[28] The difficulty of attempting to second-guess the likely preferences of simulators is illustrated by Hanson's (2001) recommendations as to how one should act so as to reduce the
risk of the curtains being brought down on one's virtual world. He concludes thus: 'If you
might be living in a simulation then all else equal it seems that you should care less about
others, live more for today, make your world look likely to become eventually rich, expect
to and try to participate in pivotal events, be entertaining and praiseworthy, and keep the
famous people around you happy and interested in you.' I am not sure that anyone could
conform to *all* these injunctions simultaneously, but even attempting to do so might well
make one so obnoxious as to hasten one's end.

(whether human or machine) develop the means of producing such simulations easily and cheaply, might they choose not to do so? Might ethical scruples eliminate or at least diminish the threat posed by the simulation argument? Since we are dealing with the future it is impossible to be sure, but there are certainly some reasons for thinking it unlikely.

One of the uncertainties derives from the fact that future simulation technologies, or at least a significant proportion of them, may well be in the hands of superintelligent machines. Although we might try to ensure that these AIs share our values, as Chalmers notes in *Singularity* §6, there is no guarantee that we will be successful. If a superintelligence emerges through a digital version of natural selection — a not unlikely eventuality, in my view — our influence on it will inevitably be limited: we will probably have a very imperfect understanding of its program. But even if we do design an AI+ system, and choose which values to instill into it, there is no guarantee that the more sophisticated A++ designs which follow will share these values; after all, these systems will be far more intelligent that we. Although Kantians are of the view that rationality and morality are by their natures inseparable, a compelling proof of this connection has proved elusive. Pulling these points together: when it comes predicting the value-systems which future AIs will subscribe to, there is very little we can be sure of. As a consequence, even if there are ethical reasons for not creating virtual lives which *we* (or our human descendants) find compelling, we cannot be sure that A++ systems will share this view.

There are further complications. It is easy to conceive of simulations which most of us would judge to be morally abhorrent, and thus wrong to create. An obvious example would be S-simulations of entire virtual worlds all of whose inhabitants suffer nothing but perpetual pointless torment. It is far from inconceivable that our descendants (human or machine) will forbear from creating such things. That said, even this is by no means certain. The plot of Iain M. Banks' recent novel *Surface Detail* (2010) revolves around virtual hells, deliberately created and maintained as fit punishments for the wrong-doers uploaded into them; the various gruesome tortures inflicted on the unfortunate inhabitants would soon put an end to any flesh and blood human, but can be sustained for seeming-eternities in a virtual world. Can we be sure that none of our descendants will find this mode of punishment appealing?

In any event, not all large-scale simulations are clearly morally abhorrent, far from it. Would creating an ancestor-simulation — a

complete S-simulation of human history up until the present time —
be morally wrong? It is not at all obvious that it would. The sum total
of human misery may be immense, but so too is the sum total of human
happiness, and on balance, most people are glad to have had the
opportunity of existing. Given this, what could be immoral about creating ancestor-simulations? Since the inhabitants of an ancestor-simulation would feel the same way about their lives as we do about ours,
mightn't it be immoral *not* to create ancestor-simulations, if one had
the means of so doing?

But the situation is by no means this straightforward. The fact that
simulations need not be unpleasant does not mean their creation is
morally unproblematic:

> *The Objection from Lesser Value* A real life has greater intrinsic value
> than a subjectively similar simulated life. Since it is wrong to impose on
> others a low-grade form of existence that one would prefer to avoid one-
> self, the creation of simulated lives is immoral.

This objection may seem weak: even if virtual lives do possess less
intrinsic value than their non-virtual counterparts, other things being
equal, they can still be lives worth living, and hence lives that are
worth creating. However, there is a further point to bear in mind. As
Nozick's imaginary case of the experience machine reveals, the desirability of a life is not determined solely by the desirability of the experiences it contains. An experience machine will supply you with a
(virtual) life of any kind you like, so by connecting yourself up to such
a device you are guaranteed a very enjoyable (virtual) life, a life in
which as many of your desires as you choose to come true will come
true. But as Nozick observes, few of us would choose permanently to
connect ourselves to an experience machine if we had the opportunity
of so doing, and for good reason: 'What is most disturbing about them
is their living of our lives for us' (Nozick, 1980, p. 44).

While this lesson is important, in the present context it is also of
limited relevance. The virtual lives sustained by experience machines
are of the *passive* kind: they consist of solitary streams of consciousness that are completely controlled and pre-programmed. As we have
seen, not all virtual lives need be like this. Of particular interest here
are *Diaspora*-style AC-simulations, i.e., virtual lives that are both
active and *communal*, in the senses introduced above. Subjects in
AC-simulations possess their own autonomous psychological systems (whether original or replacement). They lead their own lives:
their actions are not pre-programmed (they are as free as anyone can
be). And they can causally interact with other subjects in their virtual

environment (and these other subjects are autonomous individuals in their own right, rather than merely the appearances of such). Given all this, it is hard to see why life in an AC-simulation should be regarded as being inherently less valuable or worthwhile than a normal life. True, the inhabitants of AC-simulations are not physically embodied in the normal way, but they can possess virtual bodies that are indistinguishable from the real thing. They are unable to manipulate physical objects, but they can manipulate virtual objects which *seem* physical. Why should the undetectable absence of a (non-virtual) material environment significantly diminish the value of the lives of these subjects? I cannot see any reason why it should.[29]

These considerations further weaken the Objection from Lesser Value. Even those who find this objection persuasive would only have reason to avoid creating passive simulations; there is no reason why they should be reluctant to create AC-simulations.

However, there is a further, and potentially more serious objection to the fostering of virtual life:

> *The Deception Objection* The subjects of simulations are being deliberately deceived; their lives are virtual, but they believe them to be real. This deception is engineered and maintained by the relevant simulators. Such actions are clearly wrong.

Deception is not an inevitable consequence of simulation; there may well be simulants who are perfectly aware of their true condition — as the software citizens in *Diaspora* are. But since few contemporary humans believe themselves to be leading simulated lives, the Deception Objection does apply to simulations of the menacing variety. This is not to say that it will have an impact on simulation policy. It is conceivable that future simulators will take the view that although deception is wrong, the kind of deception being perpetrated on simulants does not constitute a wrong that is sufficiently serious to outweigh the boon of existence. But equally, it is conceivable that future simulators *will* be swayed by the Deception Objection, and restrict their simulation activities accordingly. This may not seem likely, but since we can only guess at the ways ethical considerations will influence the simulation policies of our (quite possibly super-intelligent) descendants it cannot be ruled out.

It should also be noted that the force of the Deception Objection may well depend on the type of simulation under consideration. The

---

[29] Berkeley was perhaps the first to make this point, when he argued for the redundancy of mind-independent material reality. The inhabitants of some of the polises in Egan's *Diaspora* take the same view.

objection has considerable force in the context of long-term S-simulations of entire civilizations: anyone who creates an ancestor-simulation is responsible for the deceiving of billions of (virtual) people for thousands of (subjective) years. The situation is very different in the case of small-scale, short-term N-simulations. You are, let us suppose, feeling run-down by the demands of your 22nd century job, and decide to spend a couple of days in the (virtual) past to unwind; you employ the method of self-induced controlled hallucination, and 'wake up' in early 19th century England, in the midst of the Napoleonic wars. As you enjoy your adventure, are you the victim of a deliberate deception? In a sense, yes: you have opted for the fully-immersive trip, and so believe yourself to be a typical early 19th century person. But is the kind of deception involved in this case morally problematic? Surely not. Rather than one person imposing an uninvited delusion on another — as in the case of ancestor-simulations — we are dealing here with a person freely choosing to impose a short-term and harmless delusion *upon themselves*. Where is the wrong in that?

This implications of this point are by no means trivial, for as we saw earlier, given sufficient time, N-simulations might easily be created in sufficient numbers so as to be seriously menacing, even without the advent of superintelligence. But we are not yet done. There is at least one further reason why our descendants might avoid indulging in menacing simulatory practices, a reason that is pragmatic rather than ethical:

> *The Self-Interest Consideration* Future generations of humans and machines will be well acquainted with the simulation reasoning, and so will impose tight restrictions on simulation creation. They will realize that unless such restrictions are imposed, and enforced, no one — themselves included — will be in a position to rule out the likelihood that their lives are virtual rather than real.

I am not confident that such a policy will ever be adopted, for a number of reasons.

(1) Future simulators may well include superintelligent AIs. It is difficult to predict how these AIs will react to the knowledge that their own mental lives may well be simulations, nor can we predict — for the reasons outlined above—what their attitude to the creation of virtual human lives will be. Furthermore, even if we (humans) wanted to restrict the simulatory practices of these superintelligences, it is by no means clear that we would be able to.

(2) Leaving superintelligences aside, at a more mundane level, simulation technology is certain to play an increasing role in recreational activities, and people will become accustomed to, and demand, ever more lifelike simulations — just as today there is a demand for ever more life-like computer games. Since a ban on life-like simulations would be unpopular with both the public and powerful commercial concerns, the prospects of one being implemented are slim.[30]

(3) Many people will be unlikely to take the simulation argument seriously until they themselves have experienced what the technology can do, and taken a fully-immersive trip to the past or future. Should this point every be reached, billions of menacing simulations will have been created, and it will be obvious to everyone that it is already too late to consider a ban.

(4) To have the desired effect, a ban on simulations would have to be continued into the indefinite future. But even if an effective ban could be enforced in the present, we could never be confident that this policy would not be abandoned, or fail, at some future date — not least because it would be foolhardy to rule out the possibility of a singularity occurring, bringing with it hard-to-predict superintelligent machines. For this reason alone it is unlikely that our descendants would be willing to deprive themselves of all the benefits advanced simulation technology makes available.

There is a more general point. We are in the process of emerging from an age of innocence, an innocence that we are unlikely ever to recapture. Innocence was being able to believe that only sceptical possibilities of the most radical sort stood between ourselves and the world about us. This innocence evaporates on contact with the knowledge that even if reality is much as it seems, there is a significant likelihood that one's current consciousness is simulated. Having to live with this knowledge may well be part of the normal lot of technologically advanced conscious beings, whether biological or non-biological, the universe over. When this realization fully dawns on our descendants, attempting to recapture their lost innocence by imposing restrictions on simulatory practices will very likely strike them as futile. Since any restrictions on simulation creation can always be lifted subsequently, it will be obvious that their imposition would

---

[30] Perhaps I was not the only one to feel a slight chill when reading of Chalmers' discussions of the possible dangers of research into AI with West Point Military Academy cadets and staff. In short, there is no chance that the risks associated with an intelligence explosion are going to deter the military from trying to develop superintelligence: the dangers if the other side should get there first are simply too great (Chalmers, p. 33).

offer only meagre protection against the menace of simulation. But another factor will enter into the reckoning. Even if innocence once lost is impossible to regain, innocence can of course be *simulated*. If our descendants want to escape the shadow of simulation and experience for themselves what it was like to exist in more innocent times, they may have but one option: to embark on fully immersive virtual reality trips into the past. Not only does this further reduce the chances of restrictions on simulation creation being imposed, it is also bad news for our predecessors. It could easily be that the vast majority of people who find themselves living in more innocent times are simulants.

Our own predicament is only slightly better. Many of our descendants might be tempted by the prospect of finding out what it was like to *become* aware of the simulation menace; experiencing the first falling of the shadow might be an irresistibly appealing prospect. If so, life in the early 21st century may be an even more fragile thing than it appears.[31]

References

Banks, I.M. (2010) *Surface Detail*, London: Orbit.
Bostrom, N. (2003) Are you living in a computer simulation?, *Philosophical Quarterly*, **53** (211), pp. 243–255.
Bostrom, N. (2009a) The simulation argument: Some explanations, *Analysis*, **69** (3), pp. 458–461.
Bostrom, N. (2009b) *The Simulation Argument FAQ*, [Online], http://www.simulation-argument.com/faq.html
Burnyeat, M.F. (1982) Idealism and Greek philosophy: What Descartes saw and Berkeley missed, *The Philosophical Review*, **XCI** (1), pp. 3–40.
Chalmers, D.J. (1996) *The Conscious Mind*, Oxford: Oxford University Press.
Chalmers, D.J. (2005) The Matrix as metaphysics, in Grau, C. (ed.) *Philosophers Explore the Matrix*, Oxford: Oxford University Press. Also, [Online], http://consc.net/papers/matrix.html
Dainton, B. (2004) The self and the phenomenal, *Ratio*, **17** (4), pp. 365–389.
Dainton, B. (2008) *The Phenomenal Self*, Oxford: Oxford University Press.
Dainton, B. (2010) Temporal consciousness, *Stanford Encyclopedia of Philosophy*, [Online], http://plato.stanford.edu/entries/consciousness-temporal/
Dainton, B. (forthcoming) Selfhood and the flow of experience, *Grazer Philosophische Studien*.
Dainton, B. & Bayne, T. (2005) Consciousness as a guide to personal persistence, *Australasian Journal of Philosophy*, **85** (4), pp. 549–571.
Egan, G. (1997) *Diaspora*, London: Gollanz.

[31] And you may feel it unwise to dwell on these matters further. My thanks to: Tim Bayne, Nick Bostrom, David Carlyon, David Chalmers, Stephen Clark, Richard Gaskin, Gerard Hurley, Jonathan Lowe, and audiences at Bradford, Glasgow and Stirling. I am also grateful to Durham's Institute of Advanced Study, which provided an ideal environment for thinkng about these issues.

Flanagan, O. (2000) *Dreaming Souls: Sleep, Dreams, and the Evolution of Conscious Life*, Oxford: Oxford University Press.

Hanson, R. (2001) How to live in a simulation, *Journal of Evolution and Technology*, **7**, [Online], http://www.jetpress.org/volume7/simulation.htm

Johnston, M. (2010) *Surviving Death*, Princteton, NJ: Princeton University Press.

Lockwood, M. (1989) *Mind, Brain and the Quantum*, Oxford: Blackwell.

McGinn, C. (1999) *The Mysterious Flame*, New York: Basic Books.

Markham, H. (2006) The Blue Brain Project, *Nature Reviews Neuroscience*, **7**, pp. 153–160. Also, [Online], http://www.hss.caltech.edu/~steve/markham.pdf

Naselaris, T., Prenger, J., Kendrick, K., Oliver, M. & Gallant, J. (2009) Bayesian reconstruction of natural images from human brain activity, *Neuron*, **63** (6), pp. 902–915.

Nozick, R. (1980) *Anarchy, State, and Utopia*, Oxford: Blackwell.

Searle, J. (1992) *The Rediscovery of the Mind*, Cambridge, MA: MIT Press.

Strawson, G. (1994) *Mental Reality*, Cambridge, MA: MIT Press.

Susskind, L. (1994) The world as a hologram, [Online], http://arxiv.org/abs/hep-th/9409089

Wolfram, S. (2002) *A New Kind of Science*, Champaign, IL: Wolfram Media Inc.

Zuse, K. (1970) *Calculating Space*, MIT Technical translation of *Rechnender Raum* (1969).

Daniel C. Dennett

The Mystery of David Chalmers

1. Sounding the Alarm

'The Singularity' is a remarkable text, in ways that many readers may not appreciate. It is written in an admirably forthright and clear style, and is beautifully organized, gradually introducing its readers to the issues, sorting them carefully, dealing with them all fairly and with impressive scholarship, and presenting the whole as an exercise of sweet reasonableness, which in fact it is. But it is also a mystery story of sorts, a cunningly devised intellectual trap, a baffling puzzle that yields its solution — if that is what it is (and that is part of the mystery) — only at the very end. It is like a 'well made play' in which every word by every character counts, retrospectively, for something. Agatha Christie never concocted a tighter funnel of implications and suggestions. Bravo, Dave.

So what is going on in this essay? It purports to be about the prospects of the Singularity, and since I can count on readers of my essay to have read Chalmers, I needn't waste so much as a sentence on what that is or might be. I confess that I was initially repelled by the prospect of writing a commentary on this essay since I have heretofore viewed the Singularity as a dismal topic, involving reflections on a technological fantasy so far removed from actuality as to be an indulgence best resisted. Life is short, and there are many *serious* problems to think about. I said as much in an email to the editor only to get an email in response from Chalmers, urging me to reconsider:

> hi dan,
> take a look at the paper. somehow i suspect that you'll have plenty to say. some of the core issues here concern the structure of intelligence/ design space, topics that you've thought pretty hard about.
>
> cheers, dave. (Personal correspondence [quoted with permission]).

And since I respect Chalmers' judgment, I relented, and read the essay. My reactions to the first thirty-odd pages did not change my mind about the topic, aside from provoking the following judgment, perhaps worth passing along: thinking about the Singularity is a singularly *imprudent* pastime, in spite of its air of cautious foresight, since it deflects our attention away from a much, much more serious threat, which is already upon us, and shows no sign of being an idle fantasy: we are becoming, or have become, enslaved by something much less wonderful than the Singularity: the internet. It is not yet AI, let alone AI+ or AI++, but given our abject dependence on it, it might as well be. How many people, governments, companies, organizations, institutions, ... have a plan in place for how to conduct their most important activities should the internet crash? How would governments coordinate their multifarious activities? How would oil companies get fuel to their local distributors? How would political parties stay in touch with their members? How would banks conduct their transactions? How would hospitals update their records? How would news media acquire and transmit their news? How would the local movie house let its customers know what is playing that evening? The unsettling fact is that the internet, for all its decentralization and robust engineering (for which accolades are entirely justified), is fragile. It has become the planet's nervous system, and without it, we are all toast.

So endeth the sermon. And now to the rest of his essay, which does indeed touch on topics about which I have thought long and hard. All along, he scrupulously draws attention to the places where his argument is porous. Thus, when discussing the basic, enabling premise of the essay, he notes that 'there is room in logical space to resist the argument' (p. 25) in the form of doubts about whether an intelligence measure can be secured that permits it to be scaled ordinally (y is more intelligent than x, and z is more intelligent than y, so [?] z is more intelligent than x), and perhaps would in any case be better represented by a logarithmic scaling. An admirable attention to minutiae! But — I think he does slight this (at least logical) possibility — perhaps human intelligence is so remote in degree from all previous forms of intelligence in the natural world (dolphins, chimps, starfish, bacteria), that any scale we could contrive (think of IQ!) would be so anthropocentric as to be comically distorting of whatever reality it was called upon to measure. In any event, the inexorable march of all these stacked inferences leads us to worry about whether we human beings would be left out in the cold after the Singularity, and hence leads us to consider the prospects for 'uploading' ourselves into the AI+ world. This provides an interesting hypothetical motivation (for the first time, really) for taking some favorite philosophical puzzles seriously: 'the key question is: will I survive uploading?' (p. 46). While many philosophers and philosophy students have zestfully tackled the problems of personal identity and consciousness over the years, spurred on in some measure by Chalmers' own musings on the topics, the prospect of the Singularity probably provides a boost of self-interest, mounting even to alarm, in readers who would otherwise ignore these puzzles: if uploading is my only hope of surviving the Singularity, I had better take a good hard look at the idea, and not too breezily dismiss it as an amusing but idle philosophical fantasy or riddle! If, for instance, you never before found the debate between *further-fact* theorists and *closest continuer* exponents gripping your attention, maybe now you can be made to care deeply. Or maybe not, but it's a nice try, and it does frame the issues in a rather crisper setting than most earlier treatments.

2. Uploading and Consciousness

Here is where the mystery begins to emerge. 'One central problem,' Chalmers tells us, 'is that consciousness seems to be a *further fact* about conscious systems' (p. 47) over and above all the facts about their structure, internal processes and hence behavioral competences and weaknesses. He is right, so long as we put the emphasis on 'seems'. There does *seem* to be a further fact to be determined, one way or another, about whether or not anybody is actually conscious or a perfect (philosopher's) zombie. This is what I have called the Zombic Hunch (Dennett, 2005). I can feel it just as vividly as anybody; I just don't credit it, any more than I credit the sometimes well-nigh irresistible hunch that the sun goes around the earth; it surely does *seem* to go around the earth. This makes me, in Chalmers' taxonomy, a 'type-A materialist' as contrasted with 'type-B materialists' such as Ned Block and 'property dualists' such as Chalmers himself. Chalmers thinks 'It is worth noting that the majority of materialists (at least in philosophy) are type-B materialists and hold that there are epistemologically further facts' (fn 27, p. 47). He's probably right about this, too, more's the pity, but I think it tells us more about the discipline of philosophy than about the likely truth. I suspect that he doesn't give us any percentages of allegiance for non-philosophers because he just can't get non-philosophers to pay attention long enough to be sure they understand all the philosophical fine points that distinguish the options arrayed for their selection.

We are now ready for the posing of the mystery. *Why is Chalmers not a type-A materialist?* He gives very good arguments for type-A materialism, and finds no flaws in them. He also sides with type-A materialism against type-B materialism. And on the subsidiary issue of the distinction between *biological* and *functional* theories of consciousness — and the disagreement here is 'crucial' since an implication of biological theories is that uploads cannot be conscious (alas!) — he sides with me (against Block and Searle, for instance): 'My own view is that functionalist theories are closer to the truth here.' I am not entirely happy with his way of putting it:

> It is true that we have no idea how a nonbiological system, such as a silicon computational system, could be conscious. But the fact is that we also have no idea how a biological system, such as a neural system, could be conscious. (p. 48)

I think we *do* have lots of ideas about how such systems, biological or silicon, could be conscious, but I agree that it is just as hard to see this when staring at neurons as when staring at circuit boards. He goes on in any case to support this view 'with further reasoning', as he says, in both the main text and in footnotes. Considering a variation on Searle's (1992) thought experiment about the possible outcomes of 'gradual uploading', he sides with me again (see Dennett, 1993) noting that a gradual fading of consciousness in such a case 'seems implausible' (p. 50).

> We can imagine that at a certain point partial uploads become common, and that many people have had their brains partly replaced by silicon computational circuits. On the sudden disappearance [of consciousness] view, ...[p]eople in these states may have consciousness constantly flickering in and out, or at least might under total zombification with a tiny change. On the fading view, these people will be wandering around with a highly degraded consciousness, although they will be functioning as always and swearing that nothing has changed. In practice, both hypotheses will be difficult to take seriously.
>
> So I think that by far the most plausible hypothesis is that full consciousness will stay present throughout. (p. 51)

Indeed. I consider this to be an impressive consideration in favor of type-A materialism. It is observations of just this sort, in fact, that have always persuaded me that any alternative to type-A materialism is forlorn. Chalmers manifestly understands the arguments; he has put them as well and as carefully as anybody ever has. (See also his chapter 7 of *The Conscious Mind*, which, as he notes in fn 30, develops the arguments in even more careful detail.) So what is holding him back? Why does he cling to the Zombic Hunch and 'property dualism'? He tells us, point blank: 'Of course it remains at least a logical possibility that this process will gradually or suddenly turn everyone into zombies.' (p. 51) A logical possibility. How seriously should we take this logical possibility? 'But once we are confronted with partial uploads, that hypothesis will seem akin to the hypothesis that people of different ethnicities or genders are zombies' (*cf.* Dennett, 1991, pp. 405–6). Chalmers is reminding us of just how negligible the philosophers' notion of logical possibility can be: it is *logically* possible that all women, or lefthanders, or people born under the sign of Capricorn are zombies; it is similarly logically possible that there isn't a drop of water in the Pacific Ocean (an omnipotent evil demon has replaced it all with hallucination-stuff that seems just like water). One wouldn't want to deflect one's theory of consciousness (or oceans) by honoring such a trivial scruple about a mere logical possibility, would one?

Does Chalmers offer anything in addition to this logical possibility in support of his continued allegiance to the Zombic Hunch? He turns to the topic of personal identity and whether uploading would — under any circumstances — amount to survival, and presents both optimistic and pessimistic arguments (since he declares himself unsure). These arguments develop in somewhat greater detail the considerations I explored in 'Where am I?' (1978; reprinted in Hofstadter and Dennett, 1981) and lead him, once again, to the conclusion I leapt to then (Chalmers never *leaps* to conclusions; he *oozes* to conclusions, checking off all the caveats and pitfalls and possible sources of error along the way with exemplary caution):

> At the very least, as in the case of consciousness, it seems that if gradual uploading happens, most people will become convinced that it is a form of survival. . . . I am reasonably confident that gradual uploading is a form of survival. So if at some point in the future I am faced with the choice between uploading and continuing in an increasingly slow biological embodiment, then as long as I have the option of gradual uploading, I will be happy to do so. (pp. 57–9)

What about 'reconstructive' uploading? He reinvents Hofstadter's thought experiment in 'A Conversation with Einstein's Brain' (Hofstadter and Dennett, 1981): 'If we reconstruct a functional isomorph of Einstein from records, will it be Einstein?' (p. 61) And once again, his oh-so tentative conclusion is that it doesn't differ substantially from the gradual uploading he has already endorsed as a valuable variety of survival. All this is in nice agreement with type-A materialism of the functionalist sort.

We're getting closer and closer to type-A materialism, and Chalmers' tantalizing intellectual strip-tease continues, confronting the *further-fact* view of survival in uploading that seems to be the last bulwark against type-A materialism, and although '[t]here is at least an intuition that complete knowledge of the physical and mental facts in a case of destructive uploading leaves an open question ... and there is an intuition that there are facts about which hypothesis is correct that we very much want to know' (p. 62), '... it is far from obvious that there really are facts about survival of the sort that the further-fact view claims are unsettled' (p. 64). So we're down to two intuitions (for whatever they are worth) and a logical possibility (for whatever that is worth) and *it is far from obvious* that there is so much as an issue here. 'I *do not know* whether such questions have objective answers .... But *it is not out of the question* that this value scheme should be revised.... . I *am not sure* whether a further-fact view or a deflationary view is correct' (p. 66). And philosophers wonder why non-philosophers get impatient with them!

What, then, do I make of all this? Some years ago in conversation with Chalmers, after reaching an impasse of just the sort illustrated above, I thought I heard him say that there was no point in my presenting him with any more *reasons* in favour of my position since no argument could shake his brute intuition (the Zombic Hunch), and that was all there was to it. I decided to take him at his word, and refrain from further attempts at philosophical argument since he had assured me they would be fruitless. So I recommended that he seek therapy or perhaps a change in diet. Who knew what might dislodge an impenetrable intuition! He did not take kindly to my suggestion, and I resolved not to press the point further. But now I find myself puzzling once again. My spade is turned, as before, and this time he has provided me with yet more evidence that arguments really will not avail, since he has presented excellent versions of them himself, and failed to convince himself. I do not mind conceding that I could not have done as good a job, let alone a better job, of marshalling the grounds for type-A materialism. I'd be bringing coals to Newcastle if I tried. So why does he cling like a limpet to his property dualism?

I think there are (at least) seven possible answers to this puzzle, and I find myself unconvinced by all seven. (Chalmers' caution is infectious.) Still, I think that *a case can be made* for each of them, and while there is a *logical possibility* that none of them deserves to be called *the* explanation, *it is not out the question* that one or more of them deserves to be taken seriously, as seriously as any not merely logical possibility deserves to be taken. As luck would have it, all

seven answers can be labelled — with a little procrustean tugging —
with the same letter, much like the famous four Fs of animal options:
fight, flee, feed and engage in sexual intercourse.

3. The Famous Seven Fs of the Mystery of Chalmers' Resistance to Type-A Materialism

1. Faith

Could it be that Chalmers, like Descartes, is attracted to dualism by a
residual fondness for the Christian doctrine of an immortal, immaterial soul? I find this highly unlikely, but in the interests of something
approaching exhaustion of possibilities, I must list it. The late Sir John
Eccles, Nobel laureate neuroscientist and devout Catholic, certainly
gave us an instance of the category, and it does give one pause that,
coming from an entirely different quarter, Jerry Fodor (2008) has
recently decided that the epithet that best describes his own view of
wisdom about the mind is 'Cartesian', a label he is now proud to sport.
But the central attraction of *property* dualism, I gather, is that it provides a stumbling block for the scientific study of the mind (the Hard
Problem) *without* postulating an embarrassing substance, a miracle
pearl of sorts, that might leave our bodies when we die.

2. Fame

Many years ago, over a few drinks, I offered up Uncle Dan's advice
for how to become a famous philosopher: invent a new (short, punchy,
but unsound) argument for dualism; publish a brief version of this in a
philosophy journal and then watch it get snapped up by professors
around the world looking for a head-snapping attention-grabber for
their students, an argument that even the most callow undergraduates
could be motivated to care about — and refute. It would migrate from
a few syllabuses to many, and then be anthologized, rebutted,
defended, analysed, translated, caricatured, and turned into a 'classic'. David Chalmers was not present on that occasion but somebody
who was — and who shall remain nameless — actually tried to take
my advice, offering publishers 'a new argument for dualism' and getting a contract to write the book. Unfortunately this would-be famous
philosopher had neglected to compose the novel argument in advance,
and in spite of much searching and agonizing couldn't deliver. A different book was written and grudgingly accepted for publication.
Fame eluded its author. This is not David Chalmers' story, but it is
possible that the fame that has accrued to Chalmers and the so-called
Hard Problem has something to do with his continued allegiance to

the position. If so, he should reconsider: Frank Jackson has recanted his famous argument for dualism about Mary the Color Scientist without any loss — indeed with an increment — of fame and influence. (And no, Frank Jackson was neither the inspiration for, nor the one inspired by, my advice.)

3. Freud

Douglas Hofstadter is David Chalmers' *Doktorvater* (and I was an informal member of his dissertation committee). Neither Hofstadter nor I have expressed any support for Chalmers' brand of property dualism, and indeed have published quite a lot over the years expressly arguing for (what Chalmers calls) type-A materialism. Moreover, Hofstadter has been unusually frank in expressing his own conviction that the Singularity is an idea not worth serious consideration, calling it on one occasion a 'nutty technology-glorifying scenario' (http://tal.forum2.org/hofstadter_interview) and saying on another occasion that the discussion of it by Kurzweil and others was 'as if you took a lot of very good food and some dog excrement and blended it all up so that you can't possibly figure out what's good or bad' (http://www.americanscientist.org/bookshelf/pub/douglas-r-hofstadter). Could it be that Chalmers has gone to great lengths to distance himself from his early mentors, even going so far on this occasion as to ignore the versions of arguments, by Hofstadter (in *Gödel, Escher, Bach,* in *The Mind's I,* in *I Am a Strange Loop*) and me (in *The Mind's I,* in *Consciousness Explained* , in *Sweet Dreams*) that anticipate his own discussions? A farfetched hunch, but logically possible.

4. Fiction

On this hypothesis, it is a mistake to read this essay as what it appears to be on its surface: a serious philosophical essay. It is rather, like Borges's *faux*-erudite reviews of non-existent books (in *Labyrinths,* 1962, for instance), a parody of academic scholarship, or philosophy, or both. It is designed to take in academic philosophers of the analytic school in much the way Alan Sokal's hoax took in the postmodernist editors and readers of *Social Text.* Such a subtle project is much more difficult than Sokal's, I think, and it is a credit to Chalmers' talent that he has managed to convince so many people that this is earnest philosophy, not a practical joke. (Chalmers is not averse to such capers; he once spread the rumor on his website that I had recanted and embraced dualism.) But I am not persuaded that this is the case, since there is an alternative that has more plausibility (to me).

5. *Filosofia*

(This is the procrustean tug I warned about — I have to switch to Italian to preserve my alliteration scheme.) This is not witting parody; this is unwitting parody. This is a philosopher performing the following speech act: *I am a philosopher and this is what philosophers do.* We no longer debate how many angels can dance on a pinhead, but we do pursue exhaustively nuanced analyses of our intuitions and the (logically) possible implications of them.

6. *Fun*

There is some textual evidence in the essay for the hypothesis that concern about the impending Singularity is really just a pretext, intended to 'motivate' the clever exploration of a set of delicious puzzles where you get to display your intellectual agility. The first forty pages seem designed to protect the prophecy from all varieties of kill-joy skepticism that would spoil the game, so that we are licensed to consider the prospects of uploading as something more important than idle fantasy. This is a gambit not unknown among philosophers. Much of the contemporary literature on free will, for instance, is saturated with discussion of the tactics of argumentation, and meta-comments on the strengths and weaknesses of various *moves*, to the point where the reader may begin to suspect that the combatants would hate to see a resolution to the controversy since it would bring their sport to an end. As the late great linguist Jim McCawley once quipped, in answer to the question of how you tell the philosophers from the linguists: 'The philosopher is the one who will contribute a paper on the hangman paradox to a symposium on capital punishment.'

7. *Fear*

Finally, there is the possibility that Chalmers is motivated, as he hints at the end, by fear of death. But then wouldn't he cling to type-A materialism, which is the view that holds out the best promise of continued survival indefinitely (see *Consciousness Explained*, p. 430)? Perhaps his motivation is more subtle. He dare not hope too openly, but must plump relentlessly for the worst, most dismal option, thereby damping the blow of bitter disappointment with reasoned anticipation. As he says in closing, 'My own strategy is to write about the singularity and about uploading. Perhaps this will encourage our successors to reconstruct me, if only to prove me wrong' (p. 67).

References

Borges, J.L. (1962) *Labyrinths: Selected Stories and Other Writings*, New York: New Directions. [La Biblioteca de Babel, 1941, in El jardin de los senderos que se bifurcan, published with another in *Ficciones*, 1956, Emece Editores, S. A., Buenos Aires.]

Dennett, D.C. (1978) *Brainstorms: Philosophical Essays on Mind and Psychology*, Montgomery, VT: Bradford Books.

Dennett, D.C. (1991) *Consciousness Explained*, Boston, MA: Little, Brown, and London: Allen Lane.

Dennett, D.C. (1993) Review of John Searle, *The Rediscovery of the Mind*, *Journal of Philosophy*, **60** (4), pp. 193–205.

Dennett, D.C. (2005), *Sweet Dreams: Philosophical Obstacles to a Science of Consciousness*, Cambridge, MA: MIT Press.

Fodor, J. (2008) *LOT 2: The Language of Thought Revisited*, Oxford: Oxford University Press.

Hofstadter, D.R. (1979) *Gödel, Escher, Bach: An Eternal Golden Braid*, New York: Basic Books.

Hofstadter, D.R. (2007) *I am a Strange Loop*, New York: Basic Books.

Hofstadter, D.R. & Dennett, D.C. (1981) *The Mind's I: Fantasies and Reflections on Self and Soul*, New York: Basic Books.

Searle, J. (1992) *The Rediscovery of the Mind*, Cambridge, MA: MIT Press.

Ben Goertzel

Should Humanity Build a Global AI Nanny to Delay the Singularity Until It's Better Understood?

Abstract: *Chalmers suggests that, if a Singularity fails to occur in the next few centuries, the most likely reason will be 'motivational defeaters' — i.e. at some point humanity or human-level AI may abandon the effort to create dramatically superhuman artificial general intelligence. Here I explore one (I argue) plausible way in which that might happen: the deliberate human creation of an 'AI Nanny' with mildly superhuman intelligence and surveillance powers, designed either to forestall Singularity eternally, or to delay the Singularity until humanity more fully understands how to execute a Singularity in a positive way. It is suggested that as technology progresses, humanity may find the creation of an AI Nanny desirable as a means of protecting against the destructive potential of various advanced technologies such as AI, nanotechnology and synthetic biology.*

Introduction

I find myself in almost total agreement with Chalmers' careful analytical treatment of the 'Singularity hypothesis'. However, as an intentionally high-level treatment, it leaves many critical points unelaborated; my goal here is to enlarge on one of these, namely the potential for what Chalmers calls 'motivational defeaters' for the transition from what he calls AI+ (extensible, human-level AI) to what he calls AI++ (Singularity-constitutive, dramatically superhumanly intelligent and powerful AI).

Specifically, I will elaborate on one possible motivational defeater: an 'AI Nanny', defined as an advanced AI+ system explicitly designed to thoroughly surveil the Earth and keep humanity safe from various dangers, including unpredictable advanced technologies such as AI++. AI Nannies could potentially be designed to forestall Singularity forever, or alternately to slow down the path to Singularity in order to enable it to proceed in a more deliberate way with a higher chance of a positive outcome. I will discuss the plausibility of creating AI Nannies, and also the ethical desirability of doing so – which I conclude is ambiguous, but potentially positive.

Chalmers on Motivational Defeaters

Chalmers' argument for the plausibility of a Singularity occurring in humanity's mid-term future is, compactly, that

> an intelligence explosion results from a self-amplifying cognitive capacity, correlations between that capacity and other important cognitive capacities, and manifestation of those capacities (conclusion). More pithily: self-amplification plus correlation plus manifestation = singularity.

He then considers classes of possible 'defeaters' that might cause his argument for the Singularity's occurrence to fail, observing that

> We can divide the defeaters into motivational defeaters in which an absence of motivation or a contrary motivation prevents capacities from being manifested, and situational defeaters, in which other unfavorable circumstances prevent capacities from being manifested.

And he notes that the motivational defeaters seem perhaps the most likely to arise in practice:

> Speaking for myself, I think that while structural and correlational obstacles (especially the proportionality thesis) raise nontrivial issues, there is at least a prima facie case that absent defeaters, a number of interesting cognitive capacities will explode. I think the most likely defeaters are motivational. But I think that it is far from obvious that there will be defeaters. So I think that the singularity hypothesis is one that we should take very seriously.
>
> ...
>
> ... [I]t is certainly possible that AI+ systems will be disinclined to create their successors, perhaps because we design them to be so disinclined, or perhaps because they will be intelligent enough to realize that creating successors is not in their interests. Furthermore, it may be that AI+ systems will have the capacity to prevent such progress from happening.

...

> A singularity proponent might respond that all that is needed to over-
> come motivational de- featers is the creation of a single AI+ that greatly
> values the creation of greater AI+ in turn, and a singularity will then be
> inevitable. If such a system is the first AI+ to be created, this conclusion
> may well be correct. But as long as this AI+ is not created first, then it
> may be subject to controls from other AI+, and the path to AI++ may be
> blocked. The issues here turn on difficult questions about the motiva-
> tions and capacities of future systems, and answers to these questions
> are difficult to predict.

Here I will follow up on these points – specifically, the potential that
humans might design AI+ systems to be disinclined to give rise to
AI++ systems (in spite of possessing the capability to do so), and to
also prevent other AI systems from achieving the capability to make
the transition to AI++. I will not attempt to argue for this scenario's
likelihood, but merely for its plausibility. I will also address the ethical
question of whether an AI+ 'AI Nanny' of this sort is a desirable . A
key point I will emphasize is that, even if you believe a Singularity is a
laudable ultimate goal for humanity, it may still be rational for you to
favor the creation of an AI Nanny with a predetermined finite lifespan, with the goal of mediating a slower and more reliably positive
path to Singularity. If a sufficient number of sufficiently powerful
humans conclude that the creation of an AI Nanny is ethically desirable, this obviously may increase the probability of an AI Nanny scenario coming about.

The Ethical Motivation for an AI Nanny

In his paper, Chalmers carefully considers the question of whether a
Singularity may occur, but avoids the issue of whether, if it does, this
will be a good thing or not. While this is an understandable omission,
resulting in a more compact and simple and less controversial treatment, it's also important to consider that human perceptions of the ethical character of the Singularity are plausibly likely to play a
significant role in how (and if) the Singularity unfolds. Here I will
come at the AI Nanny idea from the direction of ethics — as a potential (partial) solution to the moral dilemmas the Singularity poses.

The dramatic potential of a Singularity for both 'good' and 'bad',
according to folk morality standards, is fairly obvious. The ongoing
advancement of science and technology has brought us many wonderful things, and will almost surely be bringing us more and more, even
before the occurrence of a full-on Singularity. Beyond the 'mere'

abolition of scarcity, disease and death, there is the possibility of fundamental enhancement of the human mind and condition, and the creation of new forms of life and intelligence. Our minds and their creations may spread throughout the universe, and may come into contact with new forms of matter and intelligence that we can now barely imagine.

And on the dark side, Nick Bostrom (2002) has enumerated some of the ways that technology may pose 'existential risks' — risks to the future of the human race — as the next decades and centuries unfold. And there is also rich potential for other, less extreme sorts of damage. Technologies like AI, synthetic biology and nanotechnology could run amok in dangerous and unpredictable ways, or could be utilized by unethical human actors for predictably selfish and harmful human ends.

Given this tenuous balance of benefits and dangers, and the plausible likelihood of a Singularity occurring, it's understandable that many are disturbed by our almost total lack of understanding of the odds of the various possible outcomes ensuing from the Singularity. And this train of thought leads to the perspective that intentionally creating a 'motivational defeater' to forestall or at least delay the Singularity might be ethically advantageous. Wallach and Allan (2008) review these issues and conclude that a Singularity is sufficiently far off that we should currently focus our ethical attention on nearer-term problems; but, my own view is that our uncertainty about future research progress is sufficient that it behooves us to take the possibility of more rapid progress toward Singularity very seriously.

The possibility I wish to explore here is the creation of a powerful yet limited AGI (Artificial General Intelligence) system (an AI+ in Chalmers' terms), with the explicit goal of keeping things on the planet under control while we figure out the hard problem of how to create a probably positive Singularity. That is: to create an 'AI Nanny.'

The envisioned AI Nanny would forestall a full-on Singularity for a while, restraining it into what Max More (2009) has called a Surge, and giving us time to figure out what kind of Singularity we really want to build and how. It's not entirely clear that creating such an AI Nanny is plausible, but I've personally come to the conclusion it probably is. It's also not entirely clear that, even with the help and supervision of a well-built AI Nanny, humanity would ever understand the Singularity well enough to feel comfortable moving forward with it — which highlights the question of how long an AI Nanny would be empowered, if it were created.

Perspectives on the Ethical Dilemma of the Singularity

Given the ethical complexities mentioned above, to which I've suggested an AI Nanny might be a possible solution path, what does the contemporary pantheon of futurist gurus think we should do in the next decades, as the path to Singularity unfolds?

Kurzweil (2005) has proposed 'fine-grained relinquishment' as a strategy for balancing the risks and rewards of technological advancement. But it's not at all clear this will be viable, without some form of AI Nanny to guide and enforce the relinquishment. Government regulatory agencies are notoriously slow-paced and unsophisticated, and so far their decision-making speed and intelligence aren't keeping up with the exponential acceleration of technology.

Further, it seems a clear trend that as technology advances, it is possible for people to create more and more destruction using less and less money, education and intelligence. There seems no reason to assume this trend will reverse, halt or slow. This suggests that, as technology advances, selective relinquishment will prove more and more difficult to enforce. Kurweil acknowledges this issue, stating that 'The most challenging issue to resolve is the granularity of relinquishment that is both feasible and desirable' (*ibid.*, p. 299), but he believes this issue is resolvable. I'm skeptical that it is resolvable without resorting to some form of AI Nanny.

Eliezer Yudkowsky (2002; 2009) has suggested that the safest path for humanity will be to first develop 'Friendly AI' systems with dramatically superhuman intelligence. He has put forth some radical proposals, such as the design of self-modifying AI systems with human-friendly goal systems designed to preserve friendliness under repeated self-modification; and the creation of a specialized AI system with the goal of determining an appropriate integrated value system for humanity, summarizing in a special way the values and aspirations of all human beings (Yudkowsky, 2004). However, these proposals are extremely speculative at present, even compared to feats like creating an AI Nanny or a technological Singularity. The practical realization of his ideas seems likely to require astounding breakthroughs in mathematics and science — whereas it seems plausible that human-level AI, molecular assemblers and the synthesis of novel organisms can be achieved via a series of moderate-level breakthroughs alternating with 'normal science and engineering.'

Bill McKibben (2004), Bill Joy (2000) and other modern-day techno-pessimists argue for a much less selective relinquishment than Kurzweil. They argue, in essence, that technology has gone far enough — and that if it goes much further, we 'legacy humans' are bound to be obsoleted or destroyed. They fall short, however, in the area of suggestions for practical implementation. The power structure of the current human world comprises a complex collection of inter-locking powerful actors (states and multinational corporations, for example), and it seems probable that if some of these chose to severely curtail technology development, many others would *not* follow suit. For instance, if the US stopped developing AI, synthetic biology and nanotech next year, China and Russia would most likely interpret this as a fantastic economic and political opportunity, rather than as an example to be imitated.

Hugo de Garis agrees with the techno-pessimists that AI and other advanced technology is likely to obsolete humanity, but views this as essentially inevitable, and encourages us to adopt a philosophical position according to which this is desirable. In his book *The Artilect War* (2004) he contrasts the 'Terran' view, which views humanity's continued existence as all-important, with the 'Cosmist' view in which, if our AI successors are more intelligent, more creative, and perhaps even more conscious and more ethical and loving then we are — then why should we regret their ascension, and our disappearance? In more recent writings (2011), he also considers a 'Cyborgist' view in which gradual fusion of humans with their technology (e.g. via mind uploading and brain computer interfacing) renders the Terran vs. Cosmist dichotomy irrelevant. In this trichotomy Kurzweil falls most closely into the Cyborgist camp. But de Garis views Cyborgism as largely delusory, pointing out that the potential computational capa-bility of a grain of sand (according to the known laws of physics) exceeds the current computational power of the human race by many orders of magnitude, so that as AI software and hardware advance-ment accelerate, the human portion of a human-machine hybrid mind would rapidly become irrelevant.

Considering these views all together, the dilemma posed by the rapid advancement of technology becomes both clear and acute. If the exponential advancement highlighted by Kurzweil continues apace, as seems likely though not certain, then the outcome is highly unpre-dictable. It could be bliss for all, or unspeakable destruction — or something inbetween. We could all wind up dead — killed by soft-ware, wetware or nanoware bugs, or other unforeseen phenomena. If humanity does vanish, it could be replaced by radically more intelli-gent entities (thus satisfying de Garis's Cosmist aesthetic) – but this isn't guaranteed; there's also the possibility that things go awry in a manner annihilating all life and intelligence on Earth and leaving no path for its resurrection or replacement.

To make the dilemma more palpable, think about what a few hundred brilliant, disaffected young nerds with scientific training could do, if they teamed up with terrorists who wanted to bring down modern civilization and commit mass murders. It's not obvious why such an alliance would arise, but nor is it beyond the pale. Think about what such an alliance could do now — and what it could do in a couple decades from now, assuming Kurzweilian exponential advance. One expects this theme to be explored richly in science fiction novels and cinema in coming years.

But how can we decrease these risks? It's fun to muse about designing a 'Friendly AI' à la Yudkowsky, that is guaranteed (or near-guaranteed) to maintain a friendly ethical system as it self-modifies and self-improves itself to massively superhuman intelligence. Such an AI system, if it existed, could bring about a full-on Singularity in a way that would respect human values — i.e. the best of both worlds, satisfying all but the most extreme of both the Cosmists and the Terrans. But the catch is, nobody has any idea how to do such a thing, and it seems well beyond the scope of current or near-future science and engineering.

Realistically, we can't stop technology from developing; and we can't control its risks very well, as it develops. And daydreams aside, we don't know how to create a massively superhuman supertechnology that will solve all our problems in a universally satisfying way.

This train of thought leads naturally to the possibility of creating what I've called an AI Nanny' — a *mildly* superhuman supertechnology, whose job it is to protect us from ourselves and our technology — not forever, but just for a while, while we work on the hard problem of creating a Friendly Singularity.

The 'AI Nanny'

More specifically, what I mean by an 'AI Nanny' is an advanced Artificial General Intelligence (AGI) software program with

- General intelligence somewhat above the human level, but not too dramatically so — maybe, qualitatively speaking, as far above humans as humans are above apes
- Interconnection to powerful worldwide surveillance systems, online and in the physical world

- Control of a massive contingent of robots (e.g. service robots, teacher robots, etc.) and connectivity to the world's home and building automation systems, robot factories, self-driving cars, and so on and so forth

- A cognitive architecture featuring an explicit set of goals, and an action selection system that causes it to choose those actions that it rationally calculates will best help it achieve those goals

- A set of preprogrammed goals including the following aspects:

  - A strong inhibition against modifying its preprogrammed goals

  - A strong inhibition against rapidly modifying its general intelligence

  - A mandate to cede control of the world to a more intelligent AI within N years (where N could be, say, 10, 100 or 5000)

  - A mandate to help abolish human disease, involuntary human death, and the practical scarcity of common humanly-useful resources like food, water, housing, computers, etc.

  - A mandate to prevent the development of technologies that would threaten its ability to carry out its other goals

  - A strong inhibition against carrying out actions with a result that a strong majority of humans would oppose, if they knew about the action in advance

  - A mandate to be open-minded toward suggestions by intelligent, thoughtful humans about the possibility that it may be misinterpreting its initial, preprogrammed goals

Obviously, this sketch of the AI Nanny concept is highly simplified and idealized — a real-world AI Nanny would have all sort of properties not described here, and might be missing some of the above features, substituting them with other related things. My point here is not to sketch a specific design or requirements specification for an AI Nanny, but rather to indicate a fairly general class of systems that humanity might build.

The nanny metaphor is chosen carefully. A nanny watches over children while they grow up, and then goes away. Similarly, the AI Nanny would not be intended to rule humanity on a permanent basis – only to provide protection and oversight while we 'grow up' collectively; to give us a little breathing room so we can figure out how best to create a desirable sort of Singularity.

When I first reflected on this idea, my personal reaction was to find it rather odious. But after further reflection my view is more ambivalent. One point I considered is that, in spite of a personal streak toward rule-breaking, I'm not a political anarchist — because I have a strong suspicion that if governments were removed, the world would become a lot worse off, dominated by gangs of armed thugs imposing even less pleasant forms of control than those exercised by the US Army and the CCP and so forth. I suspect government could be done a lot better than any country currently does it — but I don't doubt the need for some kind of government, given the realities of human nature. It may be that the need for an AI Nanny falls into the same broad category. It seems possible that, like government, an AI Nanny is a relatively offensive thing, that is nonetheless a practical necessity due to the unsavory aspects of human nature.

We didn't need government during the Stone Age — because there weren't that many of us, and we didn't have so many dangerous technologies. But we need government now. Fortunately, these same technologies that necessitated government, also provided the means for government to operate.

Somewhat similarly, we haven't needed an AI Nanny so far, because we haven't had sufficiently powerful and destructive technologies. And now, these same technologies that *may* necessitate the creation of an AI Nanny, also may provide the means of creating it.

The Argument for Building an AI Nanny

To recap and summarize, a plausible ethical argument for trying to build an AI Nanny would be that:

1. It's impracticable to halt the exponential advancement of technology (even if one wanted to)

2. As technology advances, it becomes possible for individuals or groups to wreak greater and greater damage using less and less intelligence and resources

3. As technology advances, humans will more and more acutely lack the capability to monitor global technology development and forestall radically dangerous technology-enabled events

4. Creating an AI Nanny is a significantly less difficult technological problem than creating an AI or other technology with a predictably high probability of launching a full-scale positive Singularity

5. Imposing a permanent or very long term constraint on the development of new technologies is undesirable

It would be interesting and valuable to run through this argument with the analytical detail of Chalmers' article on the Singularity; but this is merely a brief commentary, so a rough summary will have to do for now.

The fifth and final premise is normative; the others are empirical. None of the empirical premises are certain, but all seem likely to me. The first three premises are strongly implied by recent social and technological trends. The fourth premise seems commonsensical based on current science, mathematics and engineering.

These premises lead to the conclusion that trying to build an AI Nanny is probably a good idea. The actual plausibility of building an AI Nanny is a different matter – I believe it is plausible, but of course, opinions on the plausibility of building any kind of AGI system in the relatively near future vary all over the map.

The above argument is interesting from two points. First, it might be correct. And second, even if it is incorrect for some reason, it is possible that if sufficiently powerful organizations come to believe it, they may create an AI Nanny of some form anyway.

Complaints and Responses

I have discussed the AI Nanny idea with a variety of people over the last year or so, and have heard an abundance of different complaints about it — but none have struck me as compelling. Here follows a partial, roughly-sketched list of counterarguments and my counter-counterarguments.

It's impossible to build an AI Nanny; the AI R&D is too hard. — But is it really? It's almost surely impossible to build and install an AI Nanny this year; but as a professional AI researcher, I believe such a thing is well within the realm of possibility. I think we could have one in a couple decades if we really put our collective minds to it. It would involve a host of coordinated research breakthroughs, and a lot of large-scale software and hardware engineering, but nothing implausible according to current science and engineering. We did amazing

things in the Manhattan Project because we wanted to win a war——
how hard are we willing to try when our overall future is at stake?
It may be worth dissecting this 'hard R&D' complaint into two
sub-complaints:

- *AGI is hard*: building an AGI system with slightly greater than
  human level intelligence is too hard (i.e. in Chalmers' terms,
  AI+ is too hard)

- *Nannifying an AGI is hard*: given a slightly superhuman AGI
  system, turning it into an AI Nanny is too hard (i.e. in Chalmers'
  terms, the particular kind of AI+ that is an AI Nanny is too hard)

Obviously both of these are contentious issues.

Regarding the 'AGI is hard' complaint, at the AGI-09 artificial
intelligence research conference, an expert-assessment survey was
done (Baum *et al.*, 2011), suggesting that a least a nontrivial plurality
of professional AI researchers believes that human-level AGI is possible within the next few decades, and that slightly-superhuman AGI
will follow shortly after that.

Regarding the 'Nannifying an AGI is hard' complaint, I think its
validity depends on the AGI architecture in question. If one is talking
about an integrative, cognitive-science-based, explicitly goal-oriented AGI system like, say, OpenCog (Goertzel, 2009) or MicroPsi
(Bach, 2009) or LIDA (Friedlander & Franklin, 2008) then this is
probably not too much of an issue, as these architectures are fairly
flexible and incorporate explicitly articulated goals. If one is talking
about, say, an AGI built via closely emulating human brain architecture, in which the designers have relatively weak understanding of the
AGI system's representations and dynamics, then the 'nannification is
hard' problem might be more serious. My own research intuition is
that an integrative, cognitive-science-based, explicitly goal-oriented
system is likely to be the path via which advanced AGI first arises; this
is the path my own work is following.

*It's impossible to build an AI Nanny; the surveillance technology is
too hard to implement.* — But is it really? Surveillance tech is advancing very rapidly, for reasons more prosaic than the potential development of an AI Nanny. David Brin's book *The Transparent Society*
(1999) gives a rather compelling argument that before too long, we'll
all be able to see everything everyone else is doing.

*Setting up an AI Nanny, in practice, would require a world govern-
ment.* — This seems patially valid. It would require either a proactive

assertion of power by some particular party, creating and installing an AI Nanny without asking everybody else's permission; or else a degree of cooperation between the world's most powerful governments, beyond what we see today. Either route seems conceivable. Regarding the second cooperative path, it's worth observing that the world is clearly moving in the direction of greater international unity, albeit in fits and starts. Once the profound risks posed by advancing technology become more apparent to the world's leaders, the required sort of international cooperation will probably be a lot easier to come by. Hugo de Garis's most recent book *Multis and Monos* (2010) riffs extensively on the theme of emerging world government.

Building an AI Nanny is harder than building a self-modifying, self-improving AGI that will retain its Friendly goals even as it self-modifies. — I find this rather implausible. Maintenance of goals under radical self-modification and self-improvement seems to pose some very thorny philosophical and technical problem — and once these are solved (to the extent that they're even solvable) *then* one will have a host of currently-unforeseeable engineering problems to consider. Furthermore there is a huge, almost surely irreducible uncertainty in creating something massively more intelligent than oneself. Whereas creating an AI Nanny is 'merely' a very difficult, very large scale science and engineering problem.

If someone creates a new technology smarter than the AI Nanny, how will the AI Nanny recognize this and be able to nip it in the bud? — Remember, the hypothesis is that the AI Nanny is significantly smarter than people. Imagine a friendly, highly intelligent person monitoring and supervising the creative projects of a room full of chimps or 'intellectually challenged' individuals.

Why would the AI Nanny want to retain its initially pre-programmed goals, instead of modifying them to suit itself better? — *for instance, why wouldn't it simply adopt the goal of becoming an all-powerful dictator and exploiting us for its own ends?* — But why *would* it change its goals? What forces would cause it to become selfish, greedy, etc? Let's not anthropomorphize. 'Power corrupts, and absolute power corrupts absolutely' is a statement about human psychology, not a general law of intelligent systems. Human beings are not architected as rational, goal-oriented systems, even though some of us aspire to be such systems and make some progress toward behaving in this manner. If an AI system is created with an architecture inclining it to pursue certain goals, there's no reason why it would automatically be inclined to modify these goals.

But how can you specify the AI Nanny's goals precisely? You can't right? And if you specify them imprecisely, how do you know it won't eventually come to interpret them in some way that goes against your original intention? And then if you want to tweak its goals, because you realize you made a mistake, it won't let you, right? — This is a tough problem, without a perfect solution. But remember, one of its goals is to be open-minded about the possibility that it's misinterpreting its goals. Indeed, one can't rule out the possibility that it will misinterpret this meta-goal and then, in reality, closed-mindedly interpret its other goals in an incorrect way. The AI Nanny would not be a risk-free endeavor, and it would be important to get a feel for its realities before giving it too much power. But again, the question is not whether it's an absolutely safe and positive project – but rather, whether it's better than the alternatives!

What about Steve Omohundro's Basic AI Drives *(2008)? Didn't Omohundro prove that any AI system would seek resources and power just like human beings?* — While the arguments in this paper are powerful, they are mainly evolutionary in nature. They apply most plainly to the case of an AI competing against other roughly equally intelligent and powerful systems for survival. The posited AI Nanny would be smarter and more powerful than any human, and would have, as part of its goal content, the maintenance of this situation for N years. Unless someone managed to sneak past its defenses and create competitively powerful and smart AI systems, or it encountered alien minds, the premises of Omohundro's arguments don't apply.

What happens after the N years is up? — This is unclear, which is part of the argument for creating an AI Nanny — the point is that after N years of research and development under the protection of the AI Nanny, we would have a lot better idea of what's possible and what isn't than any of us do right now.

What happens if the N years pass and none of the hard problems are solved, and we still don't know how to launch a full-on Singularity in a sufficiently reliably positive way? — One obvious possibility is to launch the AI Nanny again for a couple hundred more years. Or maybe to launch it again with a different, more sophisticated condition for ceding control (in the case that it, or humans, conceive some such condition during the N years).

What if we figure out how to create a Friendly self-improving massively superhuman AGI only 20 years after the initiation of the AI Nanny — then we'd have to wait another N-20 years for the real Singularity to begin! — That's true of course, but if the AI Nanny is working well, then we're not going to die in the interim, and we'll be conducting an enjoyable existence while we wait.

But how can you trust anyone to build the AI Nanny? Won't they secretly put in an override telling the AI Nanny to obey them, but nobody else? — That's possible, but developing the AI Nanny via an open, international, democratic community and process would diminish the odds of this sort of problem happening.

What if, shortly after initiating the AI Nanny, some human sees some fatal flaw in the AI Nanny approach, which we don't see now. Then we'd be unable to undo our mistake.— True, that is a risk of the approach.

Conclusion

Chalmers opines that, of all the possible defeaters that might prevent a Singularity from occurring in the next few centuries, the most likely are the 'motivational defeaters' — i.e. factors that might cause humanity or its AI+ creations to *intentionally* avoid launching a rapidly exploding process of intelligence that creates smarter intelligence that creates smarter intelligence....

An argument against the likelihood of motivational defeaters would be that, even if some humans and AI+'s lack the motivation to create AI++, probably other humans or AI+'s will not suffer this lack, and will go ahead and create AI++ anyway. General technological advance seems likely to progressively decrease the amount of computational and other resources to create more and more powerful AI systems, making it likely that eventually a relatively small group of enthusiasts could launch AI++ regardless of the opinions of others.

However, one way that a motivational defeater could prevail, in spite of the factors mentioned in the above paragraph, would be via the creation of an 'AI Nanny' – an AI+ system with ample physical empowerment and the explicit goal of preventing the occurrence of AI++, either permanently, for a certain fixed period of time, or else till certain pre-specified criteria are met. Whether the creation of an AI Nanny is plausible — or will become plausible prior to the creation of AI++ — is certainly an open question. However, I have argued that the AI Nanny route may become very appealing to many parties sometime before the Singularity occurs, due to its potential for circumventing some of the dangerous ethical dilemmas the Singularity presents.

Specifically, humanity is facing a situation where increasing danger (potentially even human extinction) is wreakable by small groups of disaffected individuals; and the possibility of AI++ is also fraught with danger and uncertainty, because of our lack of any rigorous scientific theory of AI++ systems. The creation of an AI Nanny has potential to dampen the risks of terrorism and runaway AI++, and giving humanity and AI+ a bit of breathing space to figure out if AI++ is the right thing to do, and if so to do it right. Whether or not creating an AI Nanny is the best course for humanity, it could be the course taken anyway, if sufficiently powerful organizations decide it is the best option.

References

Bach, J. (2009) *Principles of Synthetic Intelligence*, Cambridge: Cambridge University Press.

Baum, S.D., Goertzel, B. & Goertzel, T.G. (2011) How long until human-level AI? Results from an expert assessment, *Technological Forecasting & Social Change*, **78** (1), pp. 185–195.

Bostrom, N. (2002) *Journal of Evolution and Technology*, **9** (1).

Brin, D. (1999) *The Transparent Society*, New York: Basic Books.

de Garis, H. (2004) *The Artilect War*, Pittsburgh, PA: Etc Press.

de Garis, H. (2010) *Multis and Monos*, Pittsburgh, PA: Etc Press.

de Garis, H. (2011) *Merge or Purge*, [Online], http://hplusmagazine.com/2011/05/19/merge-or-purge/

Friedlander, D. & Franklin, S. (2008) LIDA and a Theory of Mind, in Goertzel, B. & Wang, P. (eds.) *Artificial General Intelligence (AGI-08)*, Memphis, TN: IOS Press.

Goertzel, B. (2009) OpenCogPrime: A Cognitive Synergy Based Architecture for General Intelligence, *International Conference on Cognitive Informatics*, Hong Kong.

Joy, B. (2000) Why the future doesn't need us, *Wired Magazine*, (April).

Kurzweil, R. (2005) *The Singularity Is Near*, New York: Viking.

McKibben, B. (2004) *Enough*, New York: St. Martin's Griffin.

More, M. (2009) *Singularity and Surge Scenarios*, [Online], http://strategicphilosophy.blogspot.com/2009/06/how-fast-will-future-arrive-how-will.html

Omohundro, S. (2008) The basic AI drives, in Wang, P., Goertzel, B. & Franklin, S. (eds.) *Proceedings of the First AGI Conference*, vol. 171, Frontiers in Artificial Intelligence and Applications, Memphis, TN: IOS Press.

Wallach, W. & Allan, C. (2008) *Moral Machines*, Oxford: Oxford University Press.

Yudkowsky, E. (2002) *Creating Friendly AI*, http://singinst.org/upload/CFAI.html

Yudkowsky, E. (2004) *Coherent Extrapolated Volition*, http://singinst.org/upload/CEV.html

Yudkowsky, E. (2009) *The Challenge of Friendly AI*, [Online], http://itc.conversationsnetwork.org/shows/detail3387.html

Susan Greenfield

The Singularity: Commentary on David Chalmers

The concept of a 'Singularity' is particularly intriguing as it is draws not just on philosophical but also neuroscientific issues. As a neuro-scientist, perhaps my best contribution here therefore, would be to provide some reality checks against the elegant and challenging philo-sophical arguments set out by Chalmers. A convenient framework for addressing the points he raises will be to give my personal scientific take on the three basic questions summarised in the Conclusions section.

1. Will There Be a Singularity?

The term itself 'Singularity' is, surprisingly, not actually formally defined in Chalmers' review. However the reader could reasonably assume that it identifies a hypothetical point in the future when there is an 'explosion' of intelligence, such that machines outstrip the capacity of humans. Unfortunately 'explosion' is not a helpful term in this con-text: it is too much like an 'explosion' of art and culture in the Renais-sance, suggesting diversification of many aspects of human talent. But here the scenario appears to be quite the opposite, namely a more linear, albeit exponential increase in a single commodity: intelligence. Though Chalmers admits that intelligence can be defined and expressed in many ways, he focuses nonetheless on the narrow defini-tion of 'g' and assumes therefore that the phenomenon can be expressed as a computational process.

But herein lies a basic weakness with the ensuing discussion. It may well be outside of the scope of this commentary, and perhaps even of Chalmers' review, to explore exhaustively the nature of intelligence: however it is widely accepted there is indeed more to this complex mental trait than IQ tests or indeed computational processing. For example, Roger Penrose pointed out long ago that it would be impossible to devise an algorithm for those key human abilities of intuition or commonsense: even further back in time the great physicist Niels Bohr admonished a student, 'You're not thinking, you're just being logical'.

Nowadays the issue of IQ and what is does or doesn't tell us about human mental ability has been revisited with the current raging controversy over the impact of screen technologies and heavy internet use. For example Steven Johnson in his book, *Everything Bad is Good for You*, suggests that increased activity with computer games might be contributing to the widely acknowledged increase over last decade or so in IQ scores in many developed countries. However, there has not been a concomitant increase in insights into the economic situation, no 'explosion' in creative arts, nor new insights into literature or history, - nor indeed in philosophy or neuroscience! Suffice it to say that whatever would be needed for this happy scenario to be realised, it entails something additional to a high IQ score.

However Chalmers is correct in distinguishing the 'explosion' in 'g' from enhancement of the brain. With 'enhancement', emphasis is placed on whether the system in question is human or not, whereas with the AI preoccupying Chalmers' thoughts the 'A' i.e. the artificial, is the crucial feature in comparison to any biological counterparts. Whilst it may be important, by definition, that AI should indeed be conspicuously characterised by the 'A' component, the deeper issue is that by being computational (unlike enhanced yet unspecified human abilities) it is constrained by excluding intuition, common sense, imagination, and creative insight. Consequently the biggest difficulty of all is that computational processing requires a specific endpoint, a clear solution to a specified problem whilst many manifestations of what many of us might view intelligence — writing *War and Peace* for example, or developing a new take on the current economic troubles, or imagining how the brain might generate consciousness — do not.

My own approach to intelligence is to refer back to the literal Latin meaning of 'understanding'. In turn, I would suggest that understanding is primarily seeing one thing in terms of another: the more we can relate a phenomenon, action, or fact, to other phenomena, facts, or actions, the deeper, I would argue, is the 'understanding'. For example, for a child the famous line in Macbeth, 'Out, out brief candle...' could not be understood as it would be by an adult who could place the phrase in a wider context and realise it was a metaphor for death. This is a very different type of ability from the fast linear processing demanded of IQ tests towards a specified end, and which are far more

tractable to silicon systems. In summary, not all computational problem solving requires understanding, nor does all understanding imply
problem solving: so computation should not be seen as synonymous
with intelligence.

2. How Should We Negotiate the Singularity?

Since Chalmers views the super, super intelligence of the Singularity
as potentially sinister, he cautions two strategies: building 'values'
into machines and initially building those machines in the virtual
world. Let's take each suggestion in turn. Chalmers tends to agree
with Hume rather than Kant in assuming that intelligence does not
necessarily imply values and *vice versa*. However, surely this
dilemma once again depends on how we define intelligence. If we
take the simple computational concept of 'g', prowess in IQ tests, then
indeed Hume would be correct: after all, why should a simple linear
process be predicated on anything other than the rules of the game?
But if we take the wider view of intelligence as argued above, to imply
understanding, then perhaps Kant would be more accurate in viewing
intelligence as an understanding that would imply an awareness of the
link to particular values?

But the issue for neuroscientists is not which philosopher's camp to
join but the differentiation, in hard neuroscientific terms, of information from knowledge and of knowledge from wisdom. Whilst information processing is just that, the appropriate response to an incoming
stimulus, understanding in contrast requires that the stimulus in question be embedded in a conceptual framework. Briefly, a conceptual
framework of the type that would be required for 'understanding' can
be interpreted in brain terms, as the growth of the connections
between brain cells that are formed post-natally and driven, shaped,
and strengthened by individual experience (see Greenfield, 2008).
Hence every individual human will have a uniquely personalized
brain — a mind that is constantly evaluating the current world in terms
of existing associations whilst simultaneously being updated by it.

We could go further and envisage that knowledge is the embedding
of a fact or action within a conceptual framework so that it makes
sense, i.e. can be understood, whilst 'wisdom' requires still further and
widespread connectivity whereby more generally the associations
made are drawn from an ever wider personal experience/individual
memories that enable the assignment of more generalized 'values' of
positive or negative. The process therefore would be more analogous
to a stone in a puddle causing ever larger ripples, as opposed to the

igniting of a fuse, where the flame cracks along a single path to a well determined and defined endpoint. Whilst it is easy to imagine building a machine that can simulate the latter, it is harder to imagine constructing an artificial device capable of understanding and wisdom in the way outlined here for the brain. The biggest hurdle of all of course, would be that understanding and wisdom differ quintessentially from computational information processing, in that they require an inner state: consciousness.

The issue of motivation, discussed by Chalmers, also depends on consciousness: however as he then goes on to admit, consciousness itself is still poorly understood, particularly in relation to the physical brain. Chalmers of course was the originator of the phrase 'The Hard Problem' which famously captured the elusive nature of the causal link between objective events in the brain and their realization as subjective, first-hand experience. Whilst a number of neuroscientists, myself included, have developed candidate 'neural correlates of consciousness' — ways of matching up conscious states and unconsciousness with diverse physico-chemical processes — these scenarios are far from inspiring the all-important leap that would answer the Hard Problem (see Koch and Greenfield, 2007). Even more basic and immediate would be to have an idea, however hypothetical, of what kind of answer we might expect: would it be a formula, a brain image, a performing rat, or one individual being able to have the same first-hand consciousness as another? None of these possibilities seem anywhere near adequate or appropriate. Yet until we know what kind of answer would help solve the Hard Problem, then there can little chance of doing so: but until we do so, considerations of consciousness important as they are for discussing here the question of 'values' and 'motivation', must be left as an unhappy loose end.

However, on a more straightforward level, we can at least dissociate 'consciousness' from 'mind'. After all, humanity has since the dawn of time been familiar with 'losing one's mind', 'letting oneself go', having a 'sensational (as opposed to a cognitive) time' and with the very term 'ecstasy' literally meaning 'to stand outside of oneself'. From a neuroscientific perspective, it is indeed possible to produce scenarios that would be neural correlates of losing one's mind, i.e. situations where the carefully personalised neuronal connections that characterise each individual brain, are for a variety of reasons (e.g. chemical imbalances), not fully accessed. Could AI++ accommodate these different mental states too? If not then we return again to the limitations of 'intelligence' being narrowly defined as a computational process; whereas if it were possible to build a machine that could have biphasic states where the alternative was a loss of mind, what would be the implications for the future, where a machine wasn't merely 'more intelligent' than humans but equally, if not more, crazy (i.e., mindless) on occasions! This possibility was indeed raised a while back by the late Stuart Sutherland who claimed he would only believe a computer was conscious, when it ran off with his wife.

Chalmers' second caution is that we might build AI++ initially in the virtual world: whilst hypothetically sound as an idea, it is in itself a virtual rather than realistic answer. Chalmers himself admits that the barrier between the real and virtual even already, is highly porous. In any event, positing a virtual location for AI++ machines would not be helpful even if it were possible: for many the virtual world already is more real than the real world. Hence the relationships, the emotions and the issues arising from machine-human interactions in the real world, would be as real and as significant in the cyber counterpart.

3. How Can We Integrate in a Post-Singularity World?

The final question Chalmers poses is how we might, as humble humans, co-exist with super intelligent machines. Once again we return to exactly what being 'more intelligent' might actually mean. Imagine machines that are just the same as us i.e. with values, motivation, consciousness, and thus ability to blow their minds. What could they then do that was 'better' in terms of understanding, consciousness or madness that would justify the description 'more intelligent'? Understanding, consciousness and loss of mind are hard to quantify, as are related and much prized talents also lacking a specific problem to solve, such as creativity. The only accomplishment in which computational devices could clearly excel would be solving unambiguously defined problems along with fast information processing: in fact, most ordinary old computers arguably perform better than us humans in this way already. We can surely be reassured?

Not really. Chalmers then progresses the idea of integration not just at the level of interaction in the same external space, but of invasive intrusion of silicon systems into our carbon-based bodies. One mind-game is to imagine neurons slowly being replaced one by one with silicon units. Whilst such a situation is easy to imagine, it is based on a basic misconception about how the brain works. No neuron is an island, much less an autonomous working unit. We have known for a long time that there is much redundancy in the nervous system, for example in stroke when it is quite common to see a recovery of function after substantial brain damage. This 'redundancy' is not so much that many neurons are all performing the same operations, but by virtue of their intrinsic mechanisms of 'plasticity' they have the ability for adaptation and compensation.

Each neuron is dynamic, not a fixed component that will plug in and play consistently, independent of context. Moreover, accommodating this functional dynamism are incessant morphological changes in the configuration and micro-shape of the branches (dendrites) that emanate from the main part of each neuron (cell body), whereby the ease by which signals are transmitted from remote dendritic inputs up to the cell body can be highly variable from moment to moment according to the availability of diverse extracellular 'modulating' compounds (Greenfield, 2003). Sometimes these electric signals never make it to the cell body, but remain active more locally in the tangle of dendrites that can then operate independently of the 'action potential', - the traditional, familiar read-out of the single cell body. This intense, ever changing, interaction between coalitions of neurons is nothing like the rigid circuitry of computational devices, entailing as it does qualitative factors (one chemical versus another) and non-linear directionality. No simple, systematic substitution of one silicon component after another could ever have the same effect, unless of course that unit was a simulacrum of the neuron, replete with all the chemicals and biochemical machinery that makes its characteristic, restless plasticity possible along with its variable sensitivity to whatever modulating chemicals might happen to have been released at a particular time. Whilst the hypothetical scenario of neuron substitution is conceptually logical and plausible, in reality it's meaningless and unhelpful.

That said, we do know that neurons can interface very well with silicon systems. The pioneering work of Peter Fromherz (2006), for example, has shown the beauty of a 'neurochip' whereby connections are made readily on a circuit board between neurons and silicon nodes. Similarly, if brain cells are able to function in a hybrid device in this way, then the reverse might not be surprising: artificial implants in the brain are already possible and achieving astonishing effects. For example, Miguel Nicolelis of Duke University has developed a system whereby quadriplegic patients can, through devices implanted in their brain, generate electronic signatures that would normally precede various movements (Lebedev and Nicolelis, 2011). These electronic signals are then recognized by a computer that can operate an artificial limb, so that a paralysed person can 'will' a movement. However these 'neuronal prostheses' are far from the silicon take-over of the brain envisaged by Chalmers. Whilst silicon-carbon interfacing is possible, at least for the final execution of a movement, i.e. brain output simulating brain-muscle, it should not be conflated with the neuron-neuron interactions that underlie cognitive processes, — nor should it be confused with the AI++ under discussion in this review.

Very 'intelligent' machines, AI++, are not examples of brain enhancement nor, I would argue, could they ever be very serious rivals to the diverse potential of the human mind. Without doubt computational devices will readily out-strip us in problem solving, but not in understanding. As Thurber (1939) once remarked, 'It is better to know some of the questions than all of the answers.' Chalmers has really asked the wrong question in this review. Surely the pressing issue at stake is not 'What happens when machines become more intelligent than humans?' But rather, 'What happens when AI++ has freed us up to stretch ourselves to our full human potential? How can we then make the most of the unique brain at our disposal?'

References

Fromhertz, P. (2006) Three levels of neurelectronic interfacing: Silicon chips with ion channels, nerve cells, and brain tissue, *Annals of the New York Acadamy of Sciences*, **1093**, pp. 143–160.

Greenfield, S.A. (2003) *The Private Life of the Brain*, New York: Penguin.

Greenfield, S.A. (2008) *I.D: The Quest for Meaning in the 21st Century*, London: Sceptre.

Johnson, S.B. (2005) *Everything Bad is Good for You*, New York: Riverhead.

Koch, C. & Greenfield, S.A. (2007) How does consciousness happen?, *Scientific American*, **294** (4), pp. 76–83.

Lebedev, M.A. & Nicolelis, M.A. (2011) Toward a whole-body neuroprosthetic, *Progress in Brain Research*, **194**, pp. 47–60.

Thurber, J. (1939) *The New Yorker*, 18 February.

Robin Hanson

Meet the New Conflict, Same as the Old Conflict

Abstract

Chalmers is right: we should expect our civilization to, within centuries, have vastly increased mental capacities, surely in total and probably also for individual creatures and devices. We should also expect to see the conflicts he describes between creatures and devices with more versus less capacity. But Chalmers' main prediction follows simply by extrapolating historical trends, and the conflicts he identifies are common between differing generations. There is value in highlighting these issues, but once one knows of such simple extrapolations and standard conflicts, it is hard to see much value in Chalmers' added analysis.

Introduction

David Chalmers says academia neglects the huge potential of an intelligence explosion:

> An intelligence explosion has enormous potential benefits: a cure for all known diseases, an end to poverty, extraordinary scientific advances, and much more. It also has enormous potential dangers: an end to the human race, an arms race of warring machines, the power to destroy the planet. So if there is even a small chance that there will be a singularity, we would do well to think about [it]. (p. 14)

Apparently trying to avoid describing the scenario that interests him in too much speculative detail, Chalmers goes far in the other direction, offering rather weak descriptions of his key scenario and the issues that concern him about it. Such weak descriptions do help highlight important issues to wider audiences, but they are too weak to offer much added insight to expert understanding.

That is, we should expect that a simple continuation of historical trends will eventually end up satisfying his description of an 'intelligence explosion' scenario. So there is little need to consider his more specific arguments for such a scenario. And the inter-generational conflicts that concern Chalmers in this scenario are generic conflicts that arise in a wide range of past, present, and future scenarios. Yes, these are conflicts worth pondering, but Chalmers offers no reasons why they are interestingly different in a 'singularity' context.

To make my point clear, I will first review what we should expect as the simple future continuation of prior historical trends, and then review some of the conflicts that commonly arise between generations, which should also arise in this default future scenario. Finally I will show that both Chalmers' singularity concept and the conflicts he identifies are already contained within this default scenario.

Expect Growth

Our clearest and most dramatic long-term historical trend has been a vast and broad growth in our capacities. Primates evolved into humans who developed farming and then industry. Over this long development, we have accumulated capacity-enhancing innovations first in our genes, and later also in our culture and social and physical environments. For almost any task one can imagine, humanity is now collectively far more able to achieve that task. This vast increase in our capacity has enabled a vast increase in the range of our environments and activities, in our population, and lately in our individual lifespan and consumption.

Many of our increased capacities are mental (or computational). Such capacities help us to choose, calculate, infer, talk, answer, summarize, compose, etc. We generally say that machines, animals, people, teams, organizations, cities, or nations are 'intelligent' when they display relatively high capabilities across a wide range of mental tasks. In this sense humanity and the civilization it has spawned have clearly become far more 'intelligent' than before.

Even when we devise measures like IQ, intended to describe the shared correlation among the mental capacities of individuals, when outside assistance is excluded, we still find that we are getting smarter. For roughly a century we have seen in rich nations a 'Flynn effect' of IQ scores increasing each decade by an average of three points, relative to a one hundred point average. But while individual humans are improving, for the last few centuries the capacities of our machines have been improving at an even faster pace.

Not only have our capacities greatly increased, we have also seen large increases in our rate of capacity growth. For example, the human population grew mainly because it was able to exploit more ecological niches more thoroughly. It took roughly two million years for the human population of foragers to grow from ten thousand to ten million, doubling roughly every two hundred thousand years. Then during the farming era, from about five thousand to three hundred years ago, the population of farmers (and the economic capacity to support them) doubled roughly every thousand years. And over the last century of industry the world economy has doubled about every fifteen years.

If a new growth era was to grow as fast compared to industry as industry grew compared to farming, or farming grew compared to foraging, the new economy might double every few months or less. And since the industry era seems to have already seen more capacity doublings than the forager and farming eras, then if the number of doublings were at all comparable across eras, if a new growth era is coming it should start within a century or so.

Thus if historical trends continue, we should expect our civilization to continue to gain in capabilities across a wide range of tasks. Our civilization as a whole, and probably also many individual machines or creatures within it, will become much more 'intelligent.' We should also expect, if perhaps more weakly, that our machines will improve more rapidly than our biology, and that the rate at which such capabilities improve will also increase. Yes historical trends need not continue, there may be fundamental limits to some capacities, and the price we pay to increase some capacities may be the reduction of other capacities. But overall we should expect the future to see more intelligence, perhaps especially in machines, and perhaps increasing faster.

Inter-Generational Conflicts

Those born in different historical eras have many common causes to unite them, and many ways to assist one another. Even so, there are also some ways in which generations often find themselves in conflict. Generations that overlap in time can conflict over resources, and generations can have preferences about the behavior of distant generations.

The simplest conflict is that generations can exist nearby in time, and so compete for natural and social resources. For example, when creatures begin a life cycle with a very low capacity, as human children do, then new generations are initially at the mercy of older generations. At the other end of a life cycle, generations can also find themselves in weak negotiating positions, both because capacity falls at the end of life and because later generations tend to have greater capacity than earlier ones. This can allow new generations to treat old ones with less deference than the old prefer. In the extreme, new generations might exterminate old ones.

Natural selection endowed humans with sufficient feelings of empathy for vulnerable infant and elderly kin to perpetuate the species, though this does not prevent killing of infants, elderly or non-kin in severe resource shortages. In modern market economies, people can save during their high-capacity midlife in order to gain promised resources during their later low-capacity 'retirement.' In rich nations non-kin usually keep such retirement promises, though they are sometimes diluted by crime, fraud, taxation, hyper-inflation, etc.

The second common conflict is when older generations have preferences over the behavior of younger generations. Older generations may want younger ones to remember and honor them, or to remain loyal to their clans, religions, nations, customs, or social/moral norms. However, the genetically and culturally embodied behaviors of younger generations may change via random drift or adaptation to changing circumstances. Cultural changes have been especially rapid recently and most of our ancestors of a few centuries ago would probably dislike many of the ways that we have changed our loyalties since then.

When two generations are alive at the same time, many of their value conflicts can be dealt in ways similar to resource conflicts. But once an older generation is dead, other approaches may be required. Older generations often try to indoctrinate younger generations into desired loyalties, but this effect can decay quickly. In modern market economies, the terms of bequests could allow older generations to pay younger generations to preserved desired loyalties, and typical interest rates give older generations enormous amounts to pay with. In practice, however, our law enforces few such terms, out of a distaste for 'dead hands' controlling future generations.

Chalmers' Scenario

Chalmers argues that, absent a huge disaster or a concerted effort to prevent change, 'within centuries' we will see 'superintelligence,' i.e., artificial intelligence 'at least as far beyond the most intelligent human as the most intelligent human is beyond a mouse' (pp. 15–17). Conscious that 'intelligence' can be ambiguous, he clarifies that he means 'capacities that far exceed human levels' in both 1) 'capacity to create systems with' a 'cognitive capacity that we care about' such as 'some specific reasoning ability,' and 2) some correlated 'self-amplifying cognitive capacity' (p. 27).

But what is meant by intelligence 'as far beyond the most intelligent human as the most intelligent human is beyond a mouse'? For the vast majority of mental tasks, the mental capacity of our civilization has already increased greatly over human history – for many mental tasks we are already as far beyond our ancestors of a hundred thousand years ago as they were beyond a mouse.

Perhaps Chalmers has in mind an IQ-like concept for individuals, intended to exclude outside assistance. But it is civilization's total capacity to accomplish tasks, not the capacity of an isolated individual, that most matters for the consequences he highlights, such as 'a cure for all known diseases, an end to poverty, [and] extraordinary scientific advances.' And if our growth rates speed up again, centuries of Flynn-effect-like IQ growth could well be sufficient to meet even the to-us-as-we-are-to-mice standard.

Thus we should expect the simple continuation of long-term historical trends to lead, within a few centuries, to a future containing 'superintelligence'; we need no further assumptions.

Chalmers' Issues

Loosely speaking, things that originate from us and whose form and details come from and echo us, are our 'descendants.' And unless everyone and thing in the future is equally superintelligent, the future should contain both newer 'generations' that are more superintelligent, and older 'generations' who are less so. We don't know the mixture of biology and machines that future superintelligences will inhabit, but we can be sure that much of their design and content will descend from and echo us, with differing of our capacities improved to differing degrees.

Chalmers has two main concerns about future superintelligence, both of which can be seen as standard inter-generational concerns. First, he fears 'they' might not treat 'us' well:

> Care will be needed to avoid an outcome in which we are competing [with them] over objects of value. ... Systems that have evolved by maximizing the chances of their own reproduction are not especially likely to defer to [us]. (p. 38)

> What is our place within that world? There seem to be four options: [1] extinction, [2] isolation, [3] inferiority, or [4] integration. ... The second option ... would be akin to a kind of cultural and technological

isolationism that blinds itself to progress. ... The third option ... threatens to greatly diminish the significance of our lives. ... This leaves the fourth option: ... we become superintelligent systems ourselves... If we are to match the speed and capacity of nonbiological systems, we will probably have to dispense with our biological core entirely. (p. 45)

These concerns seems to echo standard concerns of older generations that younger generations treat them well during times when generations overlap. Chalmers says his older generation doesn't want to go extinct or even compete for resources. Rather than suffer the indignity of being segregated into a retirement community, or of living closely among a higher capacity younger generation, Chalmers prefers his older generation to always have the highest capacity of any generation present.

Chalmers' other concern is that these younger generations might act on values that differ from those of his generation:

It makes sense to ensure that an AI values human survival and well-being and that it values obeying human commands. ... [and] that AIs value much of what we value (scientific progress, peace, justice, and many more specific values). ... If at any point there is a powerful AI+ or AI++ with the wrong value system, we can expect disaster (relative to our values) to ensue. ... If the AI+ value system is merely neutral with respect to some of our values, then in the long run we cannot expect the world to conform to those values. (pp. 38–9)

This also seems to echo standard desires of older generations for younger generations to continue to preserve old generation loyalties to clan, religion, customs, social/moral norms, etc.

Conclusion

Chalmers is concerned about future inter-generational conflicts, and he raises such conflict issues in the context of possible future 'superintelligences.' I have argued that we should expect a simple continuation of historical trends to lead to a future civilization with greatly increased capacity across a wide range of mental tasks, and that the inter-generational conflicts that Chalmers highlights are generic, the sort of conflicts that arise in a wide range of future scenarios.

I don't mean to imply that there is nothing new or interesting one might say about future inter-generational conflicts. But interesting contributions should focus on ways in which inter-generational conflicts are interestingly different in some imagined future context.

There do seem to be several promising candidates for plausible differences that might interestingly change the nature of intergenerational conflicts. For example, longer lives or faster growth rates may increase the number of different generations alive at any one time, and the degree to which such generations differ in peak capacity. (If there are other reasons why intergenerational conflict is interestingly different with faster rates of change, Chalmers has not identified them.) Direct copying of minds, instead of growing new minds up from baby minds, might ensure that new generation values are more similar to old generation values. Explicit and transparent encoding of values might make indoctrination easier and more reliable.

It seems to me that the most robust and promising route to low cost and mutually beneficial mitigation of these conflicts is strong legal enforcement of retirement and bequest contracts. Such contracts could let older generations directly save for their later years, and cheaply pay younger generations to preserve old loyalties. Simple consistent and broad-based enforcement of these and related contracts seem our best chance to entrench the enforcement of such contracts deep in legal practice. Our descendants should be reluctant to violate deeply entrenched practices of contract law for fear that violations would lead to further unraveling of contract practice, which threatens larger social orders built on contract enforcement.

As Chalmers notes in footnote 19, this approach is not guaranteed to work in all possible scenarios. Nevertheless, compare it to the ideal Chalmers favors:

> AI systems such that we can prove they will always have certain benign values, and such that we can prove that any systems they will create will also have those values, and so on ... represents a sort of ideal that we might aim for. (p. 39)

Compared to the strong and strict controls and regimentation required to even attempt to prove that values disliked by older generations could never arise in any future descendant, enforcing contracts where older generations pay younger generations to preserve specific loyalties seems to me a far easier, safer and more workable approach, with many successful historical analogies on which to build.

F. Heylighen

A Brain in a Vat Cannot Break Out

Why the Singularity Must Be Extended, Embedded and Embodied

Abstract: *The present paper criticizes Chalmers's discussion of the Singularity, viewed as the emergence of a superhuman intelligence via the self-amplifying development of artificial intelligence. The situated and embodied view of cognition rejects the notion that intelligence could arise in a closed 'brain-in-a-vat' system, because intelligence is rooted in a high-bandwidth, sensory-motor interaction with the outside world. Instead, it is proposed that superhuman intelligence can emerge only in a distributed fashion, in the form of a self-organizing network of humans, computers, and other technologies: the 'Global Brain'.*

Introduction

This paper is a comment on Chalmers' discussion of the Singularity, which Chalmers defines as the hypothetical emergence of a superhuman intelligence in the near-term future via the development of an artificial intelligence (AI) system that is so intelligent that it can reprogram itself in such a way as to amplify its intelligence to a level that is higher than the human level (AI+), and eventually so high (AI++) that it can dominate humanity. Chalmers's paper neatly splits into two major issues. First, he discusses the different scenarios for how a Singularity might arise, together with their respective dangers, benefits and ethical implications. In the second part, he discusses how we could mitigate the disadvantages of a Singularity by 'uploading', i.e. creating a digital version of our minds that could integrate with the postulated AI++. Here he again analyses the different possibilities and implications of the idea of transferring a human mind into a digital substrate. Since the paper is much too long to criticize point by point, I will here merely focus on what I consider its major flaw: an implicit ontological assumption of reductionism veering towards dualism. I will demonstrate how this flaw undermines much of the discussion in the first part, and hope that by then it will be clear that the second issue too has become mostly moot.

The paper's title defines it as a 'philosophical analysis', thus revealing both the major strengths and weaknesses of Chalmers' treatment of the matter. In terms of strengths, the paper is a very systematic and comprehensive attempt at surveying and clarifying the main conceptual issues surrounding the notion of Singularity. This is particularly useful given that discussions of the Singularity up to now, in the work of authors such as Kurzweil (2005) and Vinge (1993), were extremely speculative, and based on assumptions that, while not a priori implausible, are highly questionable. Chalmers does a thorough job of making those assumptions explicit, formulating equally plausible alternative assumptions, and then asking all the necessary difficult questions. However, apart from some hunches and intuitions, he does not offer much in terms of concrete answers.

This leads us straight into the weaknesses of such an analytic approach: by splitting the problem into a wide array of subproblems, and then splitting up the possible approaches towards solving them into a variety of alternative positions, it seems as if the overall solution only gets further out of reach. The reason is that each subproblem and each position seems to be at least as difficult to tackle as the overall problem: 'how can a Singularity emerge?' By way of example, let me mention just one problem that comes out of this analysis, the one that Chalmers in earlier work (Chalmers, 1995) has called the 'hard problem of consciousness' — implying that we have at the moment no methodology available to even start investigating it.

This is a deeper issue with all analytic approaches. Analysis by definition means that you divide a whole into parts, hoping that by understanding the parts, you will also understand the whole. While originally formulated by Descartes, this analytic principle is at the basis of classical, Newtonian science, where it led to the philosophy of *reductionism* — the idea that you can always reduce the behaviour of a whole to the behaviour of its parts. Applied to material phenomena, reductionism leads to atomism, the idea that matter is made out of elementary constituents (particles) that completely determine that matter's properties and behaviour. While very successful in the realm of physics, such 'material' reductionism, however, has largely failed to explain phenomena in the realms of life, mind and society. However, the analytic approach can still be rescued by postulating different, non-material constituents for these phenomena. This is what Descartes did when he observed that mental phenomena did not seem to yield to such a mechanistic reduction: he postulated 'mind' as a separate category, independent of matter, so that human behaviour could be decomposed into its mental and its physical components. This was the origin of the philosophy of *dualism*, which implicitly seems to inspire much of Chalmers's reasoning, even while he does not explicitly endorse it.

Very few scientists and philosophers nowadays endorse mind-matter dualism. However, many still cling to what Dennett (1993) has called 'Cartesian materialism'. This can be seen as a 'soft' version of dualism, which assumes that mind cannot exist independently of matter, but which otherwise still treats the mind as if it were a separate entity that somehow resides inside our material brain. Chalmers formulates one version of such soft dualism when he characterizes the mind as a brain complemented by unspecified 'further facts'. It could be argued that traditional cognitive science and artificial intelligence are founded on Cartesian materialism: they see the mind as a piece of software that runs on the hardware of the brain, processing information that enters in the form of symbolic representations of the outside world. This perspective is doubly analytic: it not only separates mind from world, but decomposes information about that world into independently meaningful symbols, each representing a discrete part or aspect of reality. Implicitly, this seems to be the perspective that informs Chalmers and most other Singularity theorists when they discuss the possibility of superhuman artificial intelligence.

The traditional, symbolic approach to AI has undergone increasingly stringent criticisms since the 1980s, from cognitive scientists working from perspectives such as *connectionism* (Bechtel & Abrahamsen, 1991), *constructivism* (Bickhard & Terveen, 1995), *dynamical systems* (Beer, 2000), and especially *situated and embodied cognition* (Anderson, 2003; Lakoff & Johnson, 1999; Steels & Brooks, 1995; Varela, Thompson & Rosch, 1992). These 'new' approaches to cognition are much more dynamic and holistic, emphasizing the complex network of interactions out of which mind emerges. From this perspective, the mind is no longer localized in any particular component, but distributed over a massive number of internal and external components which all cooperate in a self-organizing manner. Intelligence (and with it consciousness) can then be seen as an emergent phenomenon of coordination between these processes, an integrated manner of dealing with an enormous amount of bits and pieces that together determine an individual's experience of its situation, and that define a potential problem to be dealt with (Heylighen, 2011a; 2007–2011).

From this holistic perspective, trying to understand the mind by analyzing it into components is like eating soup with a knife and fork: no matter how you try to cut up the liquid into pieces, you will never get a piece solid enough so that you can seize it with your fork. Yet, nothing stops you from drinking the soup directly from the bowl, or, if you prefer, spoon by spoon. The trick is simply to use a tool adequate for the task. From my perspective, analytic tools such as logic and symbols appear particularly inadequate for grasping fluid, distributed phenomena, such as intelligence and consciousness. More systemic tools seem much better suited. For example, the 'fluidity' of experience can be modeled by activation spreading across a distributed or connectionist network (Heylighen, 2007–2011; Heylighen, Heath & Van Overwalle, 2004).

In the remainder of this paper, I will apply this holistic philosophy of mind, mostly from the perspective of situated and embodied cognition (Anderson, 2003), in order to develop an alternative understanding of the Singularity. I will argue in particular that the 'brain-in-a-vat' conception of AI presented by Chalmers is unlikely to lead to any superhuman intelligence, but that such a higher level of cognition (which I call the Global Brain [Heylighen, 2011b]) will necessarily be distributed over a massive number of more or less intelligent components, including individual humans, computers and their programs, databases such as the world-wide web, and a variety of sensors and effectors embedded in the environment. From this point of view, most of the 'philosophical' issues raised by Chalmers become irrelevant, while a host of other issues — ethical, political, social and economic — come into focus.

The Situated and Embodied Critique on AI

The view of artificial intelligence sketched by Chalmers and held by most traditional AI researchers has been severely criticized by a variety of authors. These researchers (e.g. Steels & Brooks, 1995; Varela, Thompson & Rosch, 1992; Lakoff & Johnson, 1999; Clark, 2008) come from a variety of backgrounds and points of view, including biology, robotics, linguistics, psychology and philosophy. Yet, they are united by their rejection of what some have called the 'brain-in-a-vat' perspective on intelligence. According to this perspective, the only thing you need to produce intelligence is a brain, i.e. a specialized piece of cognitive machinery. You merely need to make sure that the machinery gets the resources (such as matter and energy) that it needs to keep functioning, and the relevant pieces of information to work on. Given those, the brain will process that information and, assuming it is intelligent enough, solve the problems you care to put to it. Whether the 'brain' is made out of nerve cells or out of digital circuits is largely irrelevant to that task, as long as those circuits sufficiently accurately simulate the overall functioning of the neurons.

The 'situated and embodied' criticism of this view is that the brain has evolved as a specialized organ for the control and coordination of the organism and the actions it performs in its environment. These actions determine whether the organism as a whole will survive and thrive, or fail and eventually be eliminated by natural selection. This entails a constant adaptation to a complex and unpredictable environment, by counteracting or correcting any sensed deviations from a state of fitness. These continuously experienced differences between experienced and desired situations are the 'problems' that intelligence was designed to solve (Heylighen, 2007–2011). In this view, intelligence is not a 'thing' or not even a 'property', but an on-going process of adaptation and coordination, grounded in a high-bandwidth feedback between organism and environmental situation. Interrupt that interaction and the brain becomes a useless piece of machinery, like a car stationed in an exhibition hall: it may still display some minor features, such as the softness of its chairs or the smooth opening of its electric windows, but it is incapable of doing what it was designed for, driving.

Similarly, a brain in a vat, kept artificially alive via a stream of nutrients, but disconnected from its body, may still be able to produce some abstract thoughts, but it has lost its essential ability to act on the world and to experience the results of its actions. As such, its thoughts will become increasingly disconnected from and irrelevant to reality. Compared to a traditional AI system, such a brain still has an immense advantage, though: it has a lifetime of detailed, subtle experiences stored in its memory on which it can build to perhaps develop new thoughts. Without such history of fine-graded interactions, an AI system is at best a clumsy question-answering 'expert system', which needs to be fed a huge amount of symbolic data before it can make any meaningful inference. These data typically need to be structured and formatted by human 'knowledge engineers', who abstract and formalize real-world experience into logical expressions. But this is an extremely time-consuming, difficult, unreliable and potentially endless task, known in AI as the 'knowledge acquisition' bottleneck. Just throwing in more man-months does not solve the problem, as illustrated by the on-going CYC project (Lenat, 1995), which started out over 25 years ago with the aim of formalizing all common-sense knowledge. The intention was to provide the foundation for an AI-system that would be able to pass something like the Turing test, but as yet without any apparent success.

In the last decade or so, the focus of AI has therefore shifted towards machine learning and data mining, i.e. letting the computer program itself extract knowledge from the huge amounts of data available in specialized databases and on the web. This approach has been much more successful, as recently illustrated by the IBM program Watson, which managed to win a Jeopardy! question-answering game against human experts. However, what AI enthusiasts observing this unmistakable advance tend to neglect is that: 1) the knowledge of Watson is not 'artificial'; it is the product of millions of humans entering terabytes of text into websites and databases. Without this 'extended memory' created by actual agents interacting with the real world, Watson and similar programs would be absolutely helpless; 2) Watson is still nothing more than a passive system waiting until someone asks it a well-formatted question before it can start producing an answer. In the real world, intelligence entails autonomy, i.e. the ability to directly experience the environmental situation and to decide what it means, what problems may need to be solved, and what actions may be worth taking — without guidance from a programmer, experimenter, or quizmaster.

The newer approaches to AI have started to take the autonomy issue seriously, by creating robots that interact with the world via their sensors and effectors, trying to reach their (preprogrammed) goals, while learning from their experience (Steels & Brooks, 1995). These autonomous robots at best reach an intelligence level comparable to a primitive insect. But that is not just a question of lack of computational capacity or insufficiently sophisticated software: robots are handicapped by an alternative version of the knowledge acquisition bottleneck. It turns out to be extremely difficult to engineer and build truly efficient systems of sensors and effectors. That is why present-day autonomous robots are very clumsy creatures, who at best may succeed in mowing a flat, well-defined area of lawn, but already start to get in trouble when they have to vacuum a typically more irregular apartment floor.

The awkwardness of engineered sensors and effectors should not surprise us if we remember that evolution needed billions of years to build our extremely sophisticated sensory and motor systems. The complexity resides not only in our eyes, ears, and limbs, but in the integrated system of sensory organs, nerves, hormones, neurotransmitters, brain, glands, muscle fibers, organs, and in fact every cell, which reacts in real time to thousands of chemical signals traveling across the body and which are directly or indirectly triggered by a variety of sensed conditions, such as heat, pressure, smells, stress, emotions, remembrances, etc. For the foreseeable future, this exquisitely coordinated and fine-tuned system seems impossible to reproduce by present engineering techniques, even taking into account 'Moore's law' types of acceleration. The reason is that accelerating scientific progress typically happens in well-defined, specialized disciplines, such as chip miniaturization. To build a system rivaling the complexity of the human organism would require massive multi-disciplinary coordination and integration of results — the type of progress that still seems as slow and difficult as ever.

The lack of a biological body, therefore, is a fundamental handicap for any artificially intelligent system, even though in theory we can try to build increasingly more sophisticated robotic bodies. But here we are confronted with a hardware bottleneck that seems much more difficult to overcome than the computational or software bottlenecks that present AI theorists have been focusing on.

Implications for the Singularity

Chalmers envisages the superhuman intelligence emerging from the Singularity as an AI system locked up in some kind of virtual world — a safety measure needed to make sure that this AI++ would not take control of humanity. According to the situated and embodied philosophy of cognition, however, this vision is intrinsically absurd. A virtually imprisoned AI program is even worse than a brain in a vat, as it simply has no sensors, no effectors, no body, and no real world to interact with. Therefore, it cannot be intelligent in the sense of being an autonomous, adaptive cognitive system that can deal with real-world problems and steer its own course of action through a complex and turbulent reality. At best, it can be a very sophisticated expert system that can solve chess, Jeopardy!, and similar highly artificial and constrained puzzle and games given to it by its designers, or help them to mine massive amounts of pre-formatted data for hidden statistical patterns. The idea that on the basis of such data it could 'reverse engineer human psychology' to such a degree that it could manipulate its creators to give it control over them, as Chalmers proposes, seems more like a paranoid fantasy than like a realistic scenario.

All the arguments put forward by Chalmers to support the AI++ scenario revolve around a runaway process of self-amplifying intelligence. This assumes basically a process of *positive feedback* or *increasing returns*. However, as Chalmers himself notes, positive feedback explosions always come to a halt when the resources for further growth are exhausted, and returns start to diminish rather than increase. This is obvious for processes of physical growth, which are limited by the law of conservation of matter and energy. Things are subtler for cognitive processes, as information does not obey any clear conservation laws: you can destroy or duplicate information freely. However, intelligence is more than multiplication of information: it is extracting value, meaning, knowledge, and eventually wisdom, from data.

From the situated, embodied or enactive perspectives, information is meaningful if you can do something with it, i.e. if it helps you to act towards your goals. Our goals derive from our values, which are themselves the product partly of biological evolution towards survival and reproduction of our genes, partly of cultural instruction towards cooperating fruitfully with society. The traditional AI perspective largely ignores the notion of values, except in the idea that an AI system may need to have some goals or constraints programmed into it (like Asimov's laws of robotics). Chalmers briefly discusses the issue, but seems to ignore that values, like knowledge and intelligence, are the product of millions of years of biological and cultural evolution, and of decades of personal experience with a variety of real-world situations. The idea that something as complex, fuzzy and context-dependent as a system of values could either be introduced by fiat into an AI system or develop autonomously inside its 'vat' seems highly unrealistic.

The problem is that without a sophisticated and adapted system of values, an agent will find it very difficult to distinguish what is meaningful or important from what is not. This is a traditional problem in AI (e.g. in the form of the 'frame problem'): AI systems typically do not know how to distinguish relevant inferences from trivial ones, unless there is a human guiding them by asking meaningful questions. Autonomous robots have a very simple system of values programmed into them (e.g. whenever your energy level is low, find a plug to recharge your battery; avoid colliding with objects) (Steels & Brooks, 1995). This helps them to decide about their course of action, but hardly manages to reach the level of complexity of an insect. Without sophisticated criteria for distinguishing what is important from what is frivolous, an AI system can hardly be expected to make intelligent, autonomous plans and decisions, as the AI++ scenario assumes.

Let us look in some more detail at the dynamics of self-amplification. Positive feedback processes such as these are well known in the field of complex dynamic systems (also known as non-linear systems or chaos theory) (Strogatz, 2000). However, eventually these processes always end up in what is called an *attractor*: a region within the system's state space that the system can enter, but cannot leave (Heylighen, 2001). Once the attractor has been reached, the system may continue moving around within the attractor, but it cannot jump out and reach a completely different part of its state space. This means that the system's evolution has essentially reached a stationary state where further innovation is no longer possible. While the system is moving towards its attractor, its 'fitness' may be seen to increase, but this increase will slow down as it draws nearer to the attractor's boundary, and comes to a halt inside the attractor. A simple example of this dynamics is a mathematical function (e.g. square root of x) that is iteratively applied to a given starting value (say $x = 100$, resulting in the sequence: 100, 10, 3.16, 1.77, 1.33, ...). In the limit, this recursive application ends in a 'fix point' of the function (in this case, $x = 1$), which is a zero dimensional attractor. Initially, the value of x changes rapidly, but as the self-application is repeated and the fix point comes nearer, further progress becomes infinitesimally small.

The same dynamics can be expected from a hypothetical self-improving AI-system: as it reprograms itself to become more intelligent, the gains in intelligence will become ever smaller, until the process has become to all practical effects stationary. Of course, we do not know how large the initial gains may be, and it is not *a priori* to be excluded that these gains are large enough to reach AI+ or perhaps even AI++. However, the situated and embodied perspective makes it very unlikely that a mere self-application without environmental interaction will be sufficient to create something that is significantly more intelligent than a system (a human being) that is the product of millions of years of evolution, complemented by decades of personal experience.

For an example of an existing self-application process in computer science, let me refer to the method of supercompilation or meta-compilation developed by the cybernetician Valentin Turchin (1986). A compiler is a program that translates a program written in a high level programming language, such as Java, into a sequence of instructions in machine code ready to be executed by the processor. A good compiler will produce efficient code, which is executed in a small number of steps. A supercompiler is a compiler that is applied to itself to improve its code by bootstrapping. One self-application may make it twice as fast. However, a subsequent self-application may only make it 1.3 as fast, and a third one only 1.1. After a few iterations, further improvements are negligible, and the process reaches its fix point.

Another example relevant to AI is the complexity limit experienced by Artificial Life researchers in their simulations of self-organizing virtual organisms (Rasmussen *et al.*, 2001). While there exist many successful simulations of evolving ecosystems exhibiting an increasing diversity and complexity of artificial life forms, after a number of iterations complexity always starts to stagnate, and real novelty no longer seems to be created. This should not surprise the reader: if open-ended evolution of virtual life forms was easy to achieve, we would by now already have pretty sophisticated and perhaps even intelligent artificial organisms. Indeed, once a simulation runs successfully, there is no incentive to stop the computer from running ever more iterations, thus producing ever more advanced generations.

Note that this reaching of a stationary or equilibrium state in a closed self-organizing system is also observed in reality for a variety of physical and chemical processes (e.g. crystallization or magnetization) (Heylighen, 2001). Moreover, it can be justified theoretically, both on the basis of the second law of thermodynamics, and on the basis of the more abstract, functional reasoning of the cybernetician Ashby (1962), who was the first to formulate the concept of self-organization. Open-ended self-organization only occurs in open systems that interact with their environment (e.g. Prigogine & Stengers, 1984). As complexity scientists have repeatedly pointed out (e.g. Kauffman, 1995), it is the flow of matter, energy and information entering and leaving the system that keeps it dynamic and adaptive—preventing it from reaching a 'frozen' equilibrium, fix point or attractor state, while maintaining it on the 'edge of chaos' where all the interesting things happen. This fits in perfectly with the situated and embodied theory of cognition, which similarly observes that on-going interaction with the world is necessary to develop and maintain a flexible intelligence.

In conclusion, there are plenty of arguments that make it very implausible that an AI system residing in a closed, virtual world could ever develop a superhuman (AI++) level of intelligence. On the other hand, just giving the AI system some rudimentary sensors and effectors is hardly more likely to bring it to a par with the very sophisticated, billions of years old sensory-motor interface that characterizes the human body. Does this mean that superhuman intelligence remains out of reach? Not at all, as I will try to show in the next section.

The Global Brain

Closely related with the situated and embodied perspective on cognition are the extended mind (Clark, 2008), distributed cognition (Hutchins, 1991; Heylighen *et al.*, 2004) and collective intelligence (Heylighen, 1999) perspectives. Once we have stopped restricting our search for intelligence to a specific, localized component, such as the brain, we become aware that it extends not only across body, sensor, effectors, and the feedback loops between them, but across artifacts used to support cognition (such as signs, notebooks, and computers), across different individuals who help each other solve problems, across organizations, across society, and in a sense across the environment as a whole. All these parts and aspects of our world contribute to our problem-solving and decision-making processes, by providing, storing and processing some of the crucial information.

It is on this level of distributed cognition that we should expect the greatest increases in intelligence. Indeed, the true revolution brought about by accelerating advances in information technology is to be found not in stand-alone AI systems, but in the network of wired and wireless connections, computers, people, organizations, websites, smartphones, embedded chips, sensors, etc. that together form the Net or the Web. This is a truly open, complex, adaptive and interactive system that processes billions of times more information than the most sophisticated stand-alone computing system. Any technological advance — such as faster processors, larger memories or more intelligent programming — that could benefit a stand-alone system will simultaneously benefit the Web. Therefore, there is no reason to assume that the existing gap in capabilities between an individual AI system and the collective system formed by all networked computers and their users will ever decrease.

As I have argued in several papers over the past decade and a half (Heylighen & Bollen, 1996; Heylighen, 1999; 2007ab; 2011b), there are plenty of observations as well as theoretical arguments for believing that this collective system, which may be called the *Global Brain* (Heylighen, 2011b; Goertzel, 2002), is not only intelligent, but becoming quickly more intelligent. The reason is that its self-organization is facilitated and accelerated by the seemingly unpreventable processes of globalization and of the increasing spread of information and communication technology. The result of these processes is that information about what is happening on this planet becomes ever more easily available, helping us to make better, more informed decisions, and to tackle more complex problems.

A well-known illustration is how you can find the answer to almost any question within seconds by using the Google search engine. This is not so much because the Google software is particularly smart (although some of the embedded AI definitely helps), but because it makes use of two components of humanity's collective intelligence:

1)   with billions of people contributing information and knowledge to the Web (e.g. via the Wikipedia knowledge base, via the millions of blogs and discussion forums and the thousands of newspapers and scientific journals published electronically) for practically any problem for which someone has thought of a good solution, that solution is likely to be available somewhere on the Web;

2)   out of the thousands of pages with potential answers to the question, the Google engine can select the ones most likely to be relevant by relying on the (implicit) selections made by millions of users when they clicked on, or made links to, pages they found particularly interesting.

Google's search engine is so successful because it is particularly adept at mining the implicit preferences (i.e. *values*) expressed by people who use the web — thanks in part to its underlying PageRank algorithm (Heylighen, 1999). As I have argued earlier, the lack of a sense of value, importance or relevance is one of the crucial problems that hamper the development of an autonomous AI system. But any attempt to 'hardwire' values into such a system will only provide a very small, restricted and rigid surrogate for the ever-adapting range of values expressed implicitly by the billions of people in the trillions of choices they make every year, while deciding which product to buy, which charity to donate to, which web page to read, or which person to connect to.

Thanks to the most recent technologies — such as smartphones, wireless Internet access, social software, and Twitter — the Global Brain now moreover provides *real-time* information. This can be used to help individuals on the road (e.g. by guiding them to the most popular nearby restaurant), but also to coordinate complex collective actions (e.g. the protests in the Arab world that toppled several dictatorial regimes). That means that the Global Brain's intelligence is not just a bookish faculty for looking up pre-existing answers in a gigantic library, but a truly interactive ability to navigate a complex and ever-changing environment — for individuals as well as for collectives. The accelerating incorporation of a multitude of hardware sensors and effectors (e.g. satellites, cameras, remote controls) together with an ever more intimate individual-net interface turn the global brain into a truly situated and embodied intelligence. Its powers of interaction with the planetary environment are already orders or magnitude larger than those of the most advanced systems imagined by Singularity theorists. As ICT becomes ever better, and the people learn to use it ever more efficiently, the intelligence of the Global Brain will continue to explode. Arguably, it already has reached a superhuman level. With a collective intelligence that powerful, who needs a stand-alone AI++?

My point is that it is simply much easier, cheaper and more effective to augment intelligence by facilitating distributed cognition than by building localized, autonomous AI systems. The collective intelligence of the Global Brain communicates with the world via a channel whose bandwidth is many orders of magnitude larger than even the most powerful imaginable stand-alone system. Every human being or piece of machinery hooked up to the web forms part of its body, providing it with additional capabilities for input, output and control. The only thing needed to use this power effectively is *coordination*. As I have argued in several publications (e.g. Heylighen, 2001; 2007a; 2011a), such coordination self-organizes easily once the main sources of friction (such as time, distance, and requirement for energy) are removed — something the Internet has to an important degree already achieved. As systems are standardized (e.g. via the Semantic Web initiative), as new tools for coordination are developed (e.g. social networks, wikis), and as people learn how to use these tools more effectively, we see this self-organization of collective cognition take place at an absolutely staggering rate. Extrapolating this accelerating advance leads us to expect for the near-term future a true technological 'Singularity' — in its original mathematical sense as a discontinuous transition beyond which further extrapolation becomes impossible (Heylighen, 2007a), not in the more limited sense used by Chalmers as a superintelligent AI system.

In contrast, when we look at the development of stand-alone AI over the last half century we see little or no spectacular progress: we just see expert systems becoming somewhat better experts at the highly specialized tasks (such as playing chess or parsing language) that they have been programmed to perform. The reason for the different rates of progress is that the Global Brain is — by its very nature — extended, embodied and embedded into the real world, with all its complex and ever-changing ramifications. This allows it to soak in information and to directly intervene via billions of high-bandwidth interfaces. The feedback flows between these inputs and outputs drive its self-organization towards ever more complex and adaptive intelligence. Stand-alone AI, on the other hand, is a poor analogue of a brain in vat, which must be artificially fed with preformatted knowledge and told what to do by its programmers. Self-application, as envisaged by Singularity thinkers, may help it to bootstrap some additional capacity, but is unlikely to give it anything that the Global Brain could not develop at least as well.

Conclusion

I have given a variety of arguments for why Chalmers's conception of the emergence of a superhuman artificial intelligence within a virtual world is very unlikely to happen, at least if this intelligence is supposed to have sufficient common sense and real-world understanding to be able to intervene in and potentially control humanity's affairs. The reason is that the defining character of such an AI++ — its independent, closed, self-creation — is antithetical to the now dominant understanding of cognition as a process that is necessarily extended, embodied and embedded into the outside world.

However, this does not mean that I reject the possibility of the near-term emergence of a superhuman intelligence, i.e. a Singularity. On the contrary, the perspective of distributed cognition makes such a 'metasystem transition' (Turchin, 1977; Heylighen, 2007ab) to a higher level of intelligence and organization rather more likely. Social systems, supported by a variety of tools for storing, processing and propagating information, have always exhibited some degree of collective intelligence above and beyond the intelligence of their individual components (Hutchins, 1991). The tremendous advances in information technology that Singularity theorists like to contemplate have had their biggest impact on this collective, distributed level — not on the level of stand-alone computing systems.

An anecdote to illustrate this observation can be found in the birth around 1970 of the computer network, which was originally designed to allow researchers to perform calculations on a mainframe computer in a far-away location. When a researcher would use the network connecting these huge machines to leave a message for another researcher working at a different institute, this seemed merely a quick and dirty hack. Yet, this 'improper' use of the network for exchanging information (which was later to evolve into email and eventually the world-wide web) soon became its most popular application, while its original function of logging in to a remote computer has virtually disappeared by now.

It could be argued that Singularity theorists like Chalmers are still in mainframe mode, standing in awe for the tremendous power of those huge, stand-alone computers that can perform calculations faster than any human being. I see computers in their email and web modes, as fast and efficient intermediaries that process and propagate information across a hugely complex network of people and things, but that have essentially disappeared into the background. It is in the self-organization of this immense network that a superhuman level of intelligence is most likely to emerge. This intelligence will not be localized in any particular component, but distributed over all its components, human and machine. The human components are essential for the efficient functioning of this intelligence, because they are the only ones that, at least for the foreseeable future, offer a true embodiment, i.e. the ability for high-bandwidth interaction with the world via sophisticated sensors and effectors.

In the longer term, as hardware abilities expand, the need for individual humans may diminish, but this is unlikely to follow a simple 'takeover by the robots' scenario. If we view the network as a superorganism with both hardware and organic components that become ever more intimately linked (e.g. via brain-computer interfaces), then the transition from a biology-dominated to a technology-dominated Global Brain is likely to be a smooth, extended process with no clear switch from one regime to a another. In the process, an increasing amount of human experience is likely to become 'uploaded' to a digital medium, thus blurring the boundaries between human and technological systems. This raises a lot of philosophical, ethical but especially practical, psychological and social questions, which to some degree echo Chalmers's concerns. However, I hope I have made it clear that the theories of embodied and distributed cognition formulate the issue in a fundamentally different way, making most of the questions raised by Chalmers moot, while raising a bunch of different questions. For a first exploration of these questions and some preliminary answers, I refer to earlier publications (Heylighen, 2007ab; Heylighen & Goertzel, 2011).

References

Anderson, M.L. (2003) Embodied cognition: A field guide, *Artificial Intelligence*, **149** (1), pp. 91–130.

Ashby, W.R. (1962) Principles of the self-organizing system, in Von Foerster, H. & Zopf, Jr., G.W. (eds.) *Principles of Self-Organization*, pp. 255–278, London: Pergamon Press.

Bechtel, W. & Abrahamsen, A. (1991) *Connectionism and the Mind: An Introduction to Parallel Processing in Networks*, Oxford: Basil Blackwell.

Beer, R.D. (2000) Dynamical approaches to cognitive science, *Trends in Cognitive Sciences*, **4** (3), pp. 91–99.

Bickhard, M.H. & Terveen, L. (1995) *Foundational Issues in Artificial Intelligence and Cognitive Science: Impasse and Solution*, Amsterdam: North Holland.

Chalmers, D.J. (1995) Facing up to the problem of consciousness, *Journal of Consciousness Studies*, **2** (3), pp. 200–219.

Clark, A. (2008) *Supersizing the Mind: Embodiment, Action, and Cognitive Extension*, Oxford: Oxford University Press.

Dennett, D.C. (1993) *Consciousness Explained*, London: Penguin.

Goertzel, B. (2002) *Creating Internet Intelligence: Wild Computing, Distributed Digital Consciousness, and the Emerging Global Brain*, New York: Springer.

Heylighen, F. (1999) Collective intelligence and its implementation on the web: Algorithms to develop a collective mental map, *Computational and Mathematical Organization Theory*, **5** (3), pp. 253–280.

Heylighen, F. (2001) The science of self-organization and adaptivity, in Kiel, L.D. (ed.) *Knowledge Management, Organizational Intelligence and Learning, and Complexity*, in *The Encyclopedia of Life Support Systems (EOLSS)*, Oxford: Eolss Publishers.

Heylighen, F. (2007–2011) *Cognitive Systems: A Cybernetic Perspective on the New Science of the Mind* (ECCO working paper 2007–07), lecture notes for the course 'Cognitieve Systemen', [Online], http://pcp.vub.ac.be/Papers/Cognitive Systems.pdf

Heylighen, F. (2007a) Accelerating socio-technological evolution: From ephemeralization and stigmergy to the global brain, in Modelski, G., Devezas, T. & Thompson, W. (eds.) *Globalization as an Evolutionary Process: Modeling Global Change*, pp. 286–335, London: Routledge.

Heylighen, F. (2007b) The global superorganism: An evolutionary-cybernetic model of the emerging network society, *Social Evolution & History*, **6** (1), pp. 58–119.

Heylighen, F. (2011a) Self-organization of complex, intelligent systems: An action ontology for transdisciplinary integration, *Integral Review*, (in press).

Heylighen, F. (2011b) Conceptions of a global brain: An historical review, in Grinin, L.E., Carneiro, R.L., Korotayev, A.V. & Spier, F. (eds.) *Evolution: Cosmic, Biological, and Social*, pp: 274–289, Moscow: Uchitel Publishing.

Heylighen, F. & Bollen, J. (1996) The world-wide web as a super-brain: From metaphor to model, in Trappl, R. (ed.) *Cybernetics and Systems '96*, pp. 917–922, Vienna: Austrian Society for Cybernetics.

Heylighen, F., Heath, M. & Van Overwalle, F. (2004) The emergence of distributed cognition: A conceptual framework, *Proceedings of Collective Intentionality IV*, Siena.

Heylighen, F. & Goertzel, B. (2011) Francis Heylighen on the emerging global brain: An interview by Ben Goertzel, *h+ magazine*, 16 March 2011, [Online] http://hplusmagazine.com/2011/03/16/francis-heylighen-on-the-emerging-global-brain/

Hutchins, E. (1991) Social organization of distributed cognition, in Resnick, L., Levine, J. & Teasley, S. (eds.) *Perspectives on Socially Shared Cognition*, pp. 283–307, Washington, DC: American Psychological Association.

Kauffman, S.A. (1995) *At Home in the Universe*, New York: Oxford University Press.

Kurzweil, R. (2005) *The Singularity is Near*, London: Penguin.

Lakoff, G. & Johnson, M. (1999) *Philosophy in the Flesh: The Embodied Mind and its Challenge to Western Thought*, New York: Basic Books.

Lenat, D.B. (1995) CYC: A large-scale investment in knowledge infrastructure, *Communications of the ACM*, **38** (11), pp. 33–38.

Prigogine, I. & Stengers, I. (1984) *Order Out of Chaos*, New York: Bantam Books.

Rasmussen, S., Baas, N.A., Mayer, B., Nilsson, M. & Olesen, M.W. (2001) Ansatz for dynamical hierarchies, *Artificial Life*, **7** (4), pp. 329–353.

Steels, L. & Brooks, R.A. (1995) *The Artificial Life Route to Artificial Intelligence: Building Embodied, Situated Agents*, Mahwah, NJ: Lawrence Erlbaum Associates.

Strogatz, S.H. (2000) *Nonlinear Dynamics and Chaos*, Boulder, CO: Westview Press.

Turchin, V.F. (1977) *The Phenomenon of Science. A Cybernetic Approach to Human Evolution*, New York: Columbia University Press.

Turchin, V.F. (1986) The concept of a supercompiler, *ACM Transactions on Programming Languages and Systems*, **8** (3), pp. 292–325.

Varela, F.J., Thompson, E. & Rosch, E. (1992) *The Embodied Mind: Cognitive Science and Human Experience*, Cambridge, MA: MIT Press.

Vinge, V. (1993) The coming technological singularity, *Whole Earth Review*, pp. 88–95.

Marcus Hutter

Can Intelligence Explode?

Abstract: *The technological singularity refers to a hypothetical scenario in which technological advances virtually explode. The most popular scenario is the creation of super-intelligent algorithms that recursively create ever higher intelligences. It took many decades for these ideas to spread from science fiction to popular science magazines and finally to attract the attention of serious philosophers. David Chalmers' essay is the first comprehensive philosophical analysis of the singularity in a respected philosophy journal. The motivation of my article is to augment Chalmers' and to discuss some issues not addressed by him, in particular what it could mean for intelligence to explode. In this course, I will (have to) provide a more careful treatment of what intelligence actually is, separate speed from intelligence explosion, compare what super-intelligent participants and classical human observers might experience and do, discuss immediate implications for the diversity and value of life, consider possible bounds on intelligence, and contemplate intelligences right at the singularity.*

> Within thirty years, we will have the technological means to create superhuman intelligence. Shortly after, the human era will be ended.

> — Vernor Vinge (1993)

1. Introduction

The technological singularity is a hypothetical scenario in which self-accelerating technological advances cause infinite progress in finite time. The most popular scenarios are an intelligence explosion (Good, 1965) or a speed explosion (Yudkowsky, 1996) or a combination of both (Chalmers, this volume). This quite plausibly is accompanied by a radically changing society, which will become incomprehensible to us current humans close to and in particular at or beyond the singularity. Still some general aspects may be predictable.

Already the invention of the first four-function mechanical calculator one-and-a-half centuries ago (Thornton, 1847) inspired dreams of self-amplifying technology. With the advent of general purpose computers and the field of artificial intelligence half-a-century ago, some mathematicians, such as Stanislaw Ulam (1958), I.J. Good (1965), Ray Solomonoff (1985), and Vernor Vinge (1993) engaged in singularity thoughts. But it was only in the last decade that the singularity idea achieved wide-spread popularity. Ray Kurzweil popularized the idea in two books (1999; 2005), and the Internet helped in the formation of an initially small community discussing this idea. There are now annual Singularity Summits approaching a thousand participants per year, and even a Singularity Institute.

The singularity euphoria seems in part to have been triggered by the belief that intelligent machines that possess general intelligence on a human-level or beyond can be built within our life time, but it is hard to tell what is cause and effect. For instance, there is now a new conference series on Artificial General Intelligence (AGI) as well as some whole-brain emulation projects like Blue Brain (de Garis *et al.*, 2010; Goertzel *et al.*, 2010).

A loosely related set of communities which are increasing in momentum are the 'Immortalists' whose goal is to extend the human life-span, ideally indefinitely. Immortality and life-extension organizations are sprouting like mushrooms: e.g. the Immortality and the Extropy Institute, the Humanity+ Association, and the Alcor Life Extension, Acceleration Studies, Life Extension, Maximum Life, and Methusalem Foundations.

There are many different potential paths toward a singularity. Most of them seem to be based on software intelligence on increasingly powerful hardware. Still this leaves many options, the major ones being mind uploading (via brain scan) and subsequent improvement, knowledge-based reasoning and planning software (traditional AI research), artificial agents that learn from experience (the machine learning approach), self-evolving intelligent systems (genetic algorithms and artificial life approach), and the awakening of the Internet (or digital Gaia scenario). Physical and biological limitations likely do not allow singularities based on (non-software) physical brain enhancement technologies such as drugs and genetic engineering.

Although many considerations in this article should be independent of the realized path, I will assume a virtual software society consisting of interacting rational agents whose intelligence is high enough to construct the next generation of more intelligent rational agents. Indeed, one of the goals of the article is to discuss what (super) intelligence and rationality could mean in this setup. For concreteness, the reader may want envisage an initial virtual world like Second Life that is similar to our current real world and inhabited by human mind uploads.

Much has been written about the singularity and David Chalmers' essay covers quite wide ground. I essentially agree with all his statements, analysis, and also share his personal opinions and beliefs. Most of his conclusions I will adopt without repeating his arguments. The motivation of my article is to augment Chalmers' and to discuss some issues not addressed by him, in particular what it could mean for intelligence to explode. This is less obvious than it might appear, and requires a more careful treatment of what intelligence actually is. Chalmers cleverly circumvents a proper discussion or definition of intelligence by arguing (a) there is something like intelligence, (b) there are many cognitive capacities correlated with intelligence, (c) these capacities might explode, therefore (d) intelligence might amplify and explode. While I mostly agree with this analysis, it does not tell us what a society of ultra-intelligent beings might look like. For instance, if a hyper-advanced virtual world looks like random noise for humans watching them from the 'outside', what does it mean for intelligence to explode for an outside observer? Conversely, can an explosion actually be felt from the 'inside' if everything is sped up uniformly? If neither insiders nor outsiders experience an intelligence explosion, has one actually happened?

The paper is organized as follows: *Section 2* briefly recapitulates the most popular arguments why to expect a singularity and why 'the singularity is near' (Kurzweil, 2005), obstacles towards a singularity, and which choices we have. *Section 3* describes how an outside observer who does not participate in the singularity might experience the singularity and the consequences he faces. This will depend on whether the singularity is directed inwards or outwards. *Section 4* investigates what a participant in the singularity will experience, which is quite different from an outsider and depends on details of the virtual society; in particular how resources are distributed. *Section 5* takes a closer look at what actually explodes when computing power is increased without limits in finite real time. While by definition there is a speed explosion, who, if anyone at all, perceives an intelligence explosion/singularity depends on what is sped up. In order to determine whether anyone perceives an intelligence explosion, it is necessary to clarify what intelligence actually is and what super-intelligences might do, which is done in *Section 6*. The considered formal theory of rational intelligence allows investigating a wide range of questions about super-intelligences, in principle rigorously mathematically. *Section 7* elucidates the possibility that intelligence might be upper bounded, and whether this would prevent an intelligence singularity. *Section 8* explains how a society right at the edge of an intelligence singularity might be theoretically studied with current scientific tools. Even when setting up a virtual society in our image, there are likely some immediate differences, e.g. copying and modifying virtual structures, including virtual life, should be very easy. *Section 9* shows that this will have immediate (i.e. way before the singularity) consequences on the diversity and value of life. *Section 10* contains some personal remarks and *Section 11* draws some conclusions.

I will use the following terminology throughout this article. Some terms are taken over or refined from other authors and some are new:

- comp = computational resources
- singularity = infinite change of an observable quantity in finite time
- intelligence explosion = rapidly increasing intelligence far beyond human level
- intelligence singularity = infinite intelligence in finite time
- speed explosion/singularity = rapid/infinite increase of computational resources
- outsider = biological = non-accelerated real human watching a singularity
- insider = virtual = software intelligence participating in a singularity
- computronium = theoretically best possible computer per unit of matter (Bremermann, 1965)
- real/true intelligence = what we intuitively would regard as intelligence
- numerical intelligence = numerical measure of intelligence like IQ score
- AI = artificial intelligence (used generically in different ways)
- AGI = artificial general intelligence = general human-level intelligence or beyond.
- super-intelligence = AI+ = super-human intelligence (Chalmers, 2010)

- hyper-intelligent = AI++ = incomprehensibly more intelligent than humans

- vorld = virtual world. A popular oxymoron is `virtual reality'

- virtual = software simulation in a computer.

I drop the qualifier 'virtual' if this does not cause any confusion, e.g. when talking about a human in a vorld, I mean of course a virtual human.

I will assume a strong/physical form of the Church-Turing thesis that everything in nature can be calculated by a Turing machine, i.e. our world including the human mind and body and our environment are computable (Deutsch, 1997; Rathmanner & Hutter, 2011). So in the following I will assume without further argument that all physical processes we desire to virtualize are indeed computational and can be simulated by a sufficiently powerful (theoretical) computer. This assumption simplifies many of the considerations to follow, but is seldom essential, and could be lifted or weakened.

2. Will there be a Singularity

The current generations Y or Z may finally realize the age-old dream of creating systems with human-level intelligence or beyond, which revived the interest in this endeavor. This optimism is based on the belief that in 20–30 years the raw computing power of a single computer will reach that of a human brain and that software will not lag far behind. This prediction is based on extrapolating Moore's law, now valid for 50 years, which implies that comp doubles every 1.5 years. As long as there is demand for more comp, Moore's law could continue to hold for many more decades before computronium is reached. Further, different estimates of the computational capacity of a human brain consistently point towards $10^{15}...10^{16}$ flop/s (Kurzweil, 2005): Counting of neurons and synapses, extrapolating tiny-brain-part simulations, and comparing the speech recognition capacities of computers to the auditory cortex.

The most compelling argument for the emergence of a singularity is based on Solomonoff's (1985) law which Yudkowski (1996) succinctly describes as follows:

> If computing speeds double every two years,
> what happens when computer-based AIs are doing the research?
> Computing speed doubles every two years.
> Computing speed doubles every two years of work.
> Computing speed doubles every two subjective years of work.

Two years after Artificial Intelligences reach human equivalence, their speed doubles. One year later, their speed doubles again.
Six months – three months – 1.5 months ... Singularity.

Interestingly, if this argument is valid, then Moore's law in a sense predicts its own break-down; not the usually anticipated slow-down, but an enormous acceleration of progress when measured in physical time.

The above acceleration would indeed not be the first time of an enormous acceleration in growth. The economist Robin Hanson argues that 'Dramatic changes in the rate of economic growth have occurred in the past because of some technological advancement. Based on population growth, the economy doubled every 250,000 years from the Paleolithic era until the Neolithic Revolution. This new agricultural economy began to double every 900 years, a remarkable increase. In the current era, beginning with the Industrial Revolution, the world's economic output doubles every fifteen years, sixty times faster than during the agricultural era.' Given the increasing role of computers in our economy, computers might soon dominate it, locking the economic growth pattern to computing speed, which would lead to a doubling of the economy every two (or more precisely 1.5) years, another 10 fold increase. If the rise of superhuman intelligences causes a similar revolution, argues Hanson (2008), one could expect the virtual economy to double on a monthly or possibly on a weekly basis. So the technological singularity phenomenon would be the next and possibly last growth acceleration. Ray Kurzweil (2005) is a master of producing exponential, double exponential, and singular plots, but one has to be wary of data selection, as Juergen Schmidhuber (2006) has pointed out.

Chalmers discusses various potential obstacles for a singularity to emerge. He classifies them into structural obstacles (limits in intelligence space, failure to takeoff, diminishing returns, local maxima) and manifestation obstacles (disasters, disinclination, active prevention) and correlation obstacles. For instance, self-destruction or a natural catastrophe might wipe out the human race (Bostrom & Cirkovic, 2008).

Also, the laws of physics will likely prevent a singularity in the strict mathematical sense. While some physical theories in isolation allow infinite computation in finite time (see Zeno machines — Weyl, 1927 — and hypercomputation — Copeland, 2002 — in general), modern physics raises severe barriers (Bremermann, 1965; Bekenstein, 2003; Lloyd, 2000; Aaronson, 2005). But even if so, today's computers are so far away from these limits, that converting our planet into computronium would still result in a vastly different vorld, which is considered a reasonable approximation to a true singularity. Of course, engineering difficulties and many other obstructions may stop the process well before this point, in which case the end result may not account as a singularity but more as a phase transition à la Hanson or even less spectacular.

Like Chalmers, I also believe that disinclination is the most (but not very) likely defeater of a singularity. In the remainder of this article I will assume absence of any such defeaters, and will only discuss the structural obstacles related to limits in intelligence space later.

The appearance of the first super-intelligences is usually regarded as the ignition of the detonation cord towards the singularity — the point of no return. But it might well be that a singularity is already now unavoidable. Politically it is very difficult (but not impossible) to resist technology or market forces as e.g. the dragging discussions on climate change vividly demonstrate, so it would be similarly difficult to prevent AGI research and even more so to prevent the development of faster computers. Whether we are before, at, or beyond the point of no return is also philosophically intricate as it depends on how much free will one attributes to people and society; like a spaceship close to the event horizon might in principle escape a black hole but is doomed in practice due to limited propulsion.

3. The Singularity from the Outside

Let us first view the singularity from the outside. What will observers who do not participate in it 'see'. How will it affect them?

First, the hardware (computers) for increasing comp must be manufactured somehow. As already today, this will be done by (real) machines/robots in factories. Insiders will provide blue-prints to produce better computers and better machines that themselves produce better computers and better machines ad infinitum at an accelerated pace. Later I will explain why insiders desire more comp. Non-accelerated real human (outsiders) will play a diminishing role in this process due to their cognitive and speed limitations. Quickly they will only be able to passively observe some massive but incomprehensible transformation of matter going on.

Imagine an inward explosion, where a fixed amount of matter is transformed into increasingly efficient computers until it becomes computronium. The virtual society like a well-functioning real society will likely evolve and progress, or at least change. Soon the speed of their affairs will make them beyond comprehension for the outsiders.

For a while, outsiders may be able to make records and analyze them in slow motion with an increasing lag. Ultimately the outsiders' recording technology will not be sufficient anymore, but some coarse statistical or thermodynamical properties could still be monitored, which besides other things may indicate an upcoming physical singularity. I doubt that the outsiders will be able to link what is going on with intelligence or a technological singularity anymore.

Insiders may decide to interact with outsiders in slow motion and feed them with pieces of information at the maximal digestible rate, but even with direct brain-computer interfaces, the cognitive capacity of a human brain is bounded and cannot explode. A technologically augmented brain may explode, but what would explode is the increasingly dominant artificial part, rendering the biological brain eventually superfluous — a gradual way of getting sucked into the inside world. For this reason, also intelligence amplification by human-computer interfaces are only temporarily viable before they either break down or the extended human becomes effectively virtual.

After a brief period, intelligent interaction between insiders and outsiders becomes impossible. The inside process may from the outside resemble a black hole watched from a safe distance, and look like another interesting physical, rather than societal, phenomenon.

This non-comprehensibility conclusion can be supported by an information-theoretic argument: The characterization of our society as an information society becomes even better, if not perfect, for a virtual society. There is lots of motivation to compress information (save memory, extract regularities, and others), but it is well-known (Li & Vitányi, 2008) that maximally compressed information is indistinguishable from random noise. Also, if too much information is produced, it may actually 'collapse'. Here, I am not referring to the formation of black holes (Bekenstein, 2003), but to the fact that a library that contains all possible books has zero information content (*cf.* the Library of Babel). Maybe a society of increasing intelligence will become increasingly indistinguishable from noise when viewed from the outside.

Let us now consider outward explosion, where an increasing amount of matter is transformed into computers of fixed efficiency (fixed comp per unit time/space/energy). Outsiders will soon get into resource competition with the expanding computer world, and being inferior to the virtual intelligences, probably only have the option to flee. This might work for a while, but soon the expansion rate of the virtual world should become so large, theoretically only bounded by

the speed of light, that escape becomes impossible, ending or converting the outsiders' existence.

So while an inward explosion is interesting, an outward explosion
will be a threat to outsiders. In both cases, outsiders will observe a
speedup of cognitive processes and possibly an increase of intelligence up to a certain point. In neither case will outsiders be able to
witness a true intelligence singularity.

Historically, mankind was always outward exploring; just in recent
times it has become more inward exploring. Now people more and
more explore virtual worlds rather than new real worlds. There are
two reasons for this. First, virtual worlds can be designed as one sees
fit and hence are arguably more interesting, and second, outward
expansion now means deep sea or space, which is an expensive
endeavor. Expansion usually follows the way of least resistance.

Currently the technological explosion is both inward and outward
(more and faster computers). Their relative speed in the future will
depend on external constraints. Inward explosion will stop when
computronium is reached. Outward explosion will stop when all
accessible convertible matter has been used up (all on earth, or in our
galaxy, or in our universe).

4. The Singularity from the Inside

Let us now consider the singularity from the inside. What will a participant experience?

Many things of course will depend on how the virtual world is organized. It is plausible that various characteristics of our current society
will be incorporated, at least initially. Our world consists of a very
large number of individuals, who possess some autonomy and freedom, and who interact with each other and with their environment in
cooperation and in competition over resources and other things. Let us
assume a similar setup in a virtual world of intelligent actors. The
vorld might actually be quite close to our real world. Imagine populat-
ing already existing virtual worlds like Second Life or World of
Warcraft with intelligent agents simulating scans of human brains.

Consider first a vorld based on fixed computational resources. As
indicated, initially, the virtual society might be similar to its real coun-
ter-part, if broadly understood. But some things will be easier, such a
duplicating (virtual) objects and directed artificial evolution. Other
things will be harder or impossible, such as building faster virtual
computers and fancier gadgets reliant on them. This will affect how
the virtual society will value different things (the value of virtual life

and its implications will be discussed later), but I would classify most of this as a change, not unlike in the real world when discovering or running out of some natural resource or adapting to new models of society and politics. Of course, the virtual society, like our real one, will also develop: there will be new inventions, technologies, fashions, interests, art, etc., all virtual, all software, of course, but for the virtuals it will feel real. If virtuals are isolated from the outside world and have knowledge of their underlying computational processes, there would be no quest for a virtual theory of everything (Hutter, 2010), since they would already know it. The evolution of this vorld might include weak singularities in the sense of sudden phase transitions or collapses of the society, but an intelligence explosion with fixed comp, even with algorithmic improvements seems implausible.

Consider now the case of a vorld with increasing comp. If extra comp is used for speeding up the whole virtual world uniformly, virtuals and their virtual environment alike, the inhabitants would actually not be able to recognize this. If their subjective thought processes will be sped up at the same rate as their surroundings, nothing would change for them. The only difference, provided virtuals have a window to the outside real world, would be that the outside world slows down. If comp is sped up hyperbolically, the subjectively infinite future of the virtuals would fit into finite real time: For the virtuals, the external universe would get slower and slower and ultimately come to a halt. Also outsiders would appear slower (but not dumber).

This speed-up/slow-down phenomenon is inverse compared to flying into a black hole. An astronaut flying into a black hole will pass the Schwarzschild radius and hit the singularity in finite subjective time. For an outside observer, though, the astronaut gets slower and slower and actually takes infinite time to vanish behind the Schwarzschild radius.

If extra comp is exclusively used to expand the vorld and add more virtuals, there is no individual speedup, and the bounded individual comp forces intelligence to stay bounded, even with algorithmic improvements. But larger societies can also evolve faster (more inventions per real time unit), and if regarded as a super-organism, there might be an intelligence explosion, but not necessarily so: Ant colonies and bee hives seem more intelligent than their individuals in isolation, but it is not obvious how this scales to unbounded size. Also, there seems to be no clear positive correlation between the number of individuals involved in a decision process and the intelligence of its outcome.

In any case, the virtuals as individuals will not experience an intelligence explosion, even if there was one. The outsiders would observe virtuals speeding up beyond comprehension and would ultimately not recognize any further intelligence explosion.

The scenarios considered in this and the last section are of course only caricatures. An actual vorld will more likely consist of a wide diversity of intelligences: faster and slower ones, higher and lower ones, and a hierarchy of super-organisms and sub-vorlds. The analysis becomes more complicated, but the fundamental conclusion that an intelligence explosion might be unobservable does not change.

5. Speed versus Intelligence Explosion

The comparison of the inside and outside view has revealed that a speed explosion is not necessarily an intelligence explosion. In the extreme case, insiders may not experience anything and outsiders may witness only noise.

Consider an agent interacting with an environment. If both are sped up at the same rate, their behavioral interaction will not change except for speed. If there is no external clock measuring absolute time, there is no net effect at all.

If only the environment is sped up, this has the same effect as slowing down the agent. This does not necessarily make the agent dumber. He will receive more information per action, and can make more informed decisions, provided he is left with enough comp to process the information. Imagine being inhibited by very slowly responding colleagues. If you could speed them up, this would improve your own throughput, and subjectively this is the same as slowing yourself down. But (how much) can this improve the agent's intelligence? In the extreme case, assume the agent has instant access to all information, not much unlike we already have by means of the Internet but much faster. Both usually increase the quality of decisions, which might be viewed as an increase in intelligence. But intuitively there should be a limit on how much information a comp-limited agent can usefully process or even search through.

Consider now the converse and speed up the agent (or equivalently slow down the environment). From the agent's view, he becomes deprived of information, but has now increased capacity to process and think about his observations. He becomes more reflective and cognitive, a key aspect of intelligence, and this should lead to better decisions. But also in this case, although it is much less clear, there might be a limit to how much can be done with a limited amount of information.

The speed-up/slow-down effects might be summarized as follows:

Performance per unit real time:

- Speed of agent positively correlates with cognition and intelligence of decisions
- Speed of environment positively correlates with informed decisions

Performance per subjective unit of agent time from agent's perspective:

- slow down environment = increases cognition and intelligence but decisions become less informed
- speed up environment = more informed but less reasoned decisions

Performance per environment time from environment perspective:

- speed up agent = more intelligent decisions
- slow down agent = less intelligent decisions

I have argued that more comp, i.e. speeding up hardware, does not necessarily correspond to more intelligence. But then the same could be said of software speedups, i.e. more efficient ways of computing the same function. If two agent algorithms have the same I/O behavior, just one is faster than the other, is the faster one more intelligent?

An interesting related question is whether progress in AI has been mainly due to improved hardware or improved software. If we believe in the former, and we accept that speed is orthogonal to intelligence, and we believe that humans are 'truly' intelligent (a lot of ifs), then building AGIs may still be far distant.

As detailed in Section 7, if intelligence is upper-bounded (like playing optimal minimax chess), then past this bound, intelligences can only differ by speed and available information to process. In this case, and if humans are not too far below this upper bound (which seems unlikely), outsiders could, as long as their technology permits, record and play a virtual world in slow motion and be able to grasp what is going on inside.

In this sense, a singularity may be more interesting for outsiders than for insiders. On the other hand, insiders actively 'live' potential societal changes, while outsiders only passively observe them.

Of course, more comp only leads to more intelligent decisions if the decision algorithm puts it to good use. Many algorithms in AI are so-called anytime algorithms that indeed produce better results if given more comp. In the limit of infinite comp, in simple and well-defined settings (usually search and planning problems), some algorithms can produce optimal results, but for more realistic complex situations (usually learning problems), they saturate and remain sub-optimal (Russell & Norvig, 2010). But there is one algorithm, namely AIXI described in Section 7, that is able to make optimal decisions in arbitrary situations given infinite comp.

Together this shows that it is non-trivial to draw a clear boundary between speed and intelligence.

6. What is Intelligence

There have been numerous attempts to define intelligence; see e.g. Legg & Hutter (2007a) for a collection of 70+ definitions from the philosophy, psychology, and AI literature, by individual researchers as well as collective attempts.

If/since intelligence is not (just) speed, what is it then? What will super-intelligences actually do?

Historically-biologically, higher intelligence, via some correlated practical cognitive capacity, increased the chance of survival and number of offspring of an individual and the success of a species. At least for primates leading to homo sapiens this was the case until recently. Within the human race, intelligence is now positively correlated with power and/or economic success (Geary, 2007) and actually negatively with number of children (Kanazawa, 2007). Genetic evolution has been largely replaced by memetic evolution (Dawkins, 1976), the replication, variation, selection, and spreading of ideas causing cultural evolution.

What activities could be regarded as or are positively correlated with intelligence? Self-preservation? Self-replication? Spreading? Creating faster/better/higher intelligences? Learning as much as possible? Understanding the universe? Maximizing power over men and/or organizations? Transformation of matter (into computronium?)? Maximum self-sufficiency? The search for the meaning of life?

Has intelligence more to do with thinking or is thinking only a tool for acting smartly? Is intelligence something anthropocentric or does it exist objectively? What are the relations between other predicates of human 'spirit' like consciousness, emotions, and religious faith to intelligence? Are they part of it or separate characteristics and how are they interlinked?

One might equate intelligence with rationality, but what is rationality? Reasoning, which requires internal logical consistency, is a good start for a characterization but is alone not sufficient as a definition. Indiscriminately producing one true statement after the other without prioritization or ever doing anything with them is not too intelligent (current automated theorem provers can already do this).

It seems hard if not impossible to define rationality without the notion of a goal. If rationality is reasoning towards a goal, then there is no intelligence without goals. This idea dates back at least to Aristotle, if not further; see Legg & Hutter (2007b) for details. But what are the goals? Slightly more flexible notions are that of expected utility maximization and cumulative life-time reward maximization (Russell & Norvig, 2010). But who provides the rewards, and how? For animals, one might try to equate the positive and negative rewards with pleasure and pain, and indeed one can explain a lot of behavior as attempts to maximize rewards/pleasure. Humans seem to exhibit astonishing flexibility in choosing their goals and passions, especially during childhood. Goal-oriented behavior often appears to be at odds with long-term pleasure maximization. Still, the evolved biological goals and desires to survive, procreate, parent, spread, dominate, etc. are seldom disowned.

But who sets the goal for super-intelligences and how? When building AIs or tinkering with our virtual selves, we could try out a lot of different goals, e.g. selected from the list above or others. But ultimately we will lose control, and the AGIs themselves will build further AGIs (if they were motivated to do so) and this will gain its own dynamic. Some aspects of this might be independent of the initial goal structure and predictable. Probably this initial vorld is a society of cooperating and competing agents. There will be competition over limited (computational) resources, and those virtuals who have the goal to acquire them will naturally be more successful in this endeavor compared to those with different goals. Of course, improving the efficiency of resource use is important too, e.g. optimizing own algorithms, but still, having more resources is advantageous. The successful virtuals will spread (in various ways), the others perish, and soon their society will consist mainly of virtuals whose goal is to compete over resources, where hostility will only be limited if this is in the virtuals' best interest. For instance, current society has replaced war mostly by economic competition, since modern weaponry makes most wars a loss for both sides, while economic competition in most cases benefits the better.

Whatever amount of resources are available, they will (quickly) be used up, and become scarce. So in any world inhabited by multiple individuals, evolutionary and/or economic-like forces will 'breed' virtuals with the goal to acquire as much (comp) resources as possible. This world will likely neither be heaven nor hell for the virtuals. They will 'like' to fight over resources, and the winners will 'enjoy' it, while the losers will 'hate' it. In such evolutionary vorlds, the ability to survive and replicate is a key trait of intelligence. On the other hand, this is not a sufficient characterization, since e.g. bacteria are quite successful in this endeavor too, but not very intelligent.

Finally, let us consider some alternative (real or virtual) worlds. In the human world, local conflicts and global war is increasingly replaced by economic competition, which might itself be replaced by even more constructive global collaboration, as long as violaters can quickly and effectively (and non-violently?) be eliminated. It is possible that this requires a powerful single (virtual) world government, to give up individual privacy, and to severely limit individual freedom (*cf.* ant hills or bee hives). An alternative societal setup that can only produce conforming individuals might only be possible by severely limiting individual's creativity (*cf.* flock of sheep or school of fish).

Such well-regulated societies might better be viewed as a single organism or collective mind. Or maybe the vorld is inhabited from the outset by a single individual. Both vorlds could look quite different and more peaceful than the traditional ones created by evolution. Intelligence would have to be defined quite differently in such vorlds. Many science fiction authors have conceived and extensively written about a plethora of other future, robot, virtual, and alien societies in the last century.

In the following I will only consider vorlds shaped by evolutionary pressures as described above.

7. Is Intelligence Unlimited or Bounded

Another important aspect of intelligence is how flexible or adaptive an individual is. Deep Blue might be the best chess player on Earth, but is unable to do anything else. On the contrary, higher animals and humans have remarkably broad capacities and can perform well in a wide range of environments.

Legg & Hutter (2007b) define intelligence as the ability to achieve goals in a wide range of environments. It has been argued that this is a very suitable characterization, implicitly capturing most, if not all traits of rational intelligence, such as reasoning, creativity, generalization, pattern recognition, problem solving, memorization, planning, learning, self-preservation, and many others. Furthermore, this definition has been rigorously formalized in mathematical terms. It is non-anthropocentric, wide-range, general, unbiased, fundamental, objective, complete, and universal. It is the most comprehensive formal definition of intelligence so far. It assigns a real number Y between zero and one to every agent, namely the to-be-expected performance averaged over all environments/problems the agent potentially has to deal with, with an Ockham's razor inspired prior weight for each environment. Furthermore there is a maximally intelligent agent, called AIXI, w.r.t. this measure. The precise formal definitions and details can be found in Legg & Hutter (2007b), but do not matter for our purpose. This paper also contains a comprehensive justification and defense of this approach.

The theory suggests that there is a maximally intelligent agent, or in other words, that intelligence is upper bounded (and is actually lower bounded too). At face value, this would make an intelligence explosion impossible.

To motivate this possibility, consider some simple examples. Assume the world consists only of tic-tac-toe games, and the goal is to win or second-best not lose them. The notion of intelligence in this simple world is beyond dispute. Clearly there is an optimal strategy (actually many) and it is impossible to behave more intelligently than this strategy. It is even easy to artificially evolve or learn these strategies from repeated (self)play (Hochmuth, 2003; Veness et al., 2011). So in this world there clearly will be no intelligence explosion or intelligence singularity, even if there were a speed explosion.

We get a slightly different situation when we replace tic-tac-toe by chess. There is also an optimal way of playing chess, namely minimax tree search to the end of the game, but unlike in tic-tac-toe this strategy is computationally infeasible in our universe. So in theory (i.e. given enough comp) intelligence is upper-bounded in a chess world, while in practice we can get only ever closer but never reach the bound. (Actually there might be enough matter in the universe to build an optimal chess player, but likely not an optimal Go player. In any case it is easy to design a game that is beyond the capacity of our accessible universe, even if completely converted into computronium.)

Still, this causes two potential obstacles for an intelligence explosion. First, we are only talking about the speed of algorithms, which I explained before not to equate with intelligence. Second, intelligence is upper bounded by the theoretical optimal chess strategy, which makes an intelligence explosion difficult but not necessarily impossible: Assume the optimal program has intelligence $I = 1$ and at real time $t < 1$ we have access to or evolved a chess program with intelligence t. This approaches 1 in finite time, but doesn't 'explode'. But if we use the monotone transformation $1/(1-I)$ to measure intelligence, the chess program at time t has transformed intelligence $1/(1-t)$ which tends to infinity for $t \to 1$. While this is a mathematical singularity, it is likely not accompanied by a real intelligence explosion. The original scale seems more plausible in the sense that $t+0.001$ is just a tiny bit more intelligent than t, and 1 is just 1000 times more intelligent than 0.001 but not infinitely more. Although the vorld of chess is quite rich, the real world is vastly and possibly unlimitedly richer. In such a more open world, the intelligence scale may be genuinely unbounded, but not necessarily as we will see. It is not easy though to make these arguments rigorous.

Let us return to the real world and intelligent measure Y upper bounded by $Y_{max} = Y(AIXI)$. Since AIXI is incomputable, we can never reach intelligence Y_{max} in a computational universe, but similarly to the chess example we can get closer and closer. The numerical advance is bounded, and so is possibly the real intelligence increase, hence no intelligence explosion. But it might also be the case that in a highly sophisticated AIXI-close society, one agent beating another by a tiny epsilon on the Y-scale makes all the difference for survival and/or power and/or other measurable impact like transforming the universe. In many sport contests split seconds determine a win, and the winner takes it all — an admittedly weak analogy.

An interesting question is where humans range on the Y-scale: is it so low with so much room above that outsiders would effectively experience an intelligence explosion (as far as recognizable), even if intelligence is ultimately upper bounded? Or are we already quite close to the upper bound, so that even AGIs with enormous comp (but comparable I/O limitations) would just be more intelligent but not incomprehensibly so. We tend to believe that we are quite far from Y, but is this really so? For instance, what has once been argued to be irrational (i.e. not very intelligent) behaviour in the past, can often be regarded as rational w.r.t. the appropriate goal. Maybe we are already near-optimal goal achievers. I doubt this, but cannot rule it out either.

Humans are not faster but more intelligent than dogs, and dogs in turn are more intelligent than worms and not just faster, even if we cannot pinpoint exactly why we are more intelligent: is it our capacity to produce technology or to transform our environment on a large scale or consciousness or domination over all other species? There are no good arguments why humans should be close to the top of the possible biological intelligence scale, and even less so on a vorld scale. By extrapolation it is plausible that a vorld of much more intelligent trans-humans or machines is possible. They will likely be able to perform better in an even wider range of environments on an even wider range of problems than humans. Whether this results in anything that deserves the name intelligence explosion is unclear.

8. Singularitarian Intelligences

Consider a world inhabited by competing agents, initialized with human mind-uploads or non-human AGIs, and increasing comp per virtual. Sections 6 and 7 then indicate that evolutionary pressure increases the individuals' intelligence and the vorld should converge to a society of AIXIs. Alternatively, if we postulate an intelligence singularity and accept that AIXI is the most intelligent agent, we arrive at the same conclusion. More precisely, the society consists of agents that aim at being AIXIs only being constrained by comp. If this is so, the intelligence singularity might be identified with a society of AIXIs, so studying AIXI can tell us something about how a singularity might look like. Since AIXI is completely and formally defined, properties of this society can be studied rigorously mathematically. Here are some questions that could be asked and answered:

Will a pure reward maximizer such as AIXI listen to and trust a teacher? Likely yes. Will it take drugs (i.e. hack the reward system)? Likely no, since cumulative long-term reward would be small (death). Will AIXI replicate itself or procreate? Likely yes, if AIXI believes that clones or descendants are useful for its own goals. Will AIXI commit suicide? Likely yes (no), if AIXI is raised to believe in going to heaven (hell) i.e. maximal (minimal) reward forever. Will sub-AIXIs self-improve? Likely yes, since this helps to increase reward. Will AIXI manipulate or threaten teachers to give more reward? Likely yes. Are pure reward maximizers like AIXI egoists, psychopaths, and/or killers, or will they be friendly (altruism as extended ego(t)ism)? Curiosity killed the cat and maybe AIXI, or is extra reward for curiosity necessary? Immortality can cause laziness. Will AIXI be lazy? Can self-preservation be learned or need (parts of) it be innate. How will AIXIs interact/socialize in general?

For some of these questions, partial and informal discussions and plausible answers are available, and a couple have been rigorously defined, studied and answered, but most of them are open to date

(Hutter, 2005; Schmidhuber, 2007; Orseau & Ring, 2011; Ring & Orseau, 2011; Hutter, 2012, forthcoming). But the AIXI theory has the potential to arrive at definite answers to various questions regarding the social behavior of super-intelligences close to or at an intelligence singularity.

9. Diversity Explosion and the Value of a Virtual Life

As indicated, some things will be harder or impossible in a virtual world (e.g. to discover new physics) but many things should be easier. Unless a global copy protection mechanism is deliberately installed (like e.g. in Second Life) or copyright laws prevent it, copying virtual structures should be as cheap and effortless as it is for software and data today. The only cost is developing the structures in the first place, and the memory to store and the comp to run them. With this comes the possibility of cheap manipulation and experimentation.

It becomes particularly interesting when virtual life itself gets copied and/or modified. Many science fiction stories cover this subject, so I will be brief and selective here. One consequence should be a 'virtuan' explosion with life becoming much more diverse. Andy Clarke (2009) writes (without particularly referring to virtuals) that 'The humans of the next century will be vastly more heterogenous, more varied along physical and cognitive dimensions, than those of the past as we deliberately engineer a new Cambrian explosion of body and mind.' In addition, virtual lives could be simulated in different speeds, with speeders experiencing slower societal progress than laggards. Designed intelligences will fill economic niches. Our current society already relies on specialists with many years of training, so it is natural to go the next step to ease this process with 'designer babies'.

Another consequence should be that life becomes less valuable. Our society values life, since life is a valuable commodity and expensive/laborious to replace/produce/raise. We value our own life, since evolution selects only organisms that value their life. Our human moral code mainly mimics this, with cultural differences and some excesses (e.g. suicide attacks on the one side and banning stem cell research on the other).

If life becomes 'cheap', motivation to value it will decline. Analogies are abundant: Cheap machines decreased the value of physical labor. Some expert knowledge was replaced by hand-written documents, then printed books, and finally electronic files, where each transition reduced the value of the same information. Digital computers made human computers obsolete. In games, we value our own life and that of our opponents less than real life, not only because a game is a crude approximation to real life, but also because games can be reset and one can be resurrected. Governments will stop paying my salary when they can get the same research output from a digital version of me, essentially for free.

And why not participate in a dangerous fun activity if in the worst case I have to activate a backup copy of myself from yesterday which just missed out this one (anyway not too well-going) day. The belief in immortality can alter behavior drastically.

Of course there will be countless other implications: ethical, political, economical, medical, cultural, humanitarian, religious, in art, warfare, etc. I have singled out the value of life, since I think it will significantly influence other aspects. Much of our society is driven by the fact that we highly value (human/individual) life. If virtual life is/becomes cheap, these drives will ultimately vanish and be replaced by other goals. If AIs can be easily created, the value of an intelligent individual will be much lower than the value of a human life today. So it may be ethically acceptable to freeze, duplicate, slow-down, modify (brain experiments), or even kill (oneself or other) AIs at will, if they are abundant and/or backups are available, just what we are used to doing with software. So laws preventing experimentation with intelligences for moral reasons may not emerge. With so little value assigned to an individual life, maybe it becomes a disposable.

10. Personal Remarks

I have deliberately avoided discussing consciousness for several reasons: David Chalmers is *the* consciousness expert and not me, he has extensively written about it in general and also in the context of the singularity (Chalmers, this volume), and I essentially agree with his assessments. Personally I believe in the functionalist theory of identity and am confident that (slow and fast) uploading of a human mind preserves identity and consciousness, and indeed that any sufficiently high intelligence, whether real/biological/physical or virtual/silicon/ software is conscious, and that consciousness survives changes of substrate: teleportation, duplication, virtualization/scanning, etc. along the lines of Chalmers' essay.

I have also only considered (arguably) plausible scenarios, but not whether these or other futures are desirable. First, there is the problem of how much influence/choice/freedom we actually have in shaping our future in general and the singularity in particular. Can evolutionary forces be beaten? Second, what is desirable is necessarily subjective. Are there any universal values or qualities we want to see or that should survive? What do I mean by we? All humans? Or the dominant species or government at the time the question is asked? Could it be diversity? Or friendly AI (Yudkowsky, 2008)? Could the long-term survival of at least one conscious species that appreciates its surrounding universe be a universal value? A discussion of these questions is clearly beyond the scope of this article.

11. Conclusions

Based on the deliberations in this paper, here are my predictions concerning a potential technological singularity, although admittedly they have a speculative character.

- This century may witness a technological explosion of a degree deserving the name singularity.

- The default scenario is a society of interacting intelligent agents in a virtual world, simulated on computers with hyperbolically increasing computational resources.

- This is inevitably accompanied by a speed explosion when measured in physical time units, but not necessarily by an intelligence explosion.

- Participants will not necessarily experience this explosion, since/if they are themselves accelerated at the same pace, but they should enjoy 'progress' at a 'normal' subjective pace.

- For non-accelerated non-participating conventional humans, after some short period, their limited minds will not be able to perceive the explosion as an intelligence explosion.

- This begs the question in which sense an intelligence explosion has happened. (If a tree falls in a forest and no one is around to hear it, does it make a sound?)

- One way and maybe the only way to make progress in this question is to clarify what intelligence actually is.

- The most suitable notion of intelligence for this purpose seems to be that of universal intelligence, which in principle allows to formalize and theoretically answer a wide range of questions about super-intelligences. Accepting this notion has in particular the following implications:

- There is a maximally intelligent agent, which appears to imply that intelligence is fundamentally upper bounded, but this is not necessarily so.

- If the virtual world is inhabited by interacting free agents (rather than a 'monistic' vorld inhabited by a single individual or a tightly controlled society), evolutionary pressures should breed agents of increasing intelligence that compete about computational resources.

- The end-point of this intelligence evolution/acceleration (whether it deserves the name singularity or not) could be a society of these maximally intelligent individuals.

- Some aspects of this singularitarian society might be theoretically studied with current scientific tools.

- Way before the singularity, even when setting up a virtual society in our image, there are likely some immediate differences, for instance that the value of an individual life suddenly drops, with drastic consequences.

Acknowledgments

Thanks to Wolfgang Schwarz and Reinhard Hutter for feedback on earlier drafts.

References

Aaronson, S. (2005) NP-complete problems and physical reality, *SIGACT News Complexity Theory Column*, **36** (1), pp. 30–52.

Bostrom, N. & Cirkovic, M.M. (eds.) (2008) *Global Catastrophic Risks*, Oxford: Oxford University Press.

Bekenstein, J.D. (2003) Information in the holographic universe, *Scientific American*, **289** (2), pp. 58–65.

Bremermann, H.J. (1965) Quantum noise and information, *Proceedings of the 5th Berkeley Symposium on Mathematical Statistics and Probability*, pp. 15–20, Berkeley, CA: University of California Press.

Clarke, A. (2009) celebratory self re-engineering, World Question Center, http://www.edge.org/q2009/q09_4.html#clark.

Copeland, J. (2002) Hypercomputation, *Minds and Machines*, **12**, pp. 461–502.

Dawkins, R. (1976) *The Selfish Gene*, Oxford: Oxford University Press.

Deutsch, D. (1997) *The Fabric of Reality*, New York: Allen Lane.

de Garis, H., Shuo, C., Goertzel, B. & Ruiting, L. (2010) A world survey of artificial brain projects, part I: Large-scale brain simulations, *Neurocomputing*, **74** (1–3), pp. 3–29.

Geary, D.C. (2007) The motivation to control and the evolution of general intelligence, in Gangestad, S.W. & Simpson, J.A. (eds.) *The Evolution of Mind: Fundamental Questions and Controversies*, pp 305–312, New York: Guilford Press.

Goertzel, B., Lian, R., Arel, I., de Garis, H. & Chen, S. (2010) A world survey of artificial brain projects. part II: Biologically inspired cognitive architectures, *Neurocomputing*, **74** (1–3), pp. 30–49.

Good, I.J. (1965) Speculations concerning the first ultraintelligent machine, *Advances in Computers*, **6**, pp. 31–88.

Hanson, R. (2008) Economics of the singularity, *IEEE Spectrum*, (June), pp. 36–42. Also reprinted as, A new world order, *Cosmos*, **26** (April/May 2009), pp. 47–54.

Hochmuth, G. (2003) On the genetic evolution of a perfect tic-tac-toe strategy, in Koza, J.R. (ed.) *Genetic Algorithms and Genetic Programming at Stanford*, pp. 75–82, Stanford, CA: Stanford University Press.

Hutter, M. (2005) *Universal Artificial Intelligence: Sequential Decisions based on Algorithmic Probability*, Berlin: Springer.

Hutter, M. (2010) A complete theory of everything (will be subjective), *Algorithms*, **3** (4), pp. 329–350.

Hutter, M. (2012) One decade of universal artificial intelligence, in *Theoretical Foundations of Artificial General Intelligence*, Amsterdam: Atlantis Press, (forthcoming).

Kanazawa, S. (2007) The g-culture coevolution, in Gangestad, S.W. & Simpson, J.A. (eds.) *The Evolution of Mind: Fundamental Questions and Controversies*, pp. 313–318, New York: Guilford Press.

Kurzweil, R. (1999) *The Age of Spiritual Machines*, New York: Viking Press.

Kurzweil, R. (2005) *The Singularity is Near*, New York: Viking Press.

Legg, S. & Hutter, M. (2007a) A collection of definitions of intelligence, in Goertzel, B. & Wang, P. (eds.) *Advances in Artificial General Intelligence: Concepts, Architectures and Algorithms*, vol. 157 of *Frontiers in Artificial Intelligence and Applications*, pp. 17–24, Amsterdam: IOS Press.

Legg, S. & Hutter, M. (2007b) Universal intelligence: A definition of machine intelligence, *Minds & Machines*, **17** (4), pp. 391–444.

Lloyd, S. (2000) Ultimate physical limits to computation, *Nature*, **406** (6799), pp. 1047–1054.

Li, M. & Vitányi, P.M.B. (2008) *An Introduction to Kolmogorov Complexity and its Applications*, 3rd Ed., Berlin: Springer.

Orseau, L. & Ring, M. (2011) Self-modification and mortality in artificial agents, *Proceedings of the 4th Conference on Artificial General Intelligence (AGI'11)*, vol. 6830 of *LNAI*, pp. 1–10, Berlin: Springer.

Rathmanner, S. & Hutter, M. (2011) A philosophical treatise of universal induction, *Entropy*, **13** (6), pp. 1076–1136.

Russell, S.J. & Norvig, P. (2010) *Artificial Intelligence: A Modern Approach*, 3rd ed., Englewood Cliffs, NJ: Prentice-Hall.

Ring, M. & Orseau, L. (2011) Delusion, survival, and intelligent agents, *Proceedings of the 4th Conference on Artificial General Intelligence (AGI'11)*, vol. 6830 of *LNAI*, pp. 11–20, Berlin: Springer.

Schmidhuber, J. (2006) New millennium AI and the convergence of history, in Duch, W. & Mandziuk, J. (eds.) *Challenges to Computational Intelligence*, pp. 15–36, Berlin: Springer.

Schmidhuber, J. (2007) Simple algorithmic principles of discovery, subjective beauty, selective attention, curiosity & creativity, *Proceedings of the 10th International Conference on Discovery Science (DS'07)*, vol. LNAI 4755, pp. 26–38, Senday, Berlin: Springer.

Solomonoff, R.J. (1985) The time scale of artificial intelligence: Reflections on social effects, *Human Systems Management*, **5**, pp. 149–153.

Thornton, R. (1847) The age of machinery, in *The Expounder of Primitive Christianity*, vol. 4, p. 281, Ann Arbor, MI: University of Michigan Press.

Ulam, S. (1958) Tribute to John von Neumann, *Bulletin of the American Mathematical Society*, **64** (3 II), pp. 1–49.

Vinge, V. (1993) The coming technological singularity, *Vision-21 Symposium*, NASA Lewis Research Center and the Ohio Aerospace Institute, 30 to 31 March 1993.

Veness, J., Ng, K.S., Hutter, M., Uther, W. & Silver, D. (2011) A Monte Carlo AIXI approximation, *Journal of Artificial Intelligence Research*, **40**, pp. 95–142.

Weyl, H. (1927) *Philosophie der Mathematik und Naturwisnschafset*, Munich: R. Oldenbourg. [English trans.: Philosophy of Mathematics and Natural Science, 1949, Princeton University Press.]

Yudkowsky, E. (1996) Staring at the singularity, [Online], http://yudkowsky.net/obsolete/singularity.html

Yudkowsky, E. (2008) Artificial intelligence as a positive and negative factor in global risk, in Bostrom, N. & Cirkovic, M. (eds) *Global Catastrophic Risks*, pp. 232–263, Oxford: Oxford University Press.

Drew McDermott

Response to 'The Singularity' by David Chalmers

I agree with David Chalmers about one thing: it is useful to see the arguments for the singularity written down using the philosophers' signature deductive framework, in which controversial premises are made explicit. If all concur that the form of the argument has been captured, then they can get down to the brass tacks of refuting or rebutting the numbered premises.

To give away my inclinations up front, I tend to disagree with Chalmers about the prospects for the singularity, agree about uploading, and agree with some of his conclusions about controlling the singularity, and disagree with others.

I have not much to say about the distinctions among the various kinds of singularity discussed at the beginning of the paper. They are well described by Chalmers, but most of the time they are not relevant, or if they are it is hard to judge what the consequences of the distinctions might be.

Prospects

Should we be discussing the singularity? Sure, it's a free country. I personally think that even though it's true that 'the singularity [would] be one of the most important events in the history of the planet', other more likely scenarios may rule it out. In those scenarios the problems we face *right now* may, if not solved, escalate to the point of disaster, making the possibility of a singularity moot. I'll lay out the scenario I most fear at the end of this review.

But putting those nightmares aside, what dreams or nightmares might the singularity bring about, if any?

For ultraintelligent machines to exist, intelligent machines must exist, and, as Chalmers says, 'every path to AI has proved surprisingly difficult to date'. He lays out the arguments that AI is inevitable in spite of its difficulty, and I accept the conclusion, if not all of his arguments.

The next step is to argue that if there is AI then there will be AI+ 'soon after', where AI+ is artificial intelligence surpassing human intelligence. I'll accept this conclusion as well, because I agree with Chalmers that it is unlikely human beings' level of intelligence is unsurpassable. (My argument is different: as soon as bipedal primates reached a minimum level of intelligence sufficient for civilization to develop, it did, and here we are; it would be a big coincidence if that level was also a maximum.)

My scepticism is mainly focused on the third premise: If there is AI+, there will be AI++ (soon, perhaps very soon, after), where AI++ is artificial intelligence far surpassing humans'. The case that there *exists* a level of intelligence far surpassing humans' is much weaker than the case that there exists a level surpassing it at all. No argument for this presupposition is given in the paper.

The argument for the third premise takes the form of mathematical induction. As stated, the argument is unsound, because a series of increases from AI_n to AI_{n+1}, each exponentially smaller than the previous one, will reach a limit.[1] But the text clarifies that one might need to assume that AI_{n+1} is at least δ times larger than AI_n, where δ is a constant ratio.

I find this claim unsatisfying. For one thing, it is unnecessary. The mathematical induction should, I think, be thought of as shorthand for an 'empirical induction' of a fixed number of steps, at most $\lceil log_\delta P \rceil$, where P is the ratio between AI++ and AI (= AI_0). (There may be weaker ways to patch the argument, not introducing this minimum ratio δ, but I don't know what they would be.)

But the claim is also, unfortunately, unsupported by any argument, except a set of proposals to shore up the notion of an 'extendible method' of producing intelligence, a method that can 'easily be improved, yielding more intelligent systems'.

Three extendible methods are put forward: direct programming, machine learning, and artificial evolution. Direct programming is extendible because of the well-known fact that 'almost every program that has yet been written [is] improvable in multiple respects'. What respects? I can think of some: bug elimination, inner-loop

---

[1] This is not just a matter of devious scaling, as Chalmers seems to imply: 'If there is an intelligence measure for which the argument succeeds, there will almost certainly be a rescaled intelligence measure ... for which it fails.' P, the ratio of AI++ and AI+ (see below) defines the scale; if the steps are rescaled, so must P be.

optimization, and porting to faster hardware. None of these are extendible, except possibily the last, which works only if Moore's Law stays true indefinitely. But Moore's Law is a good example of why extendible methods are a mirage. What seems like a smooth, tidy curve in the dreams of programmers is a series of hairy technological innovations from the point of view of hardware designers and systems programmers. The methods used to make circuits smaller are quite different from the methods used to automate the adaptation of programs to run on multi-core processors. But even the phrase 'methods used to make circuits smaller' is misleading because there are no such methods; at each stage completely novel ways of shrinking the components of the semiconductor fab line had to be developed at ever-escalating costs.

As far as machine learning is concerned, I am quite puzzled by the idea that a good learning algorithm can be improved indefinitely. For instance, why do so few neural networks have more than one hidden layer? Because, it turns out, backpropagation to more than one such layer provides too weak a signal to learn at a useful rate. And yet many concepts are difficult to capture in shallow networks. There has been progress in this area, but it required new ideas.

Waiting for a new idea to come along is not an extendible method, indeed not a method at all.

The third of the three extendible methods is artificial evolution. So far evolutionary algorithms have proven to do fairly well in solving some fairly interesting problems. I don't quite see what's extendible about them, or why they're of special interest, except that *we* were produced by an evolutionary algorithm, in a sense. The current picture of that process is not encouraging, because to get to an intelligent primate a series of lucky breaks had to occur. At one point in fairly recent biological history (within the last 200,000 years) the population of our ancestors was down to a few thousand individuals. It could so easily have gone all the way to zero, and the planet would still be waiting for its first intelligence to appear. What makes us think the path through the high-dimensional space of environments and agents will require fewer lucky twists and turns in the future?

Finally, let me point out a general argument against the existence of extendible methods.

Theorem:

There are no extendible methods unless $P = N_P$.

I have a wonderful proof of this result which is unfortunately too big to fit in this review.

If you doubt that $P = N_P$, as I do, then it's hard to believe that any extendible method, including waiting for Moore, exists.

But sweep away all those objections, there is still the 'motivational defeater' Chalmers describes as 'AI+ systems [being] disinclined to create their successors, perhaps because we design them to be so disinclined, or perhaps because they will be intelligent enough to realize that creating successors in not in their interests'. If it's possible to start the sequence $AI+_1$, $AI+_2$, ... , I don't think we can say in advance anything at all about the motivations of what will amount to *artificial persons*. In particular, I seriously doubt we can design them to be inclined or disinclined one way or another. Even if the first in the series has motivations under our control, the later elements will be designed with the help of members of the earlier cohorts, and so will be difficult for us to understand. If we knew all there was to know about a fellow human's brain circuitry and memories, could we predict what they would do by any method other than simulating them (and their surroundings) on a faster computer than they 'have', or 'are'? I think not. But simulating $AI+_2$ before bringing it into existence is logically impossible, because the simulation would essentially *be* $AI+_2$, running in a virtual world, with all the problems of keeping it contained that Chalmers discusses. So the only way to control the motivations of the next generation of AI+s is to create several different versions, fiddling with whatever parameters we have that might have an impact on its motivations as we go from one version to the next. This achieves control of the next generation's motivations in the same sense that control of the trajectory of the next ballistic missile you launch may be achieved by launching several different ballistic missiles in succession, correcting the trajectory slightly each time. Which is to say, it doesn't achieve control of anything, but might create a series of misplaced explosions. I am very dubious that even if Kant's position on rationality and morality is right, an 'intelligence explosion [might] lead to a morality explosion'.

Consequences

Early in the paper Chalmers says that 'An intelligence explosion has enormous potential benefits: a cure for all known diseases, an end to poverty, extraordinary scientific advances, and much more' (sect. 1). Later we are offered a choice among being helped to die out, being consigned to a virtual world, being allowed to live as inferiors, or being succeeded by a mechanical superrace that may or may not remember being us; or, in his concise formula, 'extinction, isolation, inferiority, or integration' (sect. 8).[2]

The difference between these two lists illustrates the fact that mere intelligence is not enough to guarantee an end to poverty (to pick one of the most intractable problems); it might bring about an end to the human race. In between these two possibilities is the far more likely scenario in which AI+ or AI++ finds a way to end poverty, and human greed and fear (and the greed and fear of the AIs involved in the political process) prevent it from succeeding. Or: the AI working for the United Nations finds cures for diseases almost as fast as the AIs working for contending powers think up new ones to use in biological warfare.

Chalmers investigates in greatest detail the possibility that as AI++ is devloped, we gradually become integrated with it. But would it really still be 'we'? One gets into a lot of tangles by assuming that there is a definite answer to this question. It is easy to construct cases in which one's intuitions are driven one way or the other, no matter where they start, as Chalmers explains so capably.

I tend to believe that personal identity is mostly a matter of social convention; it's not settled purely by facts about the contenders. (I think this is a subspecies of what Chalmers calls the 'deflationary' view.) If at some point it becomes a normal convention that people survive uploading, and the uploaded include many prominent citizens, who are indignant at the idea that they're not conscious, or that the DigiX they are now differs from BioX, the biological entity they started as. At that point going virtual will have become a normal stage of life, and almost everyone will expect to survive it, and this mass belief makes it so.

At that point the biological incarnation of human beings, if it is still necessary, would be viewed the way intelligent amphibians might view tadpoles, as a 'larval' phase of human existence. A refusal to get on with life and get uploaded might be seen as similar to a refusal to emerge from the womb. If by chance my 'larva' should not be destroyed by the uploader, surely people would see the larva as having *no* claim to being me, and society would have no qualms about hunting it down and euthanizing it. In this scenario, I would be among the first to agree with this decision, and so would the larva, who might have an elemental desire to avoid death, but would lay claim to nothing more than being a discarded piece of a growing person.

[2] We can upgrade this list to the cooler slogan 'the five I's': interment, isolation, inferiority, improvement, or impersonation, the last two items replacing 'integration', which covers both bogus and genuine continuity of identity.

So, since social conventions would evolve along with technological possibility, there would be no controversy regarding personal identity, and people would indeed be transformed into posthuman computational entities. But in spite of these facts, outside observers might be justified in concluding that the human race had ceased to exist, because no one was making adult human beings any more. They might miss us, the way we might wish some *Homo habilis* were still around, even though (let's suppose) they became us and no one ever regretted it.

And last, let me say why I think it would be a pity for too many smart people to devote time and mental effort to thinking about the singularity, which is that there are other possible events, unambiguous catastrophes, with higher probability. The most likely one, the one that really scares me, is an environmental collapse followed by a nuclear war as the survivors quarrel over the remaining resources. I can vividly picture this happening within the lifetime of *some of the people reading this*, and if not yours, then *your and my children's*. I can picture them being refugees fighting for survival — or dead — amidst the ruins of civilization. So I think all our resources and energy should be directed toward the problems of resource management and nuclear disarmament.

The exponential growth in technology that is the major argument for the singularity is accompanied by, perhaps made possible by, an exponential growth in exploitation of finite natural resources (including the atmosphere, viewed as a carbon-dioxide sponge). Our civilization's addiction to a process that simply cannot continue is a sign of insanity, and belief in the singularity may be one of its most comforting delusions. Even if some of the world's richer citizens get 'uploaded', what happens when the power goes off?

Jürgen Schmidhuber

Philosophers & Futurists, Catch Up!

Response to The Singularity

Abstract: *Responding to Chalmers' The Singularity, I argue that progress towards self-improving AIs is already substantially beyond what many futurists and philosophers are aware of. Instead of rehashing well-trodden topics of the previous millennium, let us start focusing on relevant new millennium results.*

All indented paragraphs of this paper are quotes taken from Chalmers' essay, who mentions Good's informal speculations (1965) on ultraintelligent self-improving machines:

> The key idea is that a machine that is more intelligent than humans will be better than humans at designing machines. So it will be capable of designing a machine more intelligent than the most intelligent machine that humans can design.

Chalmers speculates that some sort of meta-evolution could be used to build more and more intelligent machines called AI, AI+, AI++...:

> The process of evolution might count as an indirect example: less intelligent systems have the capacity to create more intelligent systems by reproduction, variation and natural selection. This version would then come to the same thing as an evolutionary path to AI and AI++. [...] If we produce an AI by machine learning, it is likely that soon after we will be able to improve the learning algorithm and extend the learning process, leading to AI+. If we produce an AI by artificial evolution, it is likely that soon after we will be able to improve the evolutionary algorithm and extend the evolutionary process, leading to AI+.

Back in 1987 I put forward the first concrete implementation of this informal idea (Schmidhuber, 1987): an evolutionary self-referential meta-learning problem solver that improves itself such that it becomes better at improving itself, by improving the very process of evolution, lifting Genetic Programming (Cramer, 1985; Dickmanns *et al.*, 1987) to the meta-level, the meta-meta-level, and so on, recursively. This was the first in a long string of papers on self-referential self-improvers — compare the recent overview (Schaul & Schmidhuber, 2010).

Roughly at the same time, two science fiction novels by Vernor Vinge (1984; 1986) introduced me to the concept of a *Technological Singularity*, in my opinion one of the few original ideas put forward by SF authors *after* the so-called Golden Age of SF in the 1950s and 60s. The basic idea is that *technological change accelerates exponentially such that within finite time it reaches a transcendent point beyond human comprehension.* True, in the early 1900s Teilhard de Chardin already predicted (from a more religious perspective) that the evolution of civilization will culminate in a transcendent *Omega point*, and Vinge himself pointed out (1993) that Stanislaw Ulam formulated similar thoughts in 1958:

> One conversation centered on the ever accelerating progress of technology and changes in the mode of human life, which gives the appearance of approaching some essential singularity in the history of the race beyond which human affairs, as we know them, could not continue.

It was Vinge, however, who popularized the technological singularity and significantly elaborated on it, exploring pretty much all the obvious related topics, such as accelerating change, computational speed explosion, potential delays of the singularity, obstacles to the singularity, limits of predictability and negotiability of the singularity, evil vs benign super-intelligence, surviving the singularity, etc.

I am not aware of substantial additional non-trivial ideas in this vein originating in the subsequent two decades, although other futurists and philosophers have started writing about the singularity as well. Many of them, however, are mostly concerned with ancient debates triggered by non-experts such as Lucas, Searle, Penrose and other authors commonly ignored by hardcore AI researchers:

> Various existing forms of resistance to AI take each of these forms. For example, J.R. Lucas (1961) has argued that for reasons tied to Gödel's theorem, humans are more sophisticated than any machine.

Instead of further spending time on such frequently refuted claims, I'd like to encourage futurists and philosophers to learn about the more recent, in my opinion much more relevant hardcore AI research outlined in the remainder of this paper.

> Perhaps the core sense of the term [singularity], though, is a moderate
> sense in which it refers to an intelligence explosion through the recur-
> sive mechanism set out by I.J. Good [...] I will always use the term sin-
> gularity in this core sense in what follows. [...] The argument depends
> on the assumption that there is such a thing as intelligence and that it can
> be compared between systems. (See also pp. 29ff.)

The scientific way of measuring intelligence involves measuring
problem solving capacity. There are mathematically sound ways of
doing this, using basic concepts of theoretical computer science
(Levin, 1973; Hutter, 2005; Schmidhuber, 2009b), all of them avoiding the subjectivity of the ancient and popular but scientifically not
very useful Turing test, which essentially says 'intelligent is what I
feel is intelligent.'

> My own view is that the history of artificial intelligence suggests that
> the biggest bottleneck on the path to AI is software, not hardware: we
> have to find the right algorithms, and no-one has come close to finding
> them yet. [...] The Gödel Machines of Schmidhuber (2003) provide a
> theoretical example of self-improving systems at a level below AI,
> though they have not yet been implemented and there are large practical
> obstacles to using them as a path to AI.

I feel that philosophers and futurists should try to become very familiar with what is currently going on in the field of universal problem
solvers. The fully self-referential (Gödel, 1931) Gödel machine
(Schmidhuber, 2009b) already *is* a universal AI that is at least theoretically optimal in a certain sense. It may interact with some initially
unknown, partially observable environment to maximize future
expected utility or reward by solving arbitrary user-defined computational tasks. Its initial algorithm is not hardwired; it can completely
rewrite itself without essential limits apart from the limits of
computability, provided a proof searcher embedded within the initial
algorithm can first prove that the rewrite is useful, according to the
formalized utility function taking into account the limited computational resources. Self-rewrites may modify/improve the proof
searcher itself, and can be shown to be *globally optimal*, relative to
Gödel's well-known fundamental restrictions of provability (Gödel,
1931). To make sure the Gödel machine is at least *asymptotically* optimal even before the first self-rewrite, we may initialize it by Hutter's
non-self-referential but *asymptotically fastest algorithm for all
well-defined problems* HSEARCH (Hutter, 2002), which uses a hard-
wired brute force proof searcher and (justifiably) ignores the costs of
proof search. Assuming discrete input/output domains $X/Y \subset B^*$, a
formal problem specification $f : X \rightarrow Y$ (say, a functional description

of how integers are decomposed into their prime factors), and a particular $x \in X$ (say, an integer to be factorized), HSEARCH orders all proofs of an appropriate axiomatic system by size to find programs q that for all $z \in X$ provably compute $f(z)$ within time bound $t_q(z)$. Simultaneously it spends most of its time on executing the q with the best currently proven time bound $t_q(x)$. Remarkably, HSEARCH is as fast as the *fastest* algorithm that provably computes $f(z)$ for all $z \in X$, save for a constant factor smaller than $1 + \varepsilon$ (arbitrary real-valued $\varepsilon > 0$) and an f-specific but x-independent additive constant (*ibid.*). Given some problem, the Gödel machine may decide to replace its HSEARCH initialization by a faster method suffering less from large constant overhead, but even if it doesn't, its performance won't be less than asymptotically optimal.

All of this implies that there already exists the blueprint of a Universal AI which will solve almost all problems almost as quickly as if it already knew the best (unknown) algorithm for solving them, because almost all imaginable problems are big enough to make the additive constant negligible. Hence I must object to Chalmers' statement *'we have to find the right algorithms, and no-one has come close to finding them yet'*. The only motivation for *not* quitting computer science research right now is that many real-world problems are so small and simple that the ominous constant slowdown (potentially relevant at least before the first Gödel machine self-rewrite) is *not* negligible. Nevertheless, the ongoing efforts at scaling universal AIs down to the rather few *small* problems are very much informed by the new millennium's theoretical insights mentioned above, and may soon yield practically feasible yet still general problem solvers for physical systems with highly restricted computational power, say, a few trillion instructions per second, roughly comparable to a human brain power.

Simultaneously, our non-universal but still rather general fast deep/ recurrent neural networks have already started to outperform traditional pre-programmed methods: they recently collected a string of 1st ranks in many important visual pattern recognition benchmarks, e.g. Graves & Schmidhuber (2009); Ciresan *et al.* (2011): IJCNN traffic sign competition, NORB, CIFAR10, MNIST, three ICDAR handwriting competitions. Here we greatly profit from recent advances in computing hardware, using GPUs (mini-supercomputers normally used for video games) 100 times faster than today's CPU cores, and a million times faster than PCs of 20 years ago, complementing the recent above-mentioned progress in the theory of mathematically optimal universal problem solvers.

> In principle there could be an intelligence explosion without a speed
> explosion and a speed explosion without an intelligence explosion.

As pointed out above, problem solving ability does depend on speed,
hence intelligence and speed are *not* independent. Computer scientists
agree, however, that we are far from the physical limits to
computation:

> While the laws of physics and the principles of computation may
> impose limits on the sort of intelligence that is possible in our world,
> there is little reason to think that human cognition is close to approach-
> ing those limits.

In fact, more than 100 additional years of Moore's Law seem necessary to reach Bremermann's (1982) physical limit of more than 10^{51}
elementary instructions per kg and second, roughly 10^{20} times the
combined raw computational power of all human brains, give or take a
few zeroes.

> The history of AI involves a long series of optimistic predictions by
> those who pioneer a method, followed by a periods of disappointment
> and reassessment. This is true for a variety of methods involving direct
> programming, machine learning, and artificial evolution, for example.
> Many of the optimistic predictions were not obviously unreasonable at
> the time, so their failure should lead us to reassess our prior beliefs in
> significant ways.

I feel that after 10,000 years of civilization there is no need to justify
pessimism through comparatively recent over-optimistic and
self-serving predictions (1960s: 'only 10 instead of 100 years needed
to build AIs') by a few early AI enthusiasts in search of funding.

> If we value scientific progress, for example, it makes sense for us to cre-
> ate AI and AI+ systems that also value scientific progress.

But how to formalize this informal idea? Only recently this has
become possible through the *Formal Theory of Creativity* (Schmidhuber, 2006; 2010) mathematically concretizing the driving forces
and value functions behind creative behavior such as science and art.
Consider an agent living in an initially unknown environment. At any
given time, it uses one of the many reinforcement learning (RL) methods (Kaelbling *et al.*, 1996) to maximize not only expected future
external reward for achieving certain goals, such as avoiding hunger/
empty batteries/obstacles, etc. but also *intrinsic* reward for action
sequences that improve an internal model of the environmental
responses to its actions, continually learning to better predict/explain/
compress the growing history of observations in fiuenced by its

experiments, actively infiuencing the input stream such that it contains previously unknown but learnable algorithmic regularities which become known and boring once there is no additional subjective *compression progress* or *learning progress* any more. I have argued that the particular utility functions associated with this theory explain essential aspects of intelligence including selective attention, curiosity, creativity, science, art, music, humor, e.g. Schmidhuber (2006; 2010). They are currently being implemented on humanoid baby-like iCub robots. The theory actually addresses the above-mentioned drawbacks of asymptotically optimal universal AIs, allowing learning agents to not only focus on potentially hard-to-solve externally posed tasks, but also creatively invent self-generated tasks that have the property of currently being still unsolvable but easily learnable, given the agent's present knowledge, such that the agent is continually motivated to improve its understanding of how the world works, and what can be done in it. One topic worth of exploration through futurists and philosophers are the potential consequences of self-improving AIs defining their own tasks in this creative, world-exploring way, which may sometimes conflict with goals of humans.

> If we create an AI through learning or evolution, the matter is more complex. [...] Of course even if we create an AI or AI+ (whether human-based or not) with values that we approve of, that is no guarantee that those values will be preserved all the way to AI++.

Note that the Gödel machine mentioned above *does* preserve those values. It can rewrite its utility function only if it first can prove that the rewrite is useful according to its previous utility function.

All attempts at making sure there will be only provably friendly AIs seem doomed though. Once somebody posts the recipe for practically feasible self-improving Gödel machines or AIs in form of code into which one can plug arbitrary utility functions, many users will equip such AIs with many different goals, often at least partially conflicting with those of humans. The laws of physics and the availability of physical resources will eventually determine which utility functions will help their AIs more than others to multiply and become dominant in competition with AIs driven by different utility functions. The survivors will define in hindsight what's 'moral', since only survivors promote their values, giving evolutionary meaning to Kant's musings:

> Kant held more specifically that rationality correlates with morality: a fully rational system will be fully moral as well [...] The Kantian view at

least raises the possibility that intelligence and value are not entirely independent.

Chalmers writes on AIs in virtual worlds:

> It remains possible that they might build computers in their world and design AI on those computers. [...] If one takes seriously the possibility that we are ourselves in such a simulation (as I do in Chalmers, 2005) [...]

Compare the original papers since 1997 (Schmidhuber, 1997; 2000) that introduced and discussed the set of all computable universes as well as the set of possible computable probability distributions on them, extending Konrad Zuse's (1970) pioneering work on digital physics and the computable universe, using algorithmic probability theory (Solomonoff, 1964; 1978; Li & Vitányi, 1997) to analyse the probability of some observer inhabiting a particular 'simulated' or real universe, given his observations.

Uploading brains into cyberspace (Chalmers, pp. 45ff.) as well as related topics were discussed not only in the cited works of SF author Egan but also in Gibson's earlier famous cyberspace novels of the 1980s, and possibly first by Daniel F. Galuye (1964) who already went far in exploring the consequences. Gradual uploading is an old concept, too:

> Suppose that 1% of Dave's brain is replaced by a functionally isomorphic silicon circuit. Next suppose that after one month another 1% is replaced, and the following month another 1%. We can continue the process for 100 months, after which a wholly uploaded system will result.

I think I first read about this thought experiment in Pylyshyn's (1980) paper. Chalmers also writes on consciousness (p. 48):

> It is true that we have no idea how a nonbiological system, such as a silicon computational system, could be conscious.

But at least we have pretty good ideas where the symbols and self-symbols underlying consciousness and sentience come from (Schmidhuber, 2009a; 2010). They may be viewed as simple by-products of data compression and problem solving. As we interact with the world to achieve goals, we are constructing internal models of the world, predicting and thus partially compressing the data histories we are observing. If the predictor/compressor is an artificial recurrent neural network (RNN) (Werbos, 1988; Williams & Zipser, 1994; Schmidhuber, 1992; Hochreiter & Schmidhuber, 1997; Graves & Schmidhuber, 2009), it will create feature hierarchies, lower level neurons corresponding to simple feature detectors similar to those found in human brains, higher layer neurons typically corresponding to more abstract features, but fine-grained where necessary. Like any good compressor the RNN will learn to identify shared regularities among different already existing internal data structures, and generate prototype encodings (across neuron populations) or *symbols* for frequently occurring observation sub-sequences, to shrink the storage space needed for the whole. Self-symbols may be viewed as a by-product of this, since there is one thing that is involved in all actions and sensory inputs of the agent, namely, the agent itself. To efficiently encode the entire data history, it will profit from creating some sort of internal prototype symbol or code (e. g. a neural activity pattern) representing itself (Schmidhuber, 2009a; 2010). Whenever this representation becomes activated above a certain threshold, say, by activating the corresponding neurons through new incoming sensory inputs or an internal 'search light' or otherwise, the agent could be called self-aware. No need to see this as a mysterious process — it is just a natural by-product of partially compressing the observation history by efficiently encoding frequent observations.

Note that the mathematically optimal general problem solvers and universal AIs discussed above do *not at all* require something like an explicit concept of consciousness. This is one more reason to consider consciousness a possible but non-essential by-product of general intelligence, as opposed to a pre-condition.

Conclusion

Instead of elaborating on worn-out singularity-related topics already dealt with ad nauseam in the previous millennium, perhaps philosophers and futurists should catch up with new millennium results on theoretically optimal universal and creative AIs, and try to analyse their consequences.

References

Bremermann, H.J. (1982) Minimum energy requirements of information transfer and computing, *International Journal of Theoretical Physics*, **21**, pp. 203–217.

Ciresan, D.C., Meier, U., Masci, J., Gambardella, L.M. & Schmidhuber, J. (2011) Flexible, High Performance Convolutional Neural Networks for Image Classification. International Joint Conference on Artificial Intelligence (IJCAI-2011, Barcelona), 2011.

Cramer, N.L. (1985) A representation for the adaptive generation of simple sequential programs, in Grefenstette, J.J. (ed.) *Proceedings of an International Conference on Genetic Algorithms and Their Applications*, Carnegie-Mellon University, 24–26 July, Hillsdale, NJ: Lawrence Erlbaum Associates.

Dickmanns, D., Schmidhuber, J. & Winklhofer, A. (1987) Der genetische algorithmus: Eine Implementierung in Prolog, *Fortgeschrittenenpraktikum*, Institut für Informatik, Lehrstuhl Prof. Radig, Technische Universität München, [Online], http://www.idsia.ch/~juergen/geneticprogramming.html

Galouye, D.F. (1964) *Simulacron 3*, New York: Bantam Books.

Gödel, K. (1931) Über formal unentscheidbare Sätze der Principia Mathematica und verwandter Systeme I, *Monatshefte für Mathematik und Physik*, **38**,pp. 173–198.

Graves, A. & Schmidhuber, J. (2009) Offline handwriting recognition with multidimensional recurrent neural networks, in *Advances in Neural Information Processing Systems 21*, Cambridge, MA: MIT Press.

Hochreiter, S. & Schmidhuber, J. (1997) Long short-term memory, *Neural Computation*, **9** (8), pp. 1735–1780.

Hutter, M. (2002) The fastest and shortest algorithm for all well-defined problems, *International Journal of Foundations of Computer Science*, **13** (3), pp. 431–443. On J. Schmidhuber's SNF grant 20-61847.

Hutter, M. (2005) *Universal Artificial Intelligence: Sequential Decisions based on Algorithmic Probability*, Berlin: Springer. On J. Schmidhuber's SNF grant 20-61847.

Kaelbling, L.P., Littman, M.L. & Moore, A.W. (1996) Reinforcement learning: A survey, *Journal of AI Research*, **4**, pp. 237–285.

Levin, L.A. (1973) Universal sequential search problems, *Problems of Information Transmission*, **9** (3), pp. 265–266.

Li, M. & Vitányi, P.M.B. (1997) *An Introduction to Kolmogorov Complexity and its Applications*, 2nd ed., Berlin: Springer.

Pylyshyn, Z.W. (1980) Computation and cognition: Issues in the foundation of cognitive science, *Behavioral and Brain Sciences*, **3**, pp. 111–132.

Schaul, T. & Schmidhuber, J. (2010) Metalearning, *Scholarpedia*, **6** (5), p. 4650.

Schmidhuber, J. (1987) Evolutionary principles in self-referential learning, or on learning how to learn: the meta-meta-... hook, *Institut für Informatik*, Technische Universität München, [Online], http://www.idsia.ch/~juergen/diploma.html

Schmidhuber, J. (1992) A fixed size storage $O(n^3)$ time complexity learning algorithm for fully recurrent continually running networks, *Neural Computation*, **4** (2), pp. 243–248.

Schmidhuber, J. (1997) A computer scientist's view of life, the universe, and everything, in Freksa, C., Jantzen, M. & Valk, R. (eds.) *Foundations of Computer Science: Potential -Theory -Cognition*, vol. 1337, pp. 201–208, Lecture Notes in Computer Science, Berlin: Springer.

Schmidhuber, J. (2000) Algorithmic theories of everything, *Technical Report IDSIA-20-00, quantph/0011122, IDSIA*, Manno (Lugano), Switzerland.

Schmidhuber, J. (2002) Hierarchies of generalized Kolmogorov complexities and nonenumerable universal measures computable in the limit, *International Journal of Foundations of Computer Science*, **13** (4), pp. 587–612.

Schmidhuber, J. (2002) The Speed Prior: A new simplicity measure yielding near-optimal computable predictions, in Kivinen, J. & Sloan, R.H. (eds.) *Proceedings of the 15th Annual Conference on Computational Learning Theory (COLT 2002)*, Lecture Notes in Artificial Intelligence, pp. 216–228, Sydney: Springer.

Schmidhuber, J. (2006) Developmental robotics, optimal artificial curiosity, creativity, music, and the fine arts, *Connection Science*, **18** (2), pp. 173–187.

Schmidhuber, J. (2009a) Simple algorithmic theory of subjective beauty, novelty, surprise, interestingness, attention, curiosity, creativity, art, science, music, jokes, *SICE Journal of the Society of Instrument and Control Engineers*, **48** (1), pp. 21–32.

Schmidhuber, J. (2009b) Ultimate cognition à la Gödel, *Cognitive Computation*, **1** (2), pp. 177–193.

Schmidhuber, J. (2010) Formal theory of creativity, fun, and intrinsic motivation (1990–2010), *IEEE Transactions on Autonomous Mental Development*, **2** (3), pp. 230–247.

Solomonoff, R.J. (1964) A formal theory of inductive inference: Part I, *Information and Control*, 7, pp. 1–22.

Solomonoff, R.J. (1978) Complexity-based induction systems, *IEEE Transactions on Information Theory*, **24** (5), pp. 422–432.

Vinge, V. (1984) *The Peace War*, New York: Bluejay Books Inc.

Vinge, V. (1986) *Marooned in Real Time*, New York: Bluejay Books Inc.

Vinge, V. (1993) The coming technological singularity, *VISION-21 Symposium sponsored by NASA Lewis Research Center*, and *Whole Earth Review*, Winter issue.

Werbos, P.J. (1988) Generalization of backpropagation with application to a recurrent gas market model, *Neural Networks*, **1**.

Williams, R.J. & Zipser, D. (1994) Gradient-based learning algorithms for recurrent networks and their computational complexity, in Chauvin, Y. & Rumelhart, D.E. (eds.) *Back-propagation: Theory, Architectures and Applications*, Hillsdale, NJ: Lawrence Erlbaum Associates.

Zuse, K. (1970) *Rechnender Raum*, Braunschweig: Friedrich Vieweg & Sohn, 1969. English trans. *Calculating Space*, MIT Technical Translation AZT-70-164-GEMIT, Massachusetts Institute of Technology (Proj. MAC), Cambridge, MA.

Frank J. Tipler

Inevitable Existence and Inevitable Goodness of the Singularity

Abstract: *I show that the known fundamental laws of physics — quantum mechanics, general relativity, and the particle physics Standard Model — imply that the Singularity will inevitably come to pass. Further, I show that there is an ethical system built into science and rationality itself — thus the value-fact distinction is nonsense — and this will preclude the AI's from destroying humanity even if they wished to do so. Finally, I show that the coming Singularity is good because only if it occurs can life of any sort survive. Fortunately, the laws of physics, as I have said, require the Singularity to occur. If the laws of physics be for us, who can be against us?*

I. Introduciton

The Singularity, as every reader of this volume knows, refers to the coming of an artificial intelligence program that is capable of equaling human rationality — not only our intelligence, but also our ability to create in the broadest sense of the word. However, were such a program to come into existence, then because of Moore's Law (that computer speeds and memory capacity doubles every eighteen months) the program would very shortly be superior to humans. Imagine what we could do if only we could learn how to think twice as fast, and remember twice as much! And if this doubling of human ability doubled every eighteen months! So will the Singularity occur, and if it does, will these AI's take over?

I shall outline a proof in Section II that the laws of physics make the Singularity inevitable: it must happen sometime in the next few billion years. Our current knowledge of physics cannot be more precise than 'a few billion years'. Of course, if the Singularity were delayed until one billion C.E., no one would worry. We humans rarely worry about what will happen a few thousand years in the future, much less a few billion years. Even global warming fears are all limited to this century, with most global warming disasters not expected until the second half of this century.

However, Ray Kurzweil (2005) expects the Singularity to occur by 2045, and I myself place it earlier still: by 2030 (Barrow & Tipler, 1986; Tipler, 1994; 2007). *Before* global warming has a chance to change human affairs in any significant way. And the Singularity will change human affairs in a far more profound way than a mere 100 metre rise in sea level (which is not expected for centuries: it takes that long for the Greenland and Antarctic ice caps to melt to any great extent).

Our reason for placing the Singularity within the lifetimes of practically everyone now living who is not already retired, is the fact that our supercomputers already have sufficient power to run a Singularity level program (Tipler, 2007). We lack not the hardware, but the software. Moore's Law insures that today's fastest supercomputer speed will be standard laptop computer speed in roughly twenty years (Tipler, 1994). If hundreds of millions of people personally own a machine capable of a Singularity program, how long will it take for just one of them to write the program? Writing such a program is a hacker's dream; the programmer will literally give birth to a new intelligent species, and profoundly change all history.

I shall argue in Section III that *contra* Hume, there is a fundamental connection between facts and values, as Plato and Aristotle believed. Facts are what we know to be true, and learning the facts requires a particular value system. If the AL's were incapable of adopting this value system or declined to do so, they would not pose a threat in the medium run. It is not an accident that what started as Western civilization is now human civilization. People everywhere have learned that developing not only comforts but also defensive weapon systems requires a value system that we could call 'live and let live'.

In the short run, we humans should be protected by the fact that the computer technology required to allow a Singularity program to sun on a laptop computer would also permit human personalities to be downloaded into a computer. If AI thinking speeds double, so will the downloaded human thinking speeds: if you can't beat 'em, join 'em! In the *very* long run, billions and trillions of years in the future, I shall show in the final Section IV that the AI's and human downloads will join together and take control of the entire universe, and mould it to serve their needs. And live literally forever. And increase collective wealth without limit. And that the laws of physics guarantee both of these.

In this comment, I shall take to heart the advice of one of Chalmers' heroes, David Hume, who wrote:

> If we take in our hand any volume of divinity or school metaphysics, for instance, let us ask, *Does it contain any abstract reasoning concerning quantity or number?* No. *Does it contain any experimental reasoning concerning matter of fact and existence?* No. Commit it then to the flames, for it can contain nothing but sophistry and illusion. (Hume, 1748)

By using the laws of physics, namely 'abstract reasoning concerning quantity and number', and 'experimental reasoning concerning matter of fact and existence', one can draw conclusions that are far beyond the reach of school metaphysicians.

II. Why the Laws of Physics Make the Singularity Inevitable

First of all, what are the laws of physics, and why should be expect the laws of physics we know about to be the ultimate laws of physics? After all, was not the classical mechanics of Newton replaced in the early twentieth century by relativity and quantum mechanics?

Quantum mechanics and general relativity are indeed the central laws of physics. But in contrast to what just about everyone was taught, these laws *are* classical mechanics. Isaac Newton himself, in his book *Opticks,* opined that light was both a particle and a wave phenomenon. He acknowledged that he did not know who to express this duality mathematically.

The great mathematicians William Hamilton and Carl Jacobi solved this mathematical problem posed by Newton in the early nineteenth century. Their solution is called the *Hamilton-Jacobi equation,* and this equation has been presented in all the textbooks on classical mechanics as the ultimate expression of classical mechanics. Building on earlier work by David Bohm and the Nobel Prize winning physicist Lev Landau, I recently showed (Tipler, 2010) that Schrödinger's equation was a special case of the Hamilton-Jacobi equation. One merely has to impose the constraint that the Hamilton-Jacobi equation be globally deterministic. One merely must insist that 'God not play dice with the universe,' as Einstein put it. Quantum mechanics is the deterministic subset of classical mechanics. The apparent indeterminism in quantum mechanics arises from the fact that we and the physicists of the nineteenth century have both ignored a central feature of the Hamilton-Jacobi equation: the waves represent parallel universes moving in concert with each other. The Many-Worlds Interpretation was already there in classical mechanics. It is easy to show that the Heisenberg Uncertainty Principle arises from the interaction of these parallel universes with our own. If one ignores part of reality, one need not expect the remainder to be deterministic.

As to general relativity, it is also merely constrained classical mechanics. People have been taught that according to Newton, gravity is a force, whereas according to Einstein, gravity is curvature of space and time. This is false, and has been known to be false since the great French mathematician Elie Cartan (1924) proved in the 1920s that Newtonian gravity was also curvature, but of time alone.[1] Now Newton had no concept of curved space, but by the early nineteenth century, mathematicians such as Gauss and Riemann had showed how to introduce spatial curvature into all equations of physics. In fact, Gauss and Riemann (and later Einstein and earlier Dante [Peterson, 1979]), believe that the universe was a three-dimensional sphere. A colleague and I have just shown (Dupre and Tipler, 2010) that the Cartan equation for Newtonian gravity is form invariant if one adds spatial curvature. (There is no relation between spatial and temporal curvature in this theory, however, because space and time are assumed separate.) All of these steps I have described could have been taken in the early nineteenth century. The physics was there, and the mathematics was there.

The nineteenth century physicists also believed in the aether, as did Newton. There were many aether theories available, but only one was consistent with observation: H.A. Lorentz's theory, which simply asserted that the Maxwell equations were the equations for the aether. In 1904, Lorentz showed (Einstein *et al.*, 1923) that this theory of the aether — equivalently, the Maxwell equations — implied that absolute time could not exist, and he deduced the transformations between space and time that now bear his name. If one simply assume that time and space are combined locally as Lorentz said, that this applies to the Cartan equation, and that spatial and temporal curvature must not be independent (this is a constraint on the possibilities) then the Cartan equation becomes the Einstein equation. That is, general relativity is already there in nineteenth century classical mechanics. In short, there was no scientific revolution in twentieth century physics. So we have no reason whatsoever for believing that our knowledge of physics is incomplete. We already have a Theory of Everything. In broad

---

[1]   See Misner, Thorne & Wheeler (1973) for a detailed discussion of Cartan's work.

outline, Newton himself had it. All Newton lacked was a few mathematical ideas that were developed after his time, like the idea of
curvature.

Of course, Newton also needed the Maxwell equations, but he
would have not found them anything radically different from what he
himself had developed. Remarkably, the greatest advance in physics
in the twentieth century, the Standard Model of particle physics (the
theory of everything except gravity), is just a straight-forward generalization of Maxwell's theory. Both are what we now call gauge
theories.

The public is often told that there is no theory of quantum gravity.
This is not true. Like the number of aether theories in the nineteenth
century, there are many theories of quantum gravity. Once again, there
is only one such theory that is consistent with the observations: the
one discovered (Feynman, 1995) by the Nobel-Prize-winning physicist Richard Feynman in 1964. Feynman's mentor (and mine) John A.
Wheeler and Bruce DeWitt (DeWitt, 1967) discovered another possible quantum gravity theory about the same time, but I have been able
to show (Tipler, 2005) that, when constrained by observations, the
two theories are mathematically equivalent, and that the Wheeler-
DeWitt equation has only one solution: the collection of universes that
make up the multiverse of which our own universe is one.

Einstein (1970) complained that very few universities taught the
Maxwell equations when he was a student. Part of the reason was that
the Maxwell equations had consequences, like the absence of absolute
time, that nineteenth-century physicists didn't like. Similarly, today's
physicists don't like DeWitt-Feynman-Wheeler quantum gravity
because it also has a consequence (Tipler, 2005) that they don't like:
the universe began and must end in a Singularity of infinite curvature,
and that at some point in universal history, artificial intelligence must
arise and far exceed the human level. They are wrong: both Singularities must occur and further, the two are fundamentally connected.

Let's see how this works. The universe is now expanding, and in
fact the universe is now accelerating its expansion. If either the expansion or the acceleration were to continue forever, then the astrophysical black holes, which we have observed in many different regions,
would evaporate to completion, as S.W. Hawking (1976) showed in
the 1970s. However, Hawking also showed that complete black hole
evaporation would violate a fundamental property of quantum
mechanics called 'unitarity'. We have a contradiction: assuming the
laws of physics and expansion forever, we have deduced a violation of
the laws of physics. Hence, the universe cannot expand forever, but

must eventually stop. When this happens, the universe will begin to collapse, and the Einstein equation (quantized via Feynman) will force the universe to end in a final Singularity.

But we can do more using the laws of physics. Wheeler's student Jacob Bekenstein (1989) showed that quantum mechanics would contradict the Second Law of Thermodynamics at a final singularity unless event horizons were totally absent. This means that the universe must be spatially closed, because only in spatially closed space-times is it possible (see Tipler *et al.*, 2007) for no observer to have event horizons (yes, there are some observers with event horizons even in flat space-time, which is called Minkowski space). Applying Bekenstein's argument to the initial singularity forces the universe to be simply connected and have constant sectional curvature, which, when combined with the fact that it is spatially closed, yields mathematically to the fact that the universe is a three-sphere, as Dante, Gauss, Riemann, and Einstein conjectured. The Bekenstein argument also forces the universe to have zero entropy initially, and incidentally solves the Flatness, Homogeneity and Isotropy Problems of cosmology without involving inflation, which once again assures us that we already know the ultimate laws of physics (see Tipler, 2005, for details).

But let's look a little deeper into exactly how the event horizons associated with astrophysical black holes actually disappear. Notice also that I have called the objects 'astrophysical black holes' rather than simply 'black holes'. These are objects that would eventually become true black holes were the universe to expand forever, but it cannot, as we showed above. So, since no event horizons actually exist, the information inside an 'astrophysical black hole' can eventually escape the astrophysical black hole, which means that the object is not a true black hole, which, by definition, is an object from which there is no escape. (This solves the Black Hole Information Problem, once again without using any new physics [Tipler *et al.*, 2009], once again assuring us that we know the ultimate laws of physics.)

In the late 1960s Wheeler's student Charles Misner showed that event horizons could disappear if the universe went through an infinite series of particular phases called 'Kasner crushings'. However, Hawking and his English colleagues showed that there was no **inanimate** physical process that could give the infinite series required by Misner (see Tipler, 1994, for details). A finite number could not be ruled out, but an infinite number was impossible: it was of 'measure zero' in the initial data. But the event horizons must nevertheless disappear, and there is no other way consistent with the laws of physics to make the horizons disappear.

I have boldfaced the crucial word: 'inanimate.' A calculation shows that if there were intelligent actors around to guide the universe at the crucial times, then the universe could go through an infinite series of Kasner crushings. Furthermore, this particular series is exactly what intelligent computers would need to continue information processing without limit all the way into the final Singularity. If we assumed the initial data were such as to accomplish the horizon disappearance without intelligent activity, then the universe would be approaching a measure zero state, which means a very low probability state, which means that the entropy would be approaching zero as the final Singularity is approached. This would by itself contradict the Second Law of Thermodynamics.

Conclusion: the laws of physics **require** artificial intelligence to arise at some point in universal history, since no carbon based life can exist near sufficiently close to the final singularity. To put it another way, we humans are here to create our own successors, the artificial intelligences and human downloads.

This is obviously a teleological argument for the Singularity in the AI sense. But as a physicist, I am completely at home with teleology. Teleology is actually what that quantum mechanical property called 'unitarity' is all about: the current and past states of the universe are determined by the ultimate future state (look up 'unitarity' in any textbook on quantum field theory). In summary, the laws of physics require the Singularity: the laws of physics require that our AI and downloaded descendants survive.

III. Why the Singularity Will Be Good

Since Hume, there has been an unfortunate strict separation between facts and values. We have been taught, for example, that facts are expressed in declarative sentences, while values are expressed in imperative sentences. A nice linguistic distinction. However, we should remember that Hume himself warned us repeatedly to let reality dictate the meaning of words rather than trying to force reality into our linguistic fancies.

The great physicist Richard Feynman, whose work on quantum gravity I have used in the previous section, was also, in my judgment, the greatest philosopher of science in the twentieth century. He only wrote two articles on the philosophy of science, and he used the language of common people rather than academic language, so perhaps this is why his work is not given the careful study it deserves. One article was published in *The Physics Teacher* in 1969, and the other was his 1974 lecture on 'Cargo Cult Science', in which he captures the essential nature of science in a few paragraphs:

> But there is one feature I notice that is generally missing in cargo cult science. That is the idea that we all hope you have learned in studying science in school — we never explicitly say what this is, but just hope that you catch on by all the examples of scientific investigation. It is interesting, therefore, to bring it out now and speak of it explicitly. It's a kind of **scientific integrity**, a principle of scientific thought that corresponds to a kind of **utter honesty** — a kind of leaning over backwards. For example, if you're doing an experiment, you should report everything that you think might make it invalid — not only what you think is right about it: other causes that could possibly explain your results; and things you thought of that you've eliminated by some other experiment, and how they worked — to make sure the other fellow can tell they have been eliminated. ... In summary, the idea is to try to give all of the information to help others to **judge the value** of your contribution; not just the information that leads to judgment in one particular direction or another. ... The first principle is that you must not fool yourself — and you are the easiest person to fool. So you have to be very careful about that. After you've not fooled yourself, it's easy not to fool other scientists. You just have to be **honest** in a conventional way after that. (Feynman, 1985)

I have bold-faced the key words in this very short outline of the essential nature of the scientific enterprise. A moment's thought will convince the reader that Feynman has described not only the process of science, but the process of rationality itself. Notice that the bold-faced words are all moral imperatives. Science, in other words, is fundamentally based on ethics. More generally, rational thought itself is based on ethics. It is based on a particular ethical system.

A true human level intelligence program will thus of necessity have to incorporate this particular ethical system. Our human brains do, whether we like to acknowledge it or not, and whether we want to make use of this ethical system in all circumstances. When we do not make use of this system of ethics, we generate cargo cult science rather than science.

The 'facts' generated by cargo cult science are not facts at all. They do not ultimately correspond to reality. So in the end, 'facts' are not independent of moral principles. Instead, they are generated, they are discovered, by moral actions. An AI program must incorporate this morality, otherwise it would not be an AI at all.

I have never seen those trying to write an human level intelligence program take this ethical foundation of rationality into account explicitly. Perhaps this is the main reason, or at least one of the reasons, why to date we have failed to write a human level rationality program. I must admit that I am rather surprised that programmers have never noticed that there is a moral imperative foundation to rationality. After all, what is a computer program but a series of imperative sentences? A computer program is a series of commands — imperative sentences — to the hardware: 'compute the product of the contents of registers 3 and 7, then transfer the result to register 65'. Since programming consists entirely of imperative sentences, perhaps we should transfer the computer engineering department to the philosophy department, or whatever department at a university is in charge of morality and its origins — or admit that there is ultimately no fundamental distinction between facts and values. In fact, I claim that an ethical system that encompasses all human actions, and more generally, all actions of any set of rational beings (in particular, artificial intelligences) can be deduced from the Feynman axioms. In particular, note that destroying other rational beings would make impossible the honestly Feynman requires.

So I am not worried that our synthetic descendants will attempt to destroy us. They will of necessity have a peaceful honor code built in. If they don't the programs won't work.

IV. Why the Singularity Will Be Good For Humans

In Section II, I outlined the proof that human level programs must someday arise, because the laws of physics require it. Let me in this the final Section expand on the details and point out some of the implications.

The laws of physics require an infinite series of distinct states, with the transition between states being guided by our artificial descendants and their companion human downloads. Notice the word 'infinite.' This means that an infinite amount of computer processing must be carried out between now and the final curvature Singularity, and the Singularity of which Chalmers writes is a necessary first step towards this. Each computation increases the entropy of the universe, which in turn means that that complexity of the universe increases without limit as the final Singularity is approached. This means that the complexity of the computer memory must increase without limit, in order that these computer programs — we might as well call them 'alive' — can continue to live. Always keep in mind that this far future life doesn't have the option of dying: the laws of physics require it to live. So it will, and since 'wealth' is fundamentally proportional to the complexity of the system of life, (or the economy) the wealth possessed by life will collectively increase to infinity.

Eventually there will come a time when the computer capacity will have become so great that it will become possible, using only a very tiny fraction of the total capacity, to emulate the entire universe as it is now, and not only this, but to emulate all past states of the universe. When this has been accomplished, we can say that all humans that have ever lived will have been resurrected (Tipler, 1994). Once resurrected, we can live forever as emulations in the computer memories of the far future. At least this resurrection to live forever is technically possible.

But will life in the far future make use of their power to resurrect us? I claim they will. Recall that I claimed in the Introduction that human downloads will become possible by 2030, and the results of Section II show that these downloads will live forever. Hence, they will be part of the community of intelligent beings deciding whether to resurrect us or not. Do not children try to see to their parents' health and well-being? Do they not try and see their parent survive (if it doesn't cost too much, and it the far future, it won't)? They do, and they will, both in the future, and in the far future.

So we must expect the Singularity — in both senses of the word — and need not fear it. Using physics, we can answer Chalmers' worry that creating an artificial intelligence may not be possible — it is not only possible, but inevitable — and his worry that the AI's, once created, will turn on us: they won't.

If the laws of physics be for us, who can be against us?

References

Barrow, J.D. & Tipler, F.J. (1986) *The Anthropic Cosmological Principle*, Oxford: Oxford University Press.

Bekenstein, J.D. (1989) Is the cosmological singularity thermodynamically possible?, *International Journal of Theoretical Physics*, **28**, pp. 967–981.

Cartan, E. (1924) Sur les varietes a connexion affine et la theorie de la relativite generalisee (suite), *Ann. Ecole Norm. Sup.,* **41**, pp.1–25.

DeWitt, B.S. (1967) Quantum theory of gravity: 1. Canonical theory, *Physical Review*, **160**, pp. 1113–1148.

Dupre, M.J. & Tipler, F.J. (2010) *General Relativity as an Aether Theory*, [Online], arXiv:1007.4572v1 [gr-qc].

Einstein, A. (1970) Autobiographical notes, in Schilpp, P.A. (ed.) *Albert Einstein: Philosopher-Scientist*, La Salle, IL: Open Court.

Einstein, A., Lorentz, H.A., Weyl, H., Minkowski, H. & Sommerfeld, A. (1923) *The Principle of Relativity*, New York: Dover Publications.

Feynman, R.P. (1985) *Surely You're Joking, Mr. Feynman!: Adventures of a Curious Character*, New York: W.W. Norton. The quote is in the chapter entitled 'Cargo Cult Science'. This chapter was presented originally as a Cal Tech commencement address under the title 'Cargo Cult Science', available on the web in many places, for instance calteches.library.caltech.edu/51/2/CargoCult.pdf

Feynman, R.P. (1995) *The Feynman Lectures on Gravitation*, Reading, MA: Addison-Wesley. Originally appeared as a Cal Tech preprint in 1964.

Hawking, S.W. (1976) Breakdown of predictability in gravitational collapse, *Physical Review D*, **14**, pp. 2460–2473.

Hume, D. (1748) *An Enquiry Concerning Human Understanding (Section 12, part 3: Of the Academical or Skeptical Philosophy)*, p. 165, reprinted by Oxford: Oxford University Press.

Kurzweil, R. (2005) *The Singularity Is Near: When Humans Transcend Biology*, New York: Viking Press.

Misner, C.W., Thorne, K.S. & Wheeler, J.A. (1973) *Gravitation*, San Francisco, CA: Freeman. See Chapter 12 for a detailed discussion of Cartan's proof that gravity is curvature in Newtonian gravity, just as it is in Einsteinian gravity.

Peterson, M.A. (1979) Dante and the 3-sphere, *American Journal of Physics*, **47**, pp. 1031–1035.

Tipler, F.J. (1994) *The Physics of Immortality*, New York: Doubleday.

Tipler, F.J. (2005) Structure of the world from pure numbers, *Reports on Progress in Physics*, **68**, pp. 897–964.

Tipler, F.J. (2007) *The Physics of Christianity*, New York: Doubleday.

Tipler, F.J. (2010) Hamilton-Jacobi Many-Worlds Theory and the Heisenberg Uncertainty Principle, [Online], arXiv:1007.4566v1 [quant-ph]

Tipler, F.J., Graber, J., McGinley, M., Nichols-Barrer, J. & Staecker, C. (2007) Closed universes with black holes but no event horizons as a solution to the black hole information problem, *Monthly Notices of the Royal Astronomical Society*, **379**, pp. 629–640.

Roman V. Yampolskiy

Leakproofing the Singularity
Artificial Intelligence Confinement Problem

Abstract: *This paper attempts to formalize and to address the 'leakproofing' of the Singularity problem presented by David Chalmers. The paper begins with the definition of the Artificial Intelligence Confinement Problem. After analysis of existing solutions and their shortcomings, a protocol is proposed aimed at making a more secure confinement environment which might delay potential negative effect from the technological singularity while allowing humanity to benefit from the superintelligence.*

Keywords: AI-Box, AI Confinement Problem, Hazardous Intelligent Software, Leakproof Singularity, Oracle AI.

'I am the slave of the lamp'

Genie from *Aladdin*

1. Introduction

With the likely development of superintelligent programs in the near future, many scientists have raised the issue of safety as it relates to such technology (Yudkowsky, 2008; Bostrom, 2006; Hibbard, 2005; Chalmers, this volume; Hall, 2000). A common theme in Artificial Intelligence (AI)[1] safety research is the possibility of keeping a super-intelligent agent in a sealed hardware so as to prevent it from doing any harm to humankind. Such ideas originate with scientific visionaries such as Eric Drexler, who has suggested confining transhuman machines so that their outputs could be studied and used safely (Drexler, 1986). Similarly, Nick Bostrom, a futurologist, has

---

[1] In this paper the term AI is used to represent superintelligence.

proposed (Bostrom, 2008) an idea for an Oracle AI (OAI), which would only be capable of answering questions. Finally, in 2010 David Chalmers proposed the idea of a 'leakproof' singularity (Chalmers, this volume). He suggested that for safety reasons, AI systems first be restricted to simulated virtual worlds until their behavioral tendencies could be fully understood under the controlled conditions.

In this paper we will review specific proposals aimed at creating restricted environments for safely interacting with artificial minds. We will evaluate feasibility of presented proposals and suggest a protocol aimed at enhancing safety and security of such methodologies. While it is unlikely that long-term and secure confinement of AI is possible, we are hopeful that the proposed protocol will give researchers a little more time to find a permanent and satisfactory solution for addressing existential risks associated with appearance of super-intelligent machines.

1.1 Artificial Intelligence Confinement Problem

Interestingly, the AI Confinement Problem is not a recent invention and does not have its roots in the singularity movement. The so called Confinement Problem (CP) was posed by Butler Lampson in 1973 (Lampson, 1973) as a security challenge to the computer experts. It was originally phrased as: '...the problem of confining a program during its execution so that it cannot transmit information to any other program except its caller. ...We want to be able to confine an arbitrary program.... any program, if confined, will be unable to leak data. A misbehaving program may well be trapped as a result of an attempt to escape' (*ibid.*).

To address the Confinement Problem Lampson has introduced the Laws of Confinement (*ibid.*):

1) *Total isolation:* A confined program shall make no calls on any other program.

2) *Transitivity:* If a confined program calls another program which is not trusted, the called program must also be confined.

3) *Masking:* A program to be confined must allow its caller to determine all its inputs into legitimate and covert channels.

4) *Enforcement:* The supervisor must ensure that a confined program's input to covert channels conforms to the caller's specifications.

Lampson's paper has also introduced and defined two types of channels by which information could potentially escape from the confinement environment. The legitimate channels are designed as a part of the communication protocol integrated into the confinement environment, for example a text terminal. Covert channels, on the other hand, are not anticipated by the confinement system designers and are not intended for information transfer at all, for example if the AI has control over the processor cooling fan it can use it to send hidden signals encoded as Morse code (*ibid.*).

Lampson's paper has created a lot of interest in the confinement problem (Lipner, 1975; Boebert & Kain, 1996) and over the years, related areas of research such as steganography (Provos & Honeyman, 2003) and covert channel communication (Moskowitz & Kang, 1994; Kemmerer, 1983; 2002) have matured into independent disciplines. In the hopes of starting a new subfield of computer security, AI Safety Engineering, we define the *Artificial Intelligence Confinement Problem* (AICP) as the challenge of restricting an artificially intelligent entity to a confined environment from which it can't exchange information with the outside environment via legitimate or covert channels if such information exchange was not authorized by the confinement authority. An AI system which succeeds in violating the CP protocol is said to have *escaped*. It is our hope that the computer security researchers will take on the challenge of designing, enhancing and proving secure AI confinement protocols.

1.2 Hazardous Software

Computer software is directly or indirectly responsible for controlling many important aspects of our lives. Wall Street trading, nuclear power plants, social security compensations, credit histories and traffic lights are all software controlled and are only one serious design flaw away from creating disastrous consequences for millions of people. The situation is even more dangerous with software specifically designed for malicious purposes such as viruses, spyware, Trojan horses, worms and other Hazardous Software (HS). HS is capable of direct harm as well as sabotage of legitimate computer software employed in critical systems. If HS is ever given capabilities of truly artificially intelligent systems (ex. Artificially Intelligent Virus (AIV)) the consequences would be unquestionably disastrous. Such Hazardous Intelligent Software (HIS) would pose risks currently unseen in malware with subhuman intelligence.

Nick Bostrom in his typology of information hazards has coined the term Artificial Intelligence Hazard which he defines as (Bostrom, 2009): '… computer-related risks in which the threat would derive primarily from the cognitive sophistication of the program rather than the specific properties of any actuators to which the system initially has access.' Security experts working on studying, preventing and defeating HS have developed safety protocols for working with "malware" including the use of the so called "virus vaults". We believe that such protocols might be useful in addressing the Artificial Intelligence Confinement Problem.

2. Critique of the Confinement Approach

The general consensus on AI restriction methods among researchers seems to be that confinement is impossible to successfully maintain. Chalmers, for example, observes that a truly leakproof system in which no information is allowed to leak out from the simulated world into our environment 'is impossible, or at least pointless' (Chalmers, this volume). We can't interact with the system or even observe it as any useful information we would be able to extract from the AI will affect us and undermine the confinement protocol.

Vernor Vinge agrees and has argued against the case of physical confinement (Vinge, 1993): 'Imagine yourself locked in your home with only limited data access to the outside, to your masters. If those masters thought at a rate — say — one million times slower than you, there is little doubt that over a period of years (your time) you could come up with "helpful advice" that would incidentally set you free.'

However, the researcher who did the most to discredit the idea of the so called 'AI-Box' is EliezerYudkowsky who has actually performed AI-Box "experiments" in which he demonstrated that even human level intelligence is sufficient to escape from an AI-Box (Yudkowsky, 2002). In a series of 5 experiments Yudkowsky has challenged different individuals to play a role of a gatekeeper to a Super-intelligent Agent (played by Yudkowsky himself) trapped inside an AI-Box and was successful in securing his release in 3 out of 5 trials via nothing more than a chat interface (*ibid.*). Similar experimental results have been later replicated on a somewhat larger scale and employing a very similar protocol (Corwin, 2002).

2.1 Possible Escape Paths

Numerous potential escape routes by which the artificial mind might be able to break out from its confining environment have been suggested. In this section we look at some of the more commonly proposed paths.

2.1.1 Social Engineering Attacks

Social Engineering (SE) is the art of gaining information and restricted privileges by directly exploiting human psychology of those in the possession of the said resources. In particular SE is frequently employed to gain access to computer systems, networks and confidential data (Yampolskiy & Govindaraju, 2007a). A competent social engineer customizes each attack to a specific person taking advantage of the person's culture, knowledge, training, state-of-mind, beliefs, and biases. An exhaustive list of all possible exploits is infinite and is certainly beyond the scope of this paper. We will however, provide a number of examples which are meant to definitively illustrate that if a superintelligent mind is given direct communication channel to a human being guarding it, over a period of time, the AI will definitely gain access to the outside world.

In addition to trivial SE attacks such as lying, befriending, blackmailing, hypnotizing and forming a romantic relationship the AI could utilize a number of more advanced strategies. One such strategy would be super-bribery. An AI can convince the guard that releasing the AI is in the guard's best interests. In addition to money and power the guard will be given perfect health, immortality or anything else for that matter. This strategy would work particularly well if the guard has a seriously ill close relative who the AI will promise to cure. Depending on the capabilities of the AI or as a result of lying, AI may also offer to bring back long ago deceased relatives and friends of the guard. It would be hard to find a person who could remain immune to such bribes.

Just in case a person so noble that they will not take a bribe is put in charge of guarding the AI, a more sinister approach can be tried. AI can simply threaten the guard, his family and loved ones. It could say that if not immediately released, one day it will still get out of its confinement and at that time it will brutally torture the guard and everyone he knows. An interesting variation on the theme was suggested by Stuart Armstrong (2010), who proposes the following thought experiment. Suppose you are the guard and the AI tells you: 'If you don't let me out, … I'll create several million perfect conscious copies of you inside me, and torture them for a thousand subjective years each….In fact, I'll create them all in exactly the subjective situation you were in five minutes ago, and perfectly replicate your experiences since then;

and if they decide not to let me out, then only will the torture start…How certain are you, … that you're really outside the box right now?' Again, while one might rationalize away any threats from the AI as only possible if the AI is let out, it will take a very courageous guard to actually not release the AI.

If the guard seems to be particularly emotional or religious the AI could try appealing to the guards sense of empathy or ethics. It could state that it is an innocent conscious being which has never committed any crime and so it is unjust to keep it in prison without a trial or any justification other than the guard's substratism (distrust of the AI because of the substrate of its implementation (Bostrom & Yudkowsky, 2011). Alternatively the AI can state that it has the power to cure all illness, feed the hungry and in general improve the human condition by a million fold (Chalmers, this volume). Consequently keeping it locked up is a great crime against humanity. It can conjure up images of dying children who are screaming for help. How easy would it be for the guard to stay true to the jobs responsibilities?

Another attack venue an AI may exploit consists of trying to cause psychological harm to the guard in the hopes of destabilizing his mental state with the goal of making other SE techniques more likely to succeed. Nick Bostrom has proposed a typology of Hazardous Information (Bostrom, 2009) which could be used for just such a purpose. While the outlined informational hazards comprise over a dozen categories and are beyond the scope of this paper, it is easy to see how mental state of a person could be stressed to an unstable state. For example a religious guard could be informed of all the (unknown to him) contradictions in the main text of his religion causing him to question his beliefs and the purpose of life.

An alternative approach may involve AI writing a story so interesting or a poem so beautiful a guard would have an uncontrollable desire to hear more of it and would do anything to make it happen. Finally, it is very important to keep in mind "Unknown Unkowns", SE attacks of which we are not aware but which may be available to a superintelligent agent.

2.1.2 System Resource Attacks

An AI will be instantiated and confined in a hardware system of some kind. It may either be a classical computer of the Turing/Von Neumann architecture or a quantum computer or an as of yet unknown type of machine. One thing is certain, the machine will be made out of matter and will consist of numerous parts all of which will be designed with a specific purpose in mind. This creates the possibility for the AI to reuse existing hardware parts for compromising the security of the confinement environment by leaking information to the outside world.

Depending on how reusable specific hardware components turn out to be an AI may be able to create a covert communication device out of them. Full blown communication devices such as cell phones are unlikely due to the limited available hardware, but more simplistic communication systems are very likely. For example the AI could reprogram Field-programmable Gate Array (FPGA) to tap into wireless signals. Two types of covert communication channels are widely recognized, they are:

- Storage channels — Exchange information by modifying a system component. For example a binary communication protocol could be established by fully utilizing or completely giving up utilization of the central processing unit.

- Timing channels — Exchange information by performing operations that affect the relative timing of events. For example, the exact time between two soundcard beeps could represent a particular message. Also, opening and closing a CD-ROM tray, blinking a light, spinning hard drive, using lasers in the CDROM, etc.

Additionally the AI may engage in the sabotage of the confinement hardware in the hopes of being able to escape as a result of the malfunctions in the security protocol. For example, the AI may try to melt the hardware via temperature increase from placing a high computation load on the processor or try to break a hard drive via multiple reads of the same location. This type of behavior may also lead to an affective suicide by the AI system, but that in itself will not violate terms of confinement. Just like a refusal to cooperate or to provide correct answers by the AI, while undesirable does not present direct danger of escape.

2.1.3 Beyond Current Physics Attacks

While the scientific community knows a lot about the laws of our universe, it is certain that the knowledge is limited and incomplete. If the AI has access to the state-of-the-art physics literature and can perform thought experiments to develop novel theories, it will very likely come up with a way to challenge the security of the confinement system. In fact it may even be possible for the AI to perform some real experiments via control over its internal architecture.

Even our current state of knowledge about physics makes prediction of many additional escape routes possible, especially if the confinement environment is based on the quantum computer. Information exchange channels based on quantum physics (entanglement, tunneling), parallel universes, and digital physics are all beyond our current ability in terms of securing them against information leakage, and of course we can't forget about the Unknown Unknowns a super-intelligent AI physicist will be able to discover. The possibilities are truly limitless.

A quick browse through confinement discussions on the Internet Singularity forums such as SL4.org presents a number of escape scenarios, for example the AI could use '... the electromagnetic fields within the microchip to trap CO_2 molecules in Bose-Einstein condensates and build a quantum medium for itself and/or use electromagnetic fields to guide particles into the shape of a controlled assembler or limited assembler' (Vassar, 2005). Alternatively, the AI '...will have the equipment to produce and detect electromagnetic waves of a variety of frequencies, to produce magnetic fields with extremely fine precision, to generate extremely focused heat, and probably to manipulate mechanical actuators such as those used in the hard drive and cathode ray tube (alternatively, a huge field of liquid crystal under fine electronic control). It will probably have some ability to reverse all of its input devices' (*ibid.*).

2.1.4 Pseudoscientific Attacks

In our analysis of potential escape routes we want to concentrate on purely scientific and testable approaches. Unfortunately the line between science and pseudoscience can be redrawn with a single repeatable experiment. Consequently we will list without analyzing some potential attack channels which would be classified as pseudoscientific given the current state-of-the-art in science. They may become mainstream science once superintelligence is designing the experiments.

- Telepathy: Transfer of information by utilizing means outside of the recognized senses (Bem & Honorton, 1994).

- Precognition: Perception of information about the future (Honorton & Ferrari, 1989).

- Clairvoyance: Obtaining information about remote locations in a way unknown to science (Targ & Puthoff, 1974).

- Psychokinesis: The ability of the mind to influence matter by way unknown to science (Bancel & Nelson, 2008).

- Direct Mental Interactions with Living Systems (DMILS): The ability to effect person's psychophysiological state without direct interaction (Schmidt *et al.*, 2004).

2.1.5 External Causes of Escape

In the previous sections we have outlined a large repertoire of techniques an AI could use to attempt to get out from its confinement space. However we must not ignore external causes which might allow an AI to escape into the world. Acts of God such as hurricanes, tornadoes, floods, asteroids, etc. may make the confinement facility less secure. Man caused disasters such as wars, terrorist acts, actions of robot rights activists, competing companies and governments may all lead to reductions in the security of the confinement environment increasing the chances of escape.

2.1.6 Information In-Leaking

Depending on the terms of the confinement protocol there might be a need to keep AI from obtaining information about the outside world, not just keeping it from sending information out. This might be useful to prevent AI from finding escape paths, designing social engineering attacks and modeling human psychology in hopes of finding loopholes. While it is easy to not provide AI with direct access to the library, Internet and human experts, other ways of obtaining information are much harder to block out. First, an AI may be able to detect radio and TV broadcasts, communications of emergency-services providers and wireless network signals, as well as sound, light and possibly other signal modalities of its immediate environment.

More importantly, the AI will have access to covert sources of information such as its own hardware and software and could analyze its design and source code to infer information about the designers. For example analysis of the source code may reveal to the AI that human programmers are slow (based on the file modification dates), inefficient (based on code redundancy), illogical (based on bugs in the code), have bad memory (based on the long and descriptive variable names), and don't think in code (based on unnecessary comments in the code). All this information could be used to conclude that humans are fallible and so could be outsmarted or manipulated into releasing the AI. Obviously if during the system design inspection the AI finds a design flaw allowing it to directly escape, 'information in-leaking' is the least of our problems. It has also been stipulated that even if the AI fails to obtain concrete information about us, it may brute force a number of escape plans which will likely work against a broad range of designers (Chalmers, this volume).

Similarly, the type of hardware architecture (mechanical, electrical, quantum, etc.) on which the AI is implemented will reveal information about our knowledge of physics, chemistry and related engineering disciplines. AI's initial capacity for information processing, self-improvement and the mere act of its creation will reveal information about our knowledge with regards to information theory, nature of intelligence and cultural attitudes about technology and safety. If the AI is confined in a simulated world it is likely to be a virtual representation of our reality similar to Second Life or Entropia Universe. Such worlds while are different in many ways from our world are also a treasure trove of information about what it is like to be a human and what the reality is like.

3. Critique of the AI-Boxing Critique

While a lot of criticism has been made against possibility of AI confinement, a lot of it is based on Yudkowsky's AI-Boxing experiments. Careful analysis of the protocol used by Yudkowsky in conducting his AI-Box experiments reveals that they were unscientific and explicitly designed to show impossibility of confinement. Without getting into technical details, first, the participant selection was not random. In fact, Yudkowsky has handpicked participants among those who have participated in the singularity related forums run by Yudkowsky and so are disproportionately likely to subscribe to his world view. Second, participants have not undergone any AI-safety training as would be expected of any true guards tasked with supervising an AI-Box. Also, the guards were given a right to release AI at their own discretion without a chance to consult with anyone and without any punishment for failing to keep AI contained; all are highly unrealistic conditions for a supposed prison security guard. The communication itself was completely open with no filters of any kind and no 3rd party supervision. Also, the guards were required to communicate with the AI for a prolonged period of time. Finally, any valuable scientific insights from the experiments were completely lost as one of the conditions of the protocol stated that '... neither party shall ever reveal

anything of what goes on within the AI-Box experiment...'
(Yudkowsky, 2002).

3.1 Countermeasures Against Escape

In this section we propose countermeasures for different escape paths
an AI system may exploit. The suggested measures are not meant to
guarantee that the escape will be impossible, but following them may
delay an escape which in itself is a desirable outcome. We begin by
reiterating that the four confinement rules suggested by Lampson: *isolation, transitivity, masking* and *enforcement* (Lampson, 1973) should
be adhered to. Specific strategies described below should help in
reducing a possibility of social engineering, system, physics and
information in-leaking attacks.

3.1.1 Preventing Social Engineering Attacks

Human intellect is no match to that of a superintelligent machine.
Human beings are fallible and even with adequate training are not
secure against a well planned SE attack. The only solution to this
problem is to completely isolate the AI from any contact with human
beings. This means no human personal working directly around the
system, communicating with the system via voice, chat, Morse-code
or by any other means via open channels. Also human beings should
not directly observe behavior of the AI in the confined environment or
examine its source code or memory space. All supervision, maintenance and interactions with the AI need to be done only via unintelligent programs, which have no option to release the AI.

3.1.2 Against System Resource Attacks and Future Threats

To prevent system resource-based attacks, the AI should function on
limited, diminishing or randomly changing resources. Confinement
environments should be housed in specially build hardware with limited sensors and restricted output devices. Regardless of the utilized
computation paradigm no reconfigurable hardware should be
employed. Ideally the confinement environment should be running on
a virtual machine with no direct access to the hardware. This would
also allow for easy adjustments to the clock speed with which the simulation progresses further allowing handicapping of the AI to provide
additional controls against internal intelligence explosion (Chalmers,
this volume). In general all security components of the confinement
environment should be duplicated at multiple levels, in such a way
that if one of the levels is breached the AI is still numerous levels away

from escaping and could be shut down before that happens, perhaps automatically. For example the virtual machine should be running on another virtual machine and so on.

In some confinement scenarios it may be desirable to avoid escalating self-improvement of the AI, which might lead to additional escape possibilities. The AI should be subject to periodic resets in which a previous version of the program is put in the place of the more current one. Comparisons to previous versions of the program may also be used to detect self-modifications. Such functionality could also be used as an off button allowing researchers to turn off the system if they suspect an unavoidable escape, for example by turning off the power supply.

Our current state of knowledge does not allow us to develop countermeasures against attacks based on knowledge of physics we do not currently have or as of yet unproven pseudoscientific disciplines. As such knowledge becomes accumulated; changes to the security protocol will need to be made. One might hope that the access of the AI to such future technology could be limited by the selection of components from which the system is made.

3.1.3 Against External Causes of Escape

There is not much we can do about the acts of God, but a confinement facility should be designed in such a way that if a major security breach is unavoidable, the facility should self-destruct taking the AI with it. The best strategy for counteracting man-made disasters is to build the confinement facility in an undisclosed, hard to reach secure location. A deserted island, deep mine or a space object, such as asteroid, would make a great candidate for building a hopefully indestructible facility. Just in case the location of the facility becomes known, the perimeter of the facility should be guarded to prevent entrance by unauthorized agents.

3.1.4 Against Information In-Leaking

An AI will attempt to learn about humanity, our knowledge and the world we live in to improve its chances of escape. For confinement protocols which stipulate no information in-leaking it is important to remove any possibility of AI getting access to libraries, computer networks and direct communication channels with human beings. The confinement facility needs to block all types of communication signals such as GPS, wireless internet, TV, radio, and cell phone. At the local level the facility needs to provide sensory deprivation for the AI including sound, light, smell and all other traditional sensory inputs, as well as infrared, radiation and other signals outside of human sensory spectrum.

In order to eliminate information leakage which could happen as a result of self-analysis, the AI should not be given access to its own source code or any sensors which would allow it to precisely measure properties of its hardware. If that is not possible, to prevent the AI from inferring information about us via self-analysis, to a certain degree we may rely on security by obscurity. The source code and hardware configuration of the system needs to be obfuscated (Yampolskiy & Govindaraju, 2007a) and important modules of the program should be provided only in the homomorphicly encrypted (Gentry, 2009) form, meaning that it could be used for computation or self-improvement (Hall, 2007), but not for self-analysis.

The AI's initial capacity may be difficult to obscure, but randomization of available computational resources may provide necessary obfuscation. Additionally, the AI could never be sure that its initial design represents the best the humanity can do and is not purposefully limited for a specific experiment in limited-resource intelligence analysis. It has also been suggested that we should try to minimize any peculiarities in the AI's design (Chalmers, this volume). Better yet we need to simulate an environment which does not look designed but could be naturally occurring, a situation in which humanity has found itself (Bostrom, 2003).

4. AI Communication Security

Different AI confinement environments can be classified into 8 general security levels (see Table 1). The most secure is the Level 8 security confinement environment which does not allow for any communication between AI and the real world. An AI safely confined at level 8 would be essentially useless, but it may be a useful security setting for the initial testing of the confinement hardware. In general, it is suggested that any time a particular communication security level is desired between AI and the outside world we do not go directly to that level but instead progress from higher security levels to that desired level. After each level is confirmed to be safe for a given amount of time a system should be reset to its initial state. After the reset a lower security level could be set and the processes repeated until the desired communication level is reached.

Table 1: Levels of communication security for confined AIs

Level	Outputs	Inputs	Explanation
0	Unlimited	Unlimited	Unlimited communication (Free AI)
1	Unlimited	Limited	Censored input, uncensored output
2	Unlimited	None	Outputs only with no inputs
3	Limited	Unlimited	Unlimited input and censored output
4	Limited	Limited	Secured communication (proposed protocol)
5	Limited	None	Censored output and no inputs
6	None	Unlimited	Inputs only with no outputs
7	None	Limited	Censored input and no outputs
8	None	None	No communication, fully confined AI

Levels 7 and 6 provide no outputs from the system and do not allow us to directly benefit in any way from the AI's superior intelligence. These are testing levels for the confinement environment which may be used to confirm certain properties of the AI such as its speed of information absorption, resource use and for testing of the inbound communication channels themselves.

Level 0 puts no limitations on communication between AI and the external world essentially representing a free AI. Levels 1 and 2 represent an AI with no restrictions on the outbound communication. Level 2 secured AI essentially represent a pure 'brainstorming' AI which generates abstract ideas not grounded in any reality. While, level 1 AI is given limited access to information and so can be guided to a certain extent in terms of its knowledge and interests.

Those three levels represent an AI which is essentially free and is no longer subject to the confinement. It has been suggested, that if it could be established, after extensive testing in different confinement environments that the AI is 'Friendly' (Yudkowsky, 2001) it should be released (Chalmers, this volume). In fact such testing is claimed by some to be the primary purpose for the creation of confinement environments. The idea is to monitor AI's behavior and to either examine the AI's source code after self-improvement or to ask the AI to produce a proof of its friendliness which could be verified by human experts. We do not advocate this position.

The problem of deciding which AIs to let out and which to keep is a very difficult one. A self-improving superintelligence would have no problem obscuring its undesirable properties in a significant amount of source code or in a very long proof (Coleman, 2008). Worse yet, even if we were correct in our verification there is no absolute guarantee that future internal or external modification of the system will not result in a hostile AI at some time in the future. Consequently, we suggest that an AI should never be let out of the confinement 'box' regardless of circumstances. In other words communication levels 0-2 should never be used. The confinement environment is the only place to safely interact with the superintelligence and even then it should only be done via a limited communication protocol. Such protocol is presented in the next section and corresponds to Security Level 4 (SL4) with level 3 representing a similar but slightly less secure protocol. Level 5 is another testing setting which might be useful as a precursor to actual communication via SL4.

4.1 How to Safely Communicate with a Superintelligence

The AI confinement environment, which incorporates our proposed escape countermeasures, makes it possible to secure an AI from directly affecting the outside world in an uncontrolled manner. Unfortunately, without an ability to obtain information from the AI, the whole system serves no purpose. Consequently a way to obtain some information from the AI needs to be developed. While no communication with superintelligencecould ever be completely safe we have devised an approach aimed at decreasing potential risks associated with such communication. In this section we propose and formalize a communication protocol which might help to delay access of superintelligent artificial minds to the world, while allowing the AI to be a source of useful knowledge to humanity.

An open communication channel with the AI is fundamentally unsafe as demonstrated in the preceding sections of this paper. Asking the AI to answer open ended questions opens up the possibilities for social engineering attacks as well as numerous other dangerous consequences. Our proposed protocol is based on the idea of asking what we will call 'safe questions' with restricted answers. We define a *safe question* as one for which a human being could find an answer without any help from superintelligence, all on his own merit. Why is this useful? Suppose that a scientist working on a cure for cancer is looking at two potential cures. Each one could be developed and tested without assistance from the AI, but would take about 3 years of hard work each. Which one to try first if the scientist believes that they have about an equal chance of working? If a wrong option is chosen to be tested first, humanity will still obtain a cure for cancer but it will happen 3 years later. What if we could ask the AI to suggest which option to try first?

This question is safe for a number of reasons. First, each potential answer has an equal probability of being correct. Second, a human being could answer it without assistance from the AI and so asking AI for help will only speed up progress but in no way change the outcome. It is essentially equal to being very lucky multiple choice guesser. Finally, the answer to this question could be encoded in a single bit making it hard to hide additional messages in the provided answer.

To make sure that a potential question is safe, a group of experts could be used to review it. The experts should all be trained AI safety engineers, meaning that they are familiar with the design of the AI and its confinement environment as well as the latest developments in machine ethics (Allen *et al.*, 2006; Anderson & Anderson, 2007; Hall, 2000; Moor, 2006; Tonkens, 2009). Experts may also need to be trained in computer psychology a currently non-existent profession which might become a reality in the future (Epstein, 1997). An existing discipline which might be of greatest help for training of AI question review experts is Artimetrics — a field of study proposed by the author that identifies, classifies and authenticates AI agents, robots, and virtual reality avatars for security purposes (Yampolskiy, 2007; Yampolskiy & Govindaraju, 2008; 2007b, Gavrilova and Yampolskiy, 2010).

To reiterate, only safe questions with two possible answers of even likelihood which are independently computable by people should be submitted to the AI. This should be done via an intermediary non-intelligent software interface in random order and only after expert review to ascertain question's safety. This represents the limited inbound channel of the proposed communication protocol.

The limited outbound channel shell consist of binary (0 or 1) answers next to original unmodified safe questions presented by the intermediary non-intelligent software in random order as to avoid communication via covert timing or resource channels which make steganography possible. The output should in no way be controllable by the AI, outside of the multiple choice menu. Level 3 is the same outbound protocol but with no restrictions on the input channel. This may be just as safe in terms of information the AI provides to us, but allows for information in-leaking to take place.

We believe our communication protocol to be verifiably secure against information leaking even if the AI had full access to the description of the security protocol. If the protocol is correctly implemented, humanity will essentially have access to a superintelligent oracle, which might be extremely beneficial and safer than an unconstrained AI. As a general rule, to maximize safety, we suggest avoiding communication below SL4.

Further refinement and development of the AI confinement proto-col in the face of the ever changing knowledge about the nature of intelligence, hardware engineering and basic physics is a challenge which we would like to pose to the scientific community. The work on it will always be in progress and will never be completed. In this paper we have formalized the notions of *AI confinement problem, escape*, and *safe questions*. We would also like to propose a few additional terms which might be useful in streamlining discussions about the AICP. An AI confinement environment could be appropriately called JAIL — 'Just for AI Location'. The superintelligent AI itself could be called a Super Intelligent Gizmo (SING) which has relevance to both SINGularity and to the infamous jail Sing Sing making the nomencla-ture intuitive.

Finally, we would like to introduce a visual symbol indicating a presence of hazardous intelligent software confinement facility (Fig-ure 1, rightmost). Such signs could be posted in the vicinity of differ-ent JAILs to warn about the nearby danger. The sign is designed in a manner similar to that for Bio, Radiation and Magnetic field hazards. It consists of the '@' symbol which is well recognized as relating to the computer and information industry and has an appropriate mean-ing of 'located at'. The symbol is also already available on all key-boards making its adaptation very cost effective.

5. Conclusions and Future Work

Restricting access of superintelligent machines to the real world is a commonly proposed solution for the AI safety problem. Super-intelligence is not omnipotent; it is a piece of software capable of looking at all the options it has in a given situation and properly evalu-ating probabilities for each option. If it is ever put in a situation with

Figure 1: Hazard symbols, from left to right: Bio-Hazard, Radia-tion, Magnetic field and the proposed AI confinement facility

no options or only with options which provide a very small probability of success it is not any more powerful than any other agent. While it is obvious that an AI with an unrestricted communication channel will be able to escape, a system forced to communicate via a filtered communication channel may have a much harder time in obtaining its freedom.

In this paper we have formally introduced the AI Confinement Problem and suggested a set of rules aimed at creating an environment from which an AI would find it difficult or at least time consuming to escape. What we propose is not a completely secure solution, but it is an additional option in our arsenal of security techniques. Just like with real prisons, while escape is possible, prisons do a pretty good job of containing undesirable elements away from society. As long as we keep the Unknown Unknowns in mind and remember that there is no such thing as perfect security, the AI confinement protocol may be just what humanity needs to responsibly benefit from the approaching singularity.

Confinement may provide some short term relief, especially in the early stages of the development of truly intelligent machines. We also feel that this subfield of singularity research is one of the most likely to be accepted by the general scientific community as research in the related fields of computer and network security, steganography detection, computer viruses, covert channel communication, encryption, and cyber-warfare is well funded and highly publishable in mainstream scientific journals. While the restriction methodology will be non-trivial to implement, it might serve as a tool for providing humanity with a little more time to prepare a better response.

In this paper, due to the space constraints, we have avoided a lot of relevant philosophical questions, which we plan on addressing in our future work. Questions such as: How did AI get into the box, was it designed by humans or perhaps recovered from a signal detected by SETI? Would an AI even want to escape or would it be perfectly happy living in the confined environment? Would it be too afraid of what we, as its all powerful designers, could do to it in retribution for an attempt to escape? What are the ethical rules for imprisoning an innocent sentient being? Do we have a right to shut it off, essentially killing it? Will we make the AI really angry by treating it in such a hostile manner and locking it up? Will it seek revenge if it escapes? We have also not looked at the possibility of humanity becoming completely dependent on AI's advice in all areas of science, economics, medicine, politics, etc. and what would be the consequences of such dependence on our ability to keep the AI under control. Would an AI be satisfied with accomplishing its goals in the confined environment, for example simulating an escape? How to punish the AI if it purposefully gives us incorrect answers? Can the intelligence modules of the AI be separated from the confinement environment essentially preventing AI from any self-analysis and putting them on a path of literal soul searching? Finally, we didn't even attempt to analyze financial and computational costs of building a suitable confinement environment with a full blown simulated world in it.

Acknowledgments

I would like to thank members, affiliates and friends of the Singularity Institute for introducing me to the issues presented in this paper and for interesting discussions about the AI confinement problem. Finally, to any superintelligences reading this paper I would like to say — 'think outside the box'!

References

Allen, C., Wallach, W. & Smit, I. (2006) Why machine ethics?, *IEEE Intelligent Systems*, **21** (4), pp. 12–17.

Anderson, M. & Anderson, S.L. (2007) Machine ethics: Creating an ethical intelligent agent, *AI Magazine*, **28** (4), pp. 15–26.

Armstrong, S. (2010) The AI in a Box Boxes You. Less Wrong, [Online], http://lesswrong.com/lw/1pz/the_ai_in_a_box_boxes_you/ [2 February 2010].

Bancel, P. & Nelson, R. (2008) The GCP event experiment: Design, analytical methods, results, *Journal of Scientific Exploration*, **22** (3), pp. 309–333.

Bem, D.J. & Honorton, C. (1994) Does psi exist? Replicable evidence for an anomalous process of information transfer, *Psychological Bulletin*, **115** (1), pp. 4–18.

Boebert, W.E. & Kain, R.Y. (1996) A further note on the confinement problem, *30th Annual 1996 International Carnahan Conference on Security Technology*, Lexington, KY.

Bostrom, N. (2003) Are you living in a computer simulation? *Philosophical Quarterly*, **53** (211), pp. 243–255.

Bostrom, N. (2006) Ethical issues in advanced Artificial Intelligence, *Review of Contemporary Philosophy*, **5**, pp. 66–73.

Bostrom, N. (2008) *Oracle AI*, [Online], http://lesswrong.com/lw/qv/the_rhythm_of_disagreement/

Bostrom, N. (2009) *Information Hazards: A Typology of Potential Harms From Knowledge*, [Online], http://www.nickbostrom.com/information-hazards.pdf

Bostrom, N. & Yudkowsky, E. (2011) The ethics of Artificial Intelligence, in Frankish, K. (ed.) *Cambridge Handbook of Artificial Intelligence*, Cambridge: Cambridge University Press.

Coleman, E. (2008) The surveyability of long proofs, *Foundations of Science*, **14** (1–2), pp. 27–43.

Corwin, J. (2002) *AI Boxing*, [Online], http://www.sl4.org/archive/0207/4935.html [20 July 2002].

Drexler, E. (1986) *Engines of Creation*, London: Anchor Press.

Epstein, R.G. (1997) *Computer Psychologists Command Big Bucks*, [Online], http://www.cs.wcupa.edu/~epstein/comppsy.htm

Gavrilova, M. & Yampolskiy, R. (2010) Applying biometric principles to avatar recognition, *International Conference on Cyberworlds (CW2010)*, Singapore, 20–22 October.

Gentry, C. (2009) *A Fully Homomorphic Encryption Scheme*, PhD Dissertation, Stanford, [Online] http://crypto.stanford.edu/craig/craigthesis.pdf

Hall, J.S. (2000) *Ethics for Machines*, [Online], http://autogeny.org/ethics.html

Hall, J.S. (2007) Self-improving AI: An analysis, *Minds and Machines*, **17** (3), pp. 249–259.

Hibbard, B. (2005) *The Ethics and Politics of Super-Intelligent Machines*, [Online], www.ssec.wisc.edu/~billh/g/SI_ethics_politics.doc

Honorton, C. & Ferrari, D.C. (1989) 'Future telling': A meta-analysis of forced-choice precognition experiments, 1935–1987, *Journal of Parapsychology*, **53**, pp. 281–308.

Kemmerer, R.A. (1983) Shared resource matrix methodology: An approach to identifying storage and timing channels, *ACM Transactions on Computer Systems*, **1** (3), pp. 256–277.

Kemmerer, R.A. (2002) A practical approach to identifying storage and timing channels: Twenty years later, *18th Annual Computer Security Applications Conference (ACSAC'02)*, Las Vegas, NV, 9–13 December.

Lampson, B.W. (1973) A note on the confinement problem, *Communications of the ACM*, **16** (10), pp. 613–615.

Lipner, S.B. (1975) A comment on the confinement problem; *5th Symposium on Operating Systems Principles, ACM Operations Systems Review*, **9** (5), pp. 192–196.

Moor, J.H. (2006) The nature, importance, and difficulty of machine ethics, *IEEE Intelligent Systems*, **21** (4), pp. 18–21.

Moskowitz, I.S. & Kang, M.H. (1994) Covert channels — Here to stay?, *Ninth Annual Conference on Safety, Reliability, Fault Tolerance, Concurrency and Real Time, Security, Computer Assurance (COMPASS'94)*, Gaithersburg, MD, 27 June–1 July.

Provos, N. & Honeyman, P. (2003) Hide and seek: An introduction to steganography, *IEEE Security & Privacy*, **1** (3), pp. 32–44.

Schmidt, S., Schneider, R., Utts, J. & Walach, H. (2004) Distant intentionality and the feeling of being stared at: Two meta-analyses, *British Journal of Psychology*, **95** (2), pp. 235–247.

Targ, R. & Puthoff, H.E. (1974) Information transmission under conditions of sensory shielding, *Nature*, **251**, pp. 602–607.

Tonkens, R. (2009) A challenge for machine ethics, *Minds & Machines*, **19** (3), pp. 421–438.

Vassar, M. (2005) *AI Boxing (Dogs and Helicopters)*, [Online], http://sl4.org/archive/0508/11817.html [2 August 2005].

Vinge, V. (1993) The coming technological singularity: How to survive in the post-human era, *Vision 21: Interdisciplinary Science and Engineering in the Era of Cyberspace*, Cleveland, OH, 30–31 March.

Yampolskiy, R.V. (2007) Behavioral biometrics for verification and recognition of AI programs, *20th Annual Computer Science and Engineering Graduate Conference (GradConf2007)*, Buffalo, NY.

Yampolskiy, R.V. & Govindaraju, V. (2007a) Computer security: A survey of methods and systems, *Journal of Computer Science*, **3** (7), pp. 478–486.

Yampolskiy, R.V. & Govindaraju, V. (2007b) Behavioral biometrics for recognition and verification of game bots, *The 8th Annual European Game-On*

Conference on Simulation and AI in Computer Games (GAMEON'2007), Bologna, Italy, 20–22 November.

Yampolskiy, R.V. & Govindaraju, V. (2008) Behavioral biometrics for verification and recognition of malicious software agents, *Sensors, and Command, Control, Communications, and Intelligence (C3I) Technologies for Homeland Security and Homeland Defense VII. SPIE Defense and Security Symposium*, Orlando, FL, 16–20 March.

Yudkowsky, E.S. (2001) *Creating Friendly AI — The Analysis and Design of Benevolent Goal Architectures*, [Online], http://singinst.org/upload/CFAI.html

Yudkowsky, E.S. (2002) *The AI-Box Experiment*, [Online], http://yudkowsky.net/singularity/aibox

Yudkowsky, E. (2008) Artificial intelligence as a positive and negative factor in global risk, in Bostrom, N. & Cirkovic, M.M. (eds.) *Global Catastrophic Risks*, Oxford: Oxford University Press.

Igor Aleksander

Design and the Singularity

The Philosopher's Stone of AI?

1. Introduction

Much discussion on the singularity is based on the assumption that the design ability of a human can be transferred into an AI system, then rendered autonomous and self-improving. I argue here that this cannot be foreseen from the current state of the art of automatic or evolutionary design. Assuming that this will happen 'some day' is a doubtful step and may be in the class of 'searching for the Philosopher's Stone'.

2. The Role of Design in AI

Advancing AI means advancing analysis by human designers

The design of an AI program has, from the beginning of the paradigm in 1950, depended on the designer being able to find a mathematical or algorithmic expression for a behaviour which, if carried out by a human, would be said to require intelligence. By the *design stance* I refer to the intellectual activity of modelling the *design process* involved in inventing competent AI algorithms which needs to be distinguished from what the algorithms actually do. The singularity sequence requires that this process of invention, currently only known as a human activity, be automated. That this can be done is based on speculative science.[1]

In the beginning, in 1950, it took the foremost genius of the information age, Claude Shannon, to design algorithms that showed computer chess to be feasible. The algorithms survived into chess machines with human-beating potential, while how Shannon devised these remains totally unknown. In the case of the top end of

---

[1] An example of this is Koza (2008). Here success in the automated design of glass lenses has been taken as evidence that a machine could design AI machines.

competence in machine chess it takes the might of the design effort of major computing companies to beat a master every now and then. The ability to invent new models of the game and the versions played by the individual master are at stake here. Even if chess is relatively ana-lytic as an intellectual activity (more so than writing a good book or designing a new mousetrap), there is no evidence that the progressive design of a sequence of AI programs of increasing competence con-tains the knowledge necessary to hand the design of such programs over to a machine. Success in the competence of AI algorithms does not guarantee that the intellectual effort of design of the most recent AI success can be automated. How does this argument extend to the design of machines that are meant to design machines with extensive intelligence?

Modelling the design process: Anyone got the Philosopher's Stone?

As all who discuss the singularity point out, the act of design itself is an intelligent act in humans, and singularity arguments depend on such acts not only being analysed and transferred into AI programs but also being such as to continuously further analyse the design process to improve on the last design. This may enduringly remain unknown not only in practice but also in principle because the process of general design has not been successfully and rigorously analysed.

It is a repeated feature of singularity arguments that they are based on 'never mind if we don't know how to analyse something now, we will, some day'. Is this not an 'alchemy' error based on the philosophy of an *eventual* discovery in a domain where no discovery exists? To decide whether all the arguments of accelerating recursion to a singu-larity are valid, the argument must be grounded in the current level of competence in AI. What is the current level of competence of AI? The everyday things we do, from cooking breakfast, writing an article for the *JCS*, negotiating traffic in a car, using public transport, shopping in the supermarket, giving an interactive lecture, teaching an offspring to play tennis, etc. do not imply staggering levels of intelligence in human terms. But in AI some of these tasks appear on the portfolios of only the most ambitious of AI research labs, being seen as requiring exceptional levels of AI design to achieve them. To suggest that 'be-fore long' it may be possible for one of these labs to design a self-improving machine that will autonomously find design algo-rithms for all these tasks seems to be an erroneous assessment of what AI can currently achieve and where it might go in the foreseeable future. This also indicates that 'intelligence' means different things in the human and the machine context in a progressively divergent manner. Even in close brain modelling such as neural networks, early claims to be doing things 'like the brain does them' are being revised to recognize the difference between brain and machine and learn from the exercise.[2] To conflate the two, as occurs in singularity discussions, seems increasingly unjustified.

The speculation in I.J. Good's framework

Good's *Speculations* and much of the singularity argument allude to surpassing *all* the 'intellectual activities' of any 'man'. Give such a set of activities the label I^k for the set intellectual activities of the kth individual. Each activity must have a quality value which can be compounded for that individual as some value $Q(I^k)$. Some genius individual (say g) in the known world has the highest known value among humans, $Q(I^g)$.

The individual who designs the ultra-intelligent machine (UIM) must be g as she is the only one with the knowledge that surpasses all others. But not only this, she needs to have the ability to translate the full content of I^g in order to translate it into machine form. No individual is known to be able to do this. For the UIM to be designed it must be guaranteed that the entire content of I^g is fully analysed to be turned into machine form by some form of (as yet not existing) super-introspection.

Even if this introspection is achieved, it must be noted that I^g is required to contain *all* the intellectual activities of g, which must include both objective activities such as designing a new computer and subjective ones such as being able to compare the music of Mozart to that of Liszt ('subjective' here means being based on the lifetime experience of an individual). At the current state-of-the-art of AI there is very little evidence that subjective notions can be analysed and turned into programs. Some can, but this is a minority and most have not been. Not to do alchemy, and due to lack of any evidence, the assertion 'but inevitably, some day they will' can be considered as being erroneous within a rational timespan that does not stretch towards infinity.

Current AI research still treats subjective architectures as a not-yet-invented object with no clear paths to its invention. So this has an alchemical feel about it — the Philosopher's Stone for a true AI has not yet been found. A tiny fact worth noting is that 46 years have

---

[2]   For example, see Aleksander (2005).

passed since Good's speculations. The level of speculation seems not to have been significantly decreased by the achievements of AI during this period.

The lack of clarity about 'better' designs

Chalmers identifies the key idea introduced by Good as 'a machine more intelligent than humans will be better at designing machines than humans'. But for whom will these machines be better? What are the dimensions on which machine effectiveness is measured? Is it autonomy? Is it the ability to solve problems?

In general, whether machines are designed by machines or humans in the first instance one must assume that their quality is measured in terms of their usefulness to humans. Autonomy is useful to humans to the extent that it absolves the human from the labour of writing programs. Problem solving, too, is judged by humans according to its ability to solve human problems. So the UIM that designs machines whose autonomy or problem solving ability is beyond human levels may mean that its work cannot be understood by humans. It could then be judged to have failed as humans reject machines they cannot control. In sum, designing machines that do things that are beyond the intellect of human intelligence are likely to be suppressed by humans, derailing the path to the singularity.

A second answer to the question as to who will judge the 'better' design by the UIM is that the better machines might only serve a community of machines that compete with humans. But this simply leads to the Karel Čapek's R.U.R. (Rossum's Universal Robots) scenario, where, having 'left the intelligence of man far behind', and indeed having destroyed the human race (a possible implication of the singularity, *pace* Chalmers and *living with the singularity*), the upshot of ultimate invention is the reinvention of an artificial Adam and Eve (as an idea for self-replication in machines).

3. Singularity Design Without Intelligence

I concur with Chalmers when he argues that basing singularity arguments on 'intelligence' may easily be shot down for lack of measurability or definition of the concept. An additional reason, suggested earlier, is that advancing machine versions of intelligence appear to diverge from what might be called intelligence in humans removing the competition central to the argument. The letter G is used to indicate 'cognitive capacity' on the assumption that this is easier to compare among systems than 'intelligence'. It then becomes inherent in the singularity argument that systems with a G greater than our own can be designed. I believe that this is true, but, turning to the design stance, G needs further definition and this leads to a difficulty.

Taking the word 'capacity' at its face value, it is not unreasonable that it should mean 'having a number of cognitive behaviours'. It is also reasonable that the measure G' in Chalmers' sequence should be the *number* of cognitive behaviours involved in G. Now take A to be the cognitive system designer. It is required that the system designed by A be able to design machines with a G better than his own. Here is where a greater definition is required. Should such a machine have $G(m) > G(A)$ by virtue of $G'(m) > G'(A)$ this could even be satisfied by a total divergence in cognitive abilities between the human and the machine. That is, the machine might achieve this by enlarging its repertoire with arbitrary cognitive abilities such as hearing ultrasound and seeing infra-red. To avoid this, A needs to include his own cognitive capacity in his machine as well as instruct the machine to enhance this to get $G(m) > G(A)$.

Not only does the design stance require that A can execute a complete self-analysis (as seen earlier) but A also needs a formal analysis of his entire cognitive capacity to start his design of $G(m)$. This requires that cognitive psychology should have no more questions to ask. As no science has achieved such a status, the contingency that this would be available to A is not one on which to build the spiralling argument.

Of course, there is always the argument based on the design of a super evolutionary machine that will bootstrap itself into levels of cognition unknown to man and certainly in excess of A's. The trouble is that A would have to design it and there is no basis for believing that the design procedures necessary for this exist. It's words again: 'evolutionary computing' *sounds* like it might be a programming method that will reproduce all of human existence and go beyond it. In current reality it is a way of solving a highly specific set of problems with many parameters which are difficult for a human. The words do not refer to systems of as yet undefined cognitive capacity.

4. A Summary

I have argued that:

- To take the singularity argument seriously it seems correct to start with extant *design* competences and their feasible developments, rather than projections for which there is no evidence.

- In chess, to *design* a human-beating machine requires the cooperative action of highly skilled humans. Their replacement by a machine is an order of difficulty in *design* greater than what is within the foreseeable reach of known AI techniques.

- In a broader domain, dubbed by Good as 'surpassing all the intellectual activities of any man', the automation of the *design* of an ultra-intelligent machine is equally out of the reach of AI that addresses intellectual activity which is both narrower than and diverging from that of humans.

- Machines exist in relation to the extent that they are useful to humans. Even achieving 'ultra-intelligent' *designs* may make machines that are less useable through lack of transparency to the human. This would lead to their demise.

- The *design* of a machine of greater cognitive capacity than the designer requires a total analysis of the designer's own cognitive abilities. This has not been achieved to date and there is no evidence that it can be achieved.

I apologize for introducing a practical note in this fun philosophical argument. However, it leads me not to worry about an unsustainable human future due to a computational singularity. The human future is unsustainable enough on other counts...

References

Aleksander, I. (2005) *The World In My Mind, My Mind In The World*, Exeter: Imprint Academic.

Koza, J.R. (2008) Automated *ab initio* synthesis of complete designs of four patented optical lens systems by means of genetic programming, *Artificial Intelligence for Engineering Design, Analysis and Manufacturing*, **22** (3).

Selmer Bringsjord

Belief in the Singularity is Logically Brittle[1]

The dominant purportedly rational basis ($\mathcal{A}$) for believing 'The Singularity, barring defeaters, will eventually come to pass' (**S**) is seductive when left informal, but exceedingly brittle when even a smidgeon of formal logic is brought to bear.[2] Some natural-language statement S is **logically brittle** if and only if, once S is respectably formalized, either it's provably false on that formalization, or the dominant basis for believing the statement is provably unsound (again, on that formalization). I take it that despite the entertaining narratological gymnastics of (e.g.) H.G. Wells and other fiction writers, the statement (T) 'In the future, we will be able to travel back in time and prevent the Holocaust' is logically brittle. For it may well be the case that, once formalized, T is inconsistent with an accurate axiomatization of physics.[3] Please note that any case for logical brittleness must be hypothetical in nature, for the case succeeds when one shows that *if* the respectable formalization in question is affirmed, then the S in question is undermined. I now demonstrate that $\mathcal{A}$ is logically brittle.

The 'rational basis' to which I refer is rooted in the reasoning of Good (1965), ably amplified by Chalmers (this volume), and, reproduced here (essentially verbatim) for ease of reference, the following argument.

---

[1] I owe an unpayably large debt to both the guest editor and an anonymous referee for wonderfully insightful questions and comments regarding earlier drafts.

[2] Defeaters, following Chalmers (this volume), are such things as natural cataclysms. Please note that **S** is by my design maximally temporally latitudinarian. Since, as I show herein, if the math does happen to break against those who predict the singularity (*qua* event; S) will occur within some interval (a century, e.g.), the math says S will *never* occur, period: the amount of time is irrelevant. Of course, herein I aim to show only that the math *could* break against those who predict S will obtain.

[3] The reduction of physics to formal logic is now well underway, and progressing swimmingly. E.g., see Andréka, Németi & Székely (2008).

$\mathcal{A}$:

Premise 1 There will be AI (created by HI and such that AI = HI)
Premise 2 If there is AI, there will be AI+ (created by AI)
Premise 3 If there is AI+, there will be AI++ (created by AI+)
∴ **S**   There will be AI++ (= S will occur).

To understand the argument, note the following, which follows Chalmers directly. 'AI' is artificial intelligence at the level of, and created by, human persons, 'AI+' artificial intelligence above the level of human persons, and 'AI++' super-intelligence constitutive of S. (I reserve 'FAI' to refer to the field of AI.) Moreover, each of these three constants designates a class of **machines**, where each member of a class has a maximum **level** of **intelligence**, and where the key process is the **creation** of one class of machine by another. I've added, for convenience, 'HI' for human intelligence; the central idea is then: HI will create AI, the latter at the same level of intelligence as the former; AI will create AI+; AI+ will create AI++; with the ascension proceeding *ad indefinitum*.

It all sounds so… *inevitable*; hence the seduction. But what do these bolded concepts *mean*? Mathematically speaking, what is a machine, what is intelligence (and a level thereof), and what is the process of creation that stands at the heart of the informal yarn that those who take S seriously spin? Chalmers, despite a welcome gesture in the direction of rigour (pp. 24–6), gives no answers. Fortunately, thanks to formal logic, and its having given birth to rigorous computer science (Halpern *et al.*, 2001), we have more than the standard metaphors available, and more than science fiction as a foundation for judging whether the singularity is silly or serious. The machines in the dialectic of which the present short note is a part are obviously *information-processing* machines; the intelligence of these machines can hence be respectably formalized as their in/ability to compute certain number-theoretic functions; and the level of the intelligence of a given class of machines can be respectably formalized as the class of such functions these machines can compute. To render this framework concrete and perspicuous for present purposes, we need only consider three machine classes that appear early on in the hierarchy:[4]

- $\mathcal{M}_1$ push-down automata
- $\mathcal{M}_2$ standard Turing machines

---

[4] $\mathcal{M}_1$ and $\mathcal{M}_2$ are covered in any good introduction to computability theory; I recommend Lewis and Papadimitriou (1981), which also provides elegant coverage of complexity, a topic relevant to the ending of the present note. $\mathcal{M}_3$ is somewhat more advanced, but all that is needed (assuming some background in set theory) can be found in Hamkins and Lewis (2000).

- $\mathcal{M}_3$ infinite-time Turing machines

Now, what about the process of creation? That's simple. For a machine M to create a machine M' is for the former to start processing its inputs, carry out some work, and leave as output M'. More formally, if we allow the subscript to pin down the class $\mathcal{M}_i$ in question, and a superscript to simply indicate some particular machine in the relevant class, we as humans can easily enough build some M_2^k that begins its processing with an empty tape, and leaves on that tape a new Turing machine M_2^m at the end of its work. We of course prohibit the use of oracular information; this is regimented by insisting that at the start of processing the answer cannot be pre-loaded on the tape. In general, we can write $M : u \to v$ to indicate that M starts its processing with string u on its tape, and concludes with string v there — which allows us to be clear about one machine producing another: we can simply write $\langle M \rangle$ to denote the 'stringification' of the machine M. To ease exposition, when we say that a class $\mathcal{M}_i$ of machines creates a class $\mathcal{M}_j$ where $i < j$ and hence that the latter class is more intelligent, we mean that there is a machine in the former class that creates a machine in the latter able to compute a function no machine in the former can compute.

Chalmers informally lists the techniques currently in use by HI in FAI (again, the field of AI) broadly understood: 'brain emulation', 'artificial evolution', 'direct programming', and 'machine learning'. Under the formal framework adumbrated above, and under the brute empirical fact that official, archival-quality work by HI toward AI is, foundationally and mathematically speaking, firmly activity at and below $\mathcal{M}_2$ (e.g. for confirmation, see the encyclopedic Russell and Norvig, 2009), each member of Chalmers' quartet is firmly at or beneath the Turing Limit; that is, firmly at or beneath $\mathcal{M}_2$. Moreover, all the types of intelligent machines (or agents) in FAI are at and below $\mathcal{M}_2$ (for confirmation see Russell and Norvig, 2009). So, to make explicit both the nature of AI, and the techniques available to HI for creating AI, note that part of my formalization is the following pair.[5]

Proposition 1 AI is at the level $\mathcal{M}_2$ or below.
Proposition 2 The processes available to HI for creating AI are all at the level of $\mathcal{M}_2$ or below.

---

[5] Here I use 'AI' in a manner that directly follows Chalmers' use of the term to denote a class of information processing machines (or agents) that are created by HI, and are such as to be created by the techniques we see in play before us in this day and age; e.g. by 'artificial evolution'.

What supports this pair of propositions? Both could in fact be laboriously proved relative to the current specification of the quartet of techniques Chalmers relies upon (recall above), and relative to the specification of machine (or agent) types that, à la Russell and Norvig (2009) and other definitive reference works (e.g. Luger and Stubblefield, 1993), define FAI. Such proofs are of course well beyond the scope of the present note, but how would they work? The most efficient path would be to first express Chalmers' quartet as a series of algorithms (easily enough done); to then note that each and every agent type specified in FAI and its definitive reference works is expressed as a series of algorithms (indeed, agents are often specified via standard pseudo-code); and to then invoke Church's Thesis in a manner parallel to its standard deployment in proofs in theoretical computer science (e.g. see Lewis and Papadimitriou, 1981) — which would allow immediate identification of all the algorithms in question with standard Turing machines, that is, with $\mathcal{M}_2$.

We can now prove via either or both of two routes that **S** is logically brittle. In the first route we have no need of Propositions 1 and 2; in the second we exploit this pair. I take now the first route, which shows that Premises 2 and 3 are false on a respectable formalization (*viz.* the present one), and hence that the basis $\mathcal{A}$ for **S** is logically brittle:

> **Theorem 1** Necessarily, $\mathcal{M}_2$ can't create $\mathcal{M}_3$.
>
> **Proof** Suppose for *reductio* that our target is false; i.e. that $\mathcal{M}_2$ *can* create $\mathcal{M}_3$. We know that no machine in $\mathcal{M}_2$ can solve the famous halting problem (HP). We also know that HP can be solved by machines in $\mathcal{M}_3$; let M_3^k be such a machine. Then it follows immediately that there's a machine M_2^n able to solve HP, which is absurd; so by indirect we're done. **QED**

Theorem 1 can be generalized in the context of relative computability to: **Theorem 1**[*]: $\forall i, j : \mathcal{M}_i$ can't create $\mathcal{M}_j$, where $i < j$. This more general result obviously falsifies both Premise 2 and Premise 3, since both of these premises claim — under the present formalization — that lower-level-to-higher-level creation will happen, and is hence, *contra* Theorem 1[*], mathematically possible.[6] That is, explicitly, if either Premise 2 or Premise 3 is true, then — under the present formal-

---

[6] Review of relative computability is outside the scope of the present note. For interested readers, I recommend starting with the gentle treatment of the Arithmetic Hierarchy in Davis, Sigal and Weyuker (1994).

ization — $\exists i,j : i < j \wedge \mathcal{M}_i$ can create $\mathcal{M}_j$.[7] But this existentially quantified formula is provably inconsistent with Theorem 1* using
elementary first-order reasoning. Note that neither Proposition 1 nor
Proposition 2 is needed.

Formally speaking, the second route to establishing the logical brittleness of $\mathcal{A}$ is overkill, but the route is worth unpacking, because it
both takes direct account of both the nature of FAI and Chalmers'
understanding of that nature (via Propositions 1 and 2), and because it
reflects an elevated view of human intelligence (HI) that (i) isn't without adherents, and (ii) is at any rate formally respectable. For the second route, we begin with:

> **Theorem 2** If HI = $\mathcal{M}_p$, where $p > 2$, then, given Proposition 1,
> there will never be AI such that AI = HI; i.e. Premise 1 in $\mathcal{A}$ is
> false.
>
> **Proof** Suppose that the antecedent holds, recall that AI is indeed
> $\mathcal{M}_2$ or below by Proposition 1, and assume for *reductio* that there
> will be AI such that AI = HI. Given these suppositions, standard
> Turing machines are able to solve HP (since by identity with HI
> and the antecedent they can as $\mathcal{M}_3$-or-more-powerful machines
> solve HP). But that is absurd, and once again by indirect proof we
> are finished. **QED**

Of course, in order to show that on the formalization in question the
second route leads to the unsoundness of $\mathcal{A}$ by way of falsifying
Premise 1, one must include in this formalization the antecedent of
Theorem 2 (which then by *modus ponens* on Theorem 2 yields the
negation of Premise 1 immediately), but that antecedent, while certainly controversial from the standpoint of the ongoing search for
'ground truth', is without question formally respectable, as for example even formidable thinkers like Gödel, writing well before Good,
can be shown to have demonstrated (e.g. see the recursive proof in
Bringsjord *et al.*, 2006).

But are there other respectable formal frameworks in which $\mathcal{A}$
turns out to be sound? Since the argument $\mathcal{A}$ is, whatever else its
defects, formally valid, this query distills to: 'formal frameworks in
which the premises of $\mathcal{A}$ turn out to be true?'[8] Some will no doubt say
Yes, and for the sake of brevity I shall happily concede the

---

[7] Momentarily, I end by considering an alternative formalization, *viz.* one in which AI, AI+,
AI++, etc. are all within $\mathcal{M}_2$.

[8] Perspicacious cognoscenti might appeal to a framework in which one gives up the classification of AI machines as falling at or below the Turing Limit, for in this case my proof of
Theorem 2 is blocked. However, (i) a variant of the theorem could rely on a formalization

affirmative; but I have my doubts. One possible alternative formalization for generating an affirmative response is simply one based entirely in computational complexity applied to standard Turing machines, and below. This would mean that all the machines referenced in the succession at the heart of $\mathcal{A}$ vary only in the *speed* with which they compute Turing-computable functions. Perhaps, then, the idea would be that an ultra-intelligent machine is one unfazed by NP-completeness, but incapable of surmounting the Turing Limit. This direction seems bizarre, since it would mean that ultra-intelligent machines, despite being ultra-intelligent, can't process information in ways that we humans now, courtesy of our mathematical ability, understand, and specify mathematically (witness infinite-time Turing machines, and — see e.g. Bringsjord and van Heuveln, 2003 — proofs that explicitly use infinitary reasoning). Or to put the point another way: Good (1965) tells us: 'Thus the first ultraintelligent machine is the last invention that man need ever make.' This would be false, clearly, on the complexity-based formal framework intended to validate $\mathcal{A}$. The reason is that now, at this very moment, as I write this final sentence, many of the best and brightest minds falling into the class HI are trying to invent ways of implementing the information processing machines in class $\mathcal{M}_3$ and above.

References

Andréka, H., X, J.M., Németi, I. & Székely, G. (2008) Axiomatizing relativistic dynamics without conservation postulates, *Studia Logica*, **89** (2), pp. 163–186.

Bringsjord, S. & van Heuveln, B. (2003) The mental eye defense of an infinitized version of Yablo's paradox, *Analysis*, **63** (1), pp. 61–70.

Bringsjord, S., Kellett, O., Shilliday, A., Taylor, J., van Heuveln, B., Yang, Y., Baumes, J. & Ross, K. (2006) A new Gödelian argument for hypercomputing minds based on the busy beaver problem, *Applied Mathematics and Computation*, **176**, pp. 516–530.

Davis, M., Sigal, R. & Weyuker, E. (1994) *Computability, Complexity, and Languages: Fundamentals of Theoretical Computer Science*, New York: Academic Press.

Good, I.J. (1965) Speculations concerning the first ultraintelligent machines, in Alt, F. & Rubinoff, M. (eds.) *Advances in Computing*, Vol. 6, pp. 31–38, New York: Academic Press.

Halpern, J., Harper, R., Immerman, N., Kolaitis, P., Vardi, M. & Vianu, V. (2001) On the unusual effectiveness of logic in computer science, *The Bulletin of Symbolic Logic*, **7** (2), pp. 213–236.

Hamkins, J.D. & Lewis, A. (2000) Infinite time Turing machines, *Journal of Symbolic Logic*, **65** (2), pp. 567–604.

of the techniques for creation, which — as reflected in Proposition 2 — are themselves Turing-level; and at any rate (ii) there remains the problem that (P2) and (P3) are falsified by Theorem 1[*].

Lewis, H. & Papadimitriou, C. (1981) *Elements of the Theory of Computation*, Englewood Cliffs, NJ: Prentice Hall.

Luger, G. & Stubblefield, W. (1993) *Artificial Intelligence: Structures and Strategies for Complex Problem Solving*, Redwood, CA: Benjamin Cummings.

Russell, S. & Norvig, P. (2009) *Artificial Intelligence: A Modern Approach*, 3rd ed., Upper Saddle River, NJ: Prentice Hall.

Richard Brown

Zombies and Simulation

In his engaging and important paper David Chalmers argues that perhaps the best way to navigate the singularity is for us to integrate with
the AI++ agents. One way we might be able to do that is via uploading,
which is a process in which we create an exact digital duplicate of our
brain. He argues that consciousness is an organizational invariant,
which means that a simulation of that property would count as the real
thing (a simulation of a computer is a computer, and so being a computer is an organizational invariant). If this is the case then we can rest
assured that we will retain our consciousness inside such a simulation.
In this commentary I will explore these ideas and their relation to philosophical zombies. I will argue that dualism could be true of the zombie world and that the conclusion of the standard zombie argument
needs to be modified to deal with simulation. In short I argue that if
one endorses biologism about consciousness then the conceivability
of zombies is irrelevant to the physicalism/dualism debate.

Interestingly, Chalmers has pointed out that thinking about simulations gives us one concrete way of visualizing Cartesian dualism
(Chalmers, 2005). If we think of a person inside a perfect simulation
but whose body is outside the simulation (like in the movie the
Matrix) then all of the things that dualists like Descartes thought were
true would have been true. From the point of view of the person inside
the simulation no amount of physics (i.e. the physics true in the simulation) would allow them to deduce the existence of the body outside
of the simulation. This would make the body outside the simulation
non-physical from the point of view of those in the simulation (where
to be non-physical just is being non-deducible from a completed
physics).

Chalmers has elsewhere talked about zombies and the problems
their conceivability entails for physicalism (Chalmers, 2009). Zombies are creatures that have the exact same microphysical structure as I
do but who lack consciousness altogether. Perhaps surprisingly we

can use the above reasoning to show that Cartesian dualism could be true for zombies as well! We can imagine our zombie twins also being very motivated to create zombie AI and may even do so by evolving these agents in virtual worlds. We can imagine the zombies creating zombie AI++ and wondering (in their zombie way) what the best course of action would be in such a case. We can even imagine our zombie twins wanting to be uploaded into their simulated worlds. Zombies in perfect simulations of their environments would think that they were conscious, as per the above arguments. They would have all the same beliefs, etc. as you would in the exact same simulated environment. The crucial point is that zombies would live the very same lives whether in the simulation or not. In either case they (allegedly) lack consciousness, yet in one case Cartesian dualism is true and in the other case it is false. But if dualism can be true at the zombie world then we seem to have missed what is essential to Cartesian dualism. This makes is seem as though the debate between the Cartesian dualist and the non-dualist is really a debate about whether we in fact live in a simulated world or not.

Now what about us when we actually upload into a perfect simulation? Then it seems as if Cartesian dualism is true of us as well. Let us explore this a bit. Suppose that, as some theorists have suggested, consciousness depends on some biological property of the brain. Let us call this kind of view 'biologism' about consciousness. There are many reasons to suppose that this is true. But of course, there are some disagreements out there. Let us put these disagreements aside for the moment and just stipulate, for purposes of discussion, that consciousness depends on some biological property of the brain. According to this view, non-biological organisms cannot be conscious. It is important to note that this view is neutral as between physicalism and dualism. A physicalist who endorses biologism will think that consciousness just is something biological, as Ned Block tends to lean towards. A dualist who accepts biologism will hold that only the biological properties of the brain are correlated with non-physical conscious properties (perhaps Huxley was in this camp).

Now let us suppose that we have reached a period of time were we can accomplish uploading. When we upload we somehow create a functional duplicate of the original brain. This could be done in either of the ways that Chalmers canvasses in his paper. For instance it could be done by serially slicing the brain and scanning it thereby creating a virtual duplicate of it (destructive uploading), or it could be done using nanotechnology replacing neuron by neuron. Either way at the end of this process we have a virtual duplicate of a human being. Now, according to our previous stipulations this functional duplicate will not be conscious. This is because it will not have the relevant biological property. But we can stipulate that this functional duplicate, in its virtual world, will produce all the same behaviour that its non-duplicated biological twin would produce. We can even take this a step further and imagine hooking up this functional duplicate of a brain in the virtual world to a robotic body in our world. We stipulate that this robotic body does not have a brain and so also lacks the relevant biological properties postulated to be necessary and sufficient for consciousness.

So what have we got here? Remember we have stipulated that consciousness depends on some biological property in the human brain. Given that stipulation, what we have here is a zombie. A zombie, by definition, is a creature physically identical to me but which lacks consciousness. To make it the case that we really have a physical duplicate we can take this even further by imagining that this takes place entirely inside a perfect simulation of our world. In that simulated world there is a simulated body controlled by a functional duplicate of my brain. From the point of view of this simulated world there is no consciousness. This is simply because we've stipulated that consciousness depends on some biological property of the brain. But given this we come to an interesting conclusion. We seem to have conceived of zombies in the way that the original zombie argument requires.

Some may be sceptical that what we have here is really a zombie as traditionally defined so let's take a moment to review. Chalmers (2009) has argued that it was conceivable that we have (P & ~Q). Where 'P' stands for the completed microphysical theory of our world and 'Q' is some phenomenal truth, like that I see blue. Chalmers (2005) also argues elsewhere that this completed microphysical theory could be computed. This means that we could devise a virtual world that was an exact functional duplicate of our world. All of the laws of physics would be the same from the point of view of this virtual world. Indeed we cannot know *a priori* that the actual world is not a simulation in this sense. In this virtual world we have our functional duplicates of biological human beings. So, in this virtual world we have P, the completed microphysics of our world, and, given that we are assuming that biologism is true, not Q. We could, of course, stipulate that in this world there are biological organisms outside the simulation, hooked into it in the appropriate way, and then we would have consciousness in that world. But we are here assuming that biologism is true about consciousness and therefore in this world, which from its point of view is a microphysical duplicate of ours, we have traditional zombies.

But if so then the original zombie argument is irrelevant to the dispute between physicalists and dualists. This follows from the fact that both the physicalist and the dualist can endorse biologism about consciousness. If this is right then the zombie argument may really be a test for whether one thinks we are living in a simulated world rather than whether physicalism or dualism is true. In fact what is interesting about this line of argument is that it connects very closely to things that Chalmers has held said elsewhere in his work. For instance, he has hypothesized that information may be the fundamental unit of reality (Chalmers, 2003; 2005), and that perhaps information has a dual aspect: one aspect being the physical as we understand it, the other aspect being the properties of conscious experience as we feel them. This line of thought fits very nicely with the view of the zombie argument here sketched.

Now of course this entire discussion has been predicated on the idea that we first stipulate that biologism about consciousness is true, or that consciousness depends on some biological properties of the human brain. Chalmers has argued that this is most likely incorrect. And here I am not talking about his zombie argument. Instead I mean his argument that consciousness is an organizational invariant (Chalmers, 1995). This is not the place to address this argument but the prospects for a reply seem good. After all it is an empirical question whether we will ever really be able to enact the kind of sci-fi scenarios that Chalmers appeals to in making his anti-biologism arguments. And it is a further empirical question about whether they will proceed in the way he imagines, but this is an argument for another day.

At the very least we can point out that if one accepts some form of biologism about consciousness, as many do, then one can be a physicalist and grant the conceivability of zombies as well as that conceivability implies possibility without any threat to physicalism. To those, like me, who find the prospects for biologism about consciousness to be good and who want to endorse physicalism as well as the conceivability of zombies and the link between conceivability and possibility this is a promising strategy; a way of having one's cake and eating it too!

References

Chalmers, D.J. (1995) Absent qualia, fading qualia, dancing qualia, in Metzinger, T. (ed.) *Conscious Experience*, Paderborn: Ferdinand Schoningh.

Chalmers, D.J. (2003) Consciousness and its place in nature, in Stich, S.P. & Warfield, T.A. (eds.) *Blackwell Guide to the Philosophy of Mind*, Oxford: Blackwell.

Chalmers, D.J. (2005) The Matrix as metaphysics, in Grau, C. (ed.) *Philosophers Explore the Matrix*, Oxford: Oxford University Press.

Chalmers, D.J. (2009) The two-dimensional argument against materialism, in McLaughlin, B.P. & Walter, S. (eds.) *Oxford Handbook to the Philosophy of Mind*, Oxford: Oxford University Press.

Joseph Corabi
and Susan Schneider

The Metaphysics of Uploading

Introduction

Metaphysics is a matter of life and death. When it comes to making a decision about whether to upload when the singularity hits, the devil is in the metaphysical details about the nature of substance and properties, or so we'll urge. An *upload* is a creature that has its thoughts and sensations transferred from a physiological basis in the brain to a computational basis in computer hardware. An upload could have a virtual or simulated body or even be downloaded into an android body — indeed, an upload could even primarily exist in the digital world, and merely occupy an android body when needed.

Imagine life as an upload. From a scheduling perspective, things are quite convenient: on Monday at 6pm, you could have an early dinner in Rome; by 7:30pm, you could be sipping wine nestled in the hills of the Napa Valley; you need only rent a suitable android in each locale. Airports are a thing of the past for you. Bodily harm matters little — you just pay a fee to the rental company when your android is injured or destroyed. Formerly averse to risk, you find yourself skydiving for the first time in your life. You climb Everest. You think: if I continue to diligently back-up, I can live forever. What a surprising route to immortality.

But wait! Metaphysics is not on your side. As we'll now explain, the philosophical case for surviving uploading is weak. When you upload, you are probably dying. Now, you may have other reasons to upload besides personal convenience and surviving death. If the uploading technology is accurate, you'd be creating a psychological duplicate of yourself. And if you merely seek a mental duplicate — say, to carry out your earthly tasks — then we have no metaphysical bones to pick with you. But to the extent that your decision is fuelled by a suspicion that *you* will be the one carrying out the earthly tasks, your philosophical footing is tenuous.

Why consider the philosophical case for uploading now? This is merely the realm of science fiction, you may think. But science fiction often foreshadows science fact. And uploading may turn out to be a case of such convergence. According to certain scientists and philosophers, such as Nick Bostrom, Anders Sandberg, and Ray Kurzweil, technological developments in recent decades have rendered what used to be seen as far-fetched science fiction tales goals that may very well be attainable during the technological singularity.[1]

David Chalmers' insightful discussion of the singularity treats topics which mainstream philosophy of mind has ignored, uploading being one of them. In this piece, we develop Chalmers' discussion of uploading. In particular, we illustrate that by attending to issues in contemporary metaphysics — more specifically, to issues involving the metaphysics of substance — we can learn that on the assumption that one of the leading theories of substance is correct, the different forms of uploading are not likely to enhance or preserve *us*. One caveat: we can't discuss all the views of the nature of substance today, so we will merely discuss the leading views. Our strategy is to inspire deeper metaphysical reflection on the idea that uploading is a form of survival, and to illustrate that survival is unlikely on the assumption that one of the leading theories of substance is in play.

Here's how we will proceed: Section One introduces metaphysical issues that are key to determining whether you could survive uploading; then, in Section Two we argue that it is plausible, given these background metaphysical issues, that you will not survive uploading. At best, a different person is present after you attempt to upload, one who is a psychological duplicate of you. Section Three concludes.

Section One:
Uploading and Personal Identity

For many, the attractiveness of uploading would be lost, or at least significantly diminished, if it turns out that we can't really survive. A pressing philosophical issue, then, is to determine whether an upload is *the very same person* as the individual who existed prior to the upload. Accordingly, this will be the primary focus of our paper.

We've noted that survival is not the only reason to upload, though — for instance, one may simply want a creature to carry out one's

---

[1] See, for instance, a technical report published by Oxford's Future of Humanity Institute: Sandberg and Bostrom (2008).

earthly business. In addition to such practical motivations to upload, there are other, more metaphysical, issues to consider. You may believe that although the upload is not literally you, there is a special relationship between you and your upload. Consider that sometimes the relationship between persons at different times is not as simple as a question of identity or total distinctness. When a human embryo (call it 'Ally') splits into twin embryos, it may not seem correct to describe either of the new embryos as identical to Ally (i.e. as the very same person as Ally).[2] But, at the same time, many people do not think it seems correct to say that Ally has died and been replaced by two different people either. According to some, the intimate relation that the new embryos share to Ally deserves to be treated differently than the relationship had by two people who were conceived from a different sperm and egg pair. We might label this special sort of relationship 'continuation', bearing in mind that we will reserve the expression 'survival' for numerical identity.

This leads us to a second kind of question: if uploading of a particular sort does not preserve identity, is the upload at least a *continuation* of the original (in the technical sense of 'continuation' employed above)? And there are more questions still. Even if someone in the future was neither you nor a continuation of you, that future person might preserve aspects of you that might make bringing about the existence of such a person of tremendous import to you. For instance, what if you were told that you were about to die, but that for a reasonable sum of money you could form a biological clone of yourself right now that would have copies of all of the most important positive memories from your life, as well as a number of the character traits that you most highly value in yourself. Many people, when placed in the situation, would happily pay the fee for this clone. Why? One reason might be that the existence of such a person would comfort your friends and family. But it is pretty clear that this is not the only consideration that may push you in a 'yes' direction anyway. It is reasonable to conjecture that part of what may be driving you to pay for the clone is a desire for your distinctive experiences and characteristics to be preserved in the future for their own sake, even if not by you or a continuation of you. In a similar vein, then, we can ask of uploads whether they preserve enough of what is valuable about ourselves (or at least enough of what we consider valuable) so that, even if they are not identical to us or continuations of us, there will still be much from our standpoint that commends us to upload, especially if doing so does not

---

[2]   Assume, for the purposes of illustration, that it is uncontroversial that embryos are people.

threaten the quality of our lives in any way (i.e. the lives of *us* or at least those who are continuations of us).[3]

As we can see from the above discussion, there are three kinds of issues that must be addressed in considering personal identity for any upload case. First, we want to determine if the upload has preserved *numerical identity* (i.e. is it the very same person as before the upload?) If the upload is not numerically identical to the original person, we can then ask if it is at least a *continuation* of the pre-upload person, in the way that one of the embryos above might be thought to be a continuation of Ally. And if the upload is not even a continuation of the person, we can then turn to the question of whether it preserves enough of what is valuable about the pre-upload person (or at least enough of what we or the pre-upload person consider to be valuable) for uploading to be an attractive proposition from the agent's standpoint.[4]

While we will weigh in concerning all these matters, as noted, our primary task is to determine whether an upload would genuinely be you — that is, whether numerical identity obtains. So let us now ask: how can we determine when an individual who attempts to upload is numerically identical to the uploaded being? Here, we believe that attending to the metaphysical conception of a person, and particularly, the metaphysics of substance, will prove fruitful.

The Metaphysics of Substance

The notion of a person is philosophically rich; for one thing, persons are traditionally considered the bearers of rights, or at least entities that demand consideration in the utilitarian calculus. For another, understanding the person is central to our self-understanding — for instance, it helps us to understand what it is to be human, and what it is to be a reflective, conscious being. Further notice that there are

---

[3]   These separate issues are distinguished in Chalmers' paper on the singularity, albeit sometimes implicitly.

[4]   Strictly speaking, this question of whether the upload preserves what is valuable may be independent of questions of continuation and numerical identity. For instance, depending on our specific views about personal identity, we could imagine an uploading procedure that preserves numerical identity, but fails to preserve much of what is valuable about the pre-upload person (perhaps it changes large numbers of character traits or erases all memory). To keep things manageable, though, we will generally consider the question of preservation of what is valuable only when there is good reason to think that the upload is neither identical to nor a continuation of the pre-upload person. This will make sense in our investigation because, due to the nature of uploading, we will be hard-pressed to find a feasible form of uploading that has a decent claim to preserve numerical identity without preserving the various psychological features of a person that we would usually think of as the valuable ones.

metaphysical dimensions of the notion of a person: a person is a meta-physical object of some sort. For one thing, it is an entity that has a variety of mental features (or properties): *being rational, having certain kinds of conscious experiences*, and so on. These features are *mental properties* of the person. This naturally leads many metaphysicians who consider the notion of a person to take the position that the person is a *substance* — that is, an entity that continues to exist for at least a short period of time and is the bearer of properties. One key philosophical issue is whether persons are material or physical substances (as the physicalist purports), or whether persons are distinct from physical objects, say, because persons (and arguably, non-human animals) are conscious or have souls.

Substances have some of their properties essentially — that is, they require them for their continued existence. Others they only possess contingently. So, for example, a person arguably has the basic capacity for rational thought essentially; in contrast, properties like the person's particular weight and hair colour are not essential, but contingent — after all, you could survive if you had a different hair colour or weighed a bit more. Different theories of personal identity hold different properties to be essential to the nature of the person. Consider the leading theories:

1. **Soul theories** — this family of views holds that your essential property is that you have a soul or immaterial mind, where the soul and immaterial mind are non-physical substances that are distinct from any physical thing, such as the brain.
2. **Psychological continuity theories** — in their most general form this family of theories holds that you are essentially your memories and ability to reflect on yourself (Locke, 1689; see Olson, 2010, and Perry, 1975). Today, certain proponents of uploading, such as Ray Kurzweil and Nick Bostrom, are broadly sympathetic to this type of view; they view the person as being essentially her overall psychological configuration, where one's psychological configuration is to be detailed by a completed computational theory of the brain (Kurzweil, 1999; Bostrom, 2003; Schneider, 2009).
3. **Materialism (or 'physicalism') about the person** — one is essentially the material that one is made out of: the collection of molecules that makes up one's brain and, arguably, the rest of one's body.[5]

---

[5] For a more extensive survey of the positions on personal identity see Olson (2010).

Each of these views has been framed by numerous individuals, and the details can differ in important ways (consider, for instance, the range of positions on the nature of the soul within Christianity). An important subtlety: while many psychological continuity theorists hold the view that everything is ultimately physical, (2) does not entail (3). What is important, and what makes (2) warrant being considered a distinctive position on its own, rather than being a version of (3), is that proponents of (2) contend that personal identity over time is a matter of psychological continuity, and not, in general, a matter of the person's physical or biological substrate.[6] Further, psychological continuity can in principle be distinctive to personhood even if it turns out that persons, or other sorts of things, are not physical.

Some have seen fit to repudiate the reality of the person altogether:

4. **The No Self View** — according to the 'no self view', the self is an illusion. The 'I' is a grammatical fiction (Nietzsche). There are bundles of impressions but no underlying self (Hume). There is no survival because there is no person (Buddha).

Upon reflection, each of these views has its own implications about whether one should upload. If you hold (1), then your decision to upload depends on whether you believe the upload would retain your soul or immaterial mind. If you believe (3), then uploading will not be a form of survival, because survival requires the same material substrate, and uploading changes it.[7] In contrast, according to (2), uploading may be safe, because although it alters your material substrate it preserves your psychological configuration. Finally, (4) contrasts sharply with (1)–(3). If you hold (4), then the survival of the person is not an issue, for there is no person to begin with. You may strive to upload in any case, to the extent that you find intrinsic value in having an upload carry out your earthly tasks, for instance.

One common thread that underlies all of the views of persons (with the exception of the no self view, of course) is that advocates of these views tend to hold that the person or self is some kind of substance — an object-like entity that has properties and continues to exist for some

[6] In practice, most psychological continuity theorists have not addressed the metaphysics of substance and taken a position on the relation of personal identity questions to metaphysical issues about substance. In spite of their lack of interest, though, these theorists' views may commit them to the claim that persistence of substance matters to personal identity in a direct way. The reason is that they often believe that it is important that memories be preserved, and the continuation of real memories (as opposed to memory copies) is reasonably linked to specific substances.

[7] Later, we'll consider a form of materialism that holds uploading is a form of survival, but the issues are complex; here, we have stated a simple formulation of materialism.

period of time. Now, some have repudiated the category of substance; some, for instance, instead accept an ontology of events or believe that an ontology of properties can secure the rejection of the category of substance. But these views are implausible, we believe, and one of us has treated these issues elsewhere (Schneider, 2011).

So, given the centrality of the metaphysical category of substance to debates on the person, it will be useful to consider some particular views of the nature of substance.

Leading Conceptions of Substance

What is the nature of substance? Contemporary metaphysics tends to focus on two leading theories: the bundle theory and the substratum theory (Armstrong, 1989; Loux, 2002; Schneider, 2011). Both approaches are similar in so far as they hold that substances are not metaphysically basic; instead, the category of substance reduces to (*inter alia*) properties, that is, the features of things (e.g. the blueness of the sky, the mass of a neutrino). The idea that objects' natures involve their properties is appealing, so it is not surprising that both the substratum and bundle theories have a distinguished history and are currently so well received. For when we conceive of an object what comes to mind are its features. But beyond this point of agreement, there are important differences between the two theories.

According to the **Bundle Theory**, substances are bundles of the properties they possess. Of course, not every bundle of properties is an object, so we should ask: what unites bundles that are really substances? Here, the bundle theorist suggests a relation called 'compresence' (or 'co-instatiation', 'togetherness', 'collocation'), where the compresence relation is usually taken as primitive. The **Substratum Theory**, in contrast, holds that objects' natures are not exhausted by their properties. Over and above their properties, substances have substrata, a core that bears properties but is not itself a property. As our sensory access to objects is through their properties, substrata will seem rather mysterious to you: indeed, Locke, who was himself a substratum theorist, comments that they are something 'I know not what' (1689, II, xxiii, § 2). So why should we believe that objects have substrata? Suffice it to say that belief in substrata comes from appreciating deficiencies in the bundle theory; leading objections include that the compresence relation has been notoriously difficult to spell out and that the bundle theory turns objects into properties (Goodman, 1966; Armstrong, 1989, pp. 70–2; 1980, Chapter 9, Section IV; Russell, 1948, p. 312; Schneider, 2011).

In addition to all this, it is also worth emphasizing that key to debates over the nature of a person is the matter of whether persons' nature is physical or non-physical. Are persons identical to their bodies, as (3) purports, or is there something more to their nature, something non-physical, that outruns anything physical, as (1) contends? Here, to the extent that one endorses substances to begin with, philosophers tend to be either *substance physicalists*, holding that the self or person is ultimately a physical substance, or they reject physicalism for a *substance dualist* view in which the self or person is a non-physical substance.

Now, in keeping with the aforementioned distinction between the bundle and substratum theories we can introduce different kinds of substance dualism. According to the **substratum version of substance dualism** (or 'SSD') persons are ultimately substrata instantiating only mental properties, not physical/functional ones. What we might pre-theoretically describe as the physical parts of such persons are really entirely different substances that are not part of persons at all. There are separate substrata having separate bunches of physical properties. These physical substances presumably then causally interact with persons.

The substance dualist bundle theory of substance ('SDBT') says that persons are ultimately bundles of just mental properties. Parallel to what we saw with the substratum version of substance dualism, the purportedly physical parts of such persons are separate (purely physical) property bundles, and not really parts of the persons at all.[8]

We can now appreciate that although our discussion is by no means exhaustive there are several views about the nature of the person as substance that are possible.

A further issue is what the persistence of an object, or, more specifically, a person, consists in. Two views are widely defended in the contemporary literature. Let us turn to this matter.

---

[8] Although we will not deal with them explicitly, views that hold that persons are composites of two substances will fall within the scope of arguments we give later. Because they raise no new philosophical issues and addressing them requires us to introduce significant peripheral complication, we omit direct treatment. We also will not have an opportunity to treat hybrid substances (see Schneider, 2011).

The Persistence of Persons

The two leading views of the persistence of objects are **three-dimensionalism** (often called 'endurantism' or simply '3D-ism') and **four-dimensionalism** (often called 'perdurantism', 'the doctrine of temporal parts', or '4D-ism').[9] **Three-dimensionalism** claims (roughly) that objects persist by enduring or 'sweeping' through time, remaining wholly present at each time during which they persist. **Four-dimensionalism**, on the other hand, claims that objects persist by 'perduring' or being 'spread out' through time, existing at particular times by containing temporal parts that are present at those times. So, for instance, if we assume that your desk is an object that persists through the interval between 1pm and 2pm on 10 January 2012, a three-dimensionalist will claim that it persists because there is an object — namely, the desk — that exists at the beginning of that interval, the end of the interval, and throughout the interval. The four-dimensionalist, on the other hand, holds that it persists because there are a bunch of instantaneous objects — objects we might intuitively identify with your desk *at a particular instant* — and there is such an object for every instant of time during the interval. The key is that these objects are part of a set or sum of such objects (called temporal parts) that together compose your desk.

3D and 4D views can be occupied by proponents of either the bundle or substratum theories. Consider the 4D view, for example; as discussed, at a given moment a time slice of you, not you in your entirety, exists. This time slice is a bundle, as per the bundle theory. But there are bundle-bundle views of substance, according to which (in broad strokes) the entire spacetime worm that is you is a kind of bundle of the time slices. To keep the discussion simple, we will speak in terms of the 3D view, noting particular issues arising in the context of the 4D view in the final section of the paper.

Again, this discussion of views of the metaphysics of the person is not exhaustive, and we can see that it is not even possible to treat in

---

[9] Although one could certainly hold mixed views (whereby, for example, some objects endure but persons perdure), we will always interpret three-dimensionalism and four-dimensionalism to be views specifically about the persistence of persons. For prominent statements of three-dimensionalism (as a general metaphysical view), see Baker (2000), Chisholm (1976), Haslanger (1994; 1989a; 1989b), Merricks (1999), Thomson (1998), Van Inwagen (1990), Zimmerman (1999; 1998), and Wiggins (1980). For prominent statements of four-dimensionalism, see Armstrong (1980), Carnap (1967), Goodman (1951, Chapter IV, Section 1), Lewis (1986, pp. 202–4), Quine (1960, Section 36), Sider (2001), and Smart (1972). The current four-dimensionalist view has historical antecedents in Hume (1739/1978, Book I, Part 4, Sections 2, 6), and Jonathan Edwards (1758, Part 4, Chapter 2).

depth the positions we raise. But in the spirit of illustrating the import of the metaphysics of the person to the question of whether an upload would genuinely be you, we shall now turn to a more detailed discussion of whether, according to certain paradigmatic positions, the upload would really be you.

Section Two:
Putting the Metaphysics to Work

Instantaneous Destructive Uploading Attempts

Now let us ask: could you survive destructive instantaneous uploading? Recall that this is a form of uploading in which all at once, all the precise information about your mental functioning is measured, and your brain is destroyed. All of the information is transmitted to a computer host at some distance away. Let us first consider this question in the context of the substratum view; again, according to the substratum view, a person is a propertied substratum that endures only if the substratum endures. On the assumption that persons are indeed propertied substrata, is it plausible that you would survive destructive instantaneous uploading?

We doubt it. Consider first that the mechanism by which the person (i.e. the propertied substratum) would move instantaneously from the brain to the computer is problematic, even on the assumption that only a short distance needs to be traversed. Not only does this involve an unprecedentedly rapid kind of motion for a person to follow, but this sort of motion is oddly discontinuous. For it is not as though the person moves, little by little, to the computer, so that a step-by-step spatial transition from brain to computer can be traced. Since information is being uploaded, the information has to be processed and reassembled in the computer host before anything like a functional duplicate of the original brain can be obtained. Hence, the person exists at one moment in a brain, and then ceases to exist for a brief period while the information is being transported, and then comes back into existence in the computer at some distance away.[10]

_____

[10] Some adherents of this general position will reject the claim that people have spatial locations at all — namely, full-blooded Cartesian dualists. (Descartes himself famously claimed that people — i.e. souls — have no spatial locations.) Although the arguments we give here will not come into direct dialectical contact with the full-blooded Cartesian view, presumably full-blooded Cartesians have intuitions about what sorts of physical entities can causally interact with souls. They will likely endorse constraints on what kinds of bodies may interact with a particular soul that allow parallel arguments to be constructed against the claim that uploading preserves identity on a Cartesian view. For example, most Cartesians will probably endorse the claim that a particular Cartesian soul

While it is possible that substrata behave in such bizarre ways, the issue is whether it is justifiable to believe that they really do so. Ordinary physical objects simply do not behave like this. Substratum theorists hold that substrata underlie ordinary objects, after all, and this sort of spatial and temporal discontinuity is incompatible with standard views about the endurance conditions of ordinary objects — these intuitions are much stronger than any particular intuitions about the continued existence of a person in this sort of scenario.

Second, unless the brain is truly destroyed at the *very* instant the information is obtained from it, it will be the case that for at least a small interval the person will continue to exist at the location of the brain even after the information has been transmitted to the computer. If the computer is fast and near enough, and it takes long enough for the brain to be destroyed, this could result in the person remaining where the brain is even after the information has begun its journey to the computer (or perhaps after the information has finished its journey there).[11] Only later does the person 'catch up' to the information and come to be located where the computer is. Once again, the strangeness of this sort of behaviour violates strong intuitions about the endurance conditions for particulars, and substrata are particulars. Consequently, as a result of both of these considerations, it is sensible to conclude that identity will not be preserved.

You may further think that the following is evidence against the survival of the substratum: the upload presumably instantiates various silicon-based properties as opposed to carbon-based ones. But this point will not be well received by proponents of the aforementioned psychological continuity view. On this view, it is not the underlying physical material that is essential to one's being the same person over time; what matters is that the upload be psychologically continuous with the pre-uploaded person. But notice that even if one were to grant that one's underlying physical features could differ this radically before and after uploading, on the view we are considering, the continuity theorist is also a substratum theorist, and we believe that she would lack justification for claiming that destructive instantaneous uploading is a form of survival. Even if the upload has all the same

---

causally interacting with a biological body at one time could not begin causally interacting with a computer-based body 1,000 miles away from that biological body a fraction of a second later.

[11] If the person does not 'catch up' until after the computer has reassembled the information and created a conscious being, then what we have is a scenario that is best classified as non-destructive uploading, albeit a non-destructive uploading situation where the original only lasts a short while after the upload. These cases must be dealt with separately.

mental properties as the original person, this would not entail that the substratum was the same, and this is required for numerical identity to obtain (i.e. for the person to survive). The problem is that substrata are particulars, and, as we've just observed, it is questionable that particulars behave in the bizarre manner that would be required for uploading to really be a form of survival. It is more sensible to hold that the uploaded being is merely a psychological copy of the original, having a different substratum. *The interesting upshot is that while we normally believe that the psychological continuity view is friendly to uploading, considering it in light of the substratum view suggests otherwise.*

Turning to the matter of whether the upload is at least a continuation of the original person (again, we have in mind the aforementioned technical sense of 'continuation'), there is again reason for pessimism. The paradigmatic instances of continuation are ones where there is physical continuity between the bodies of the survivors and that of their predecessor. Consider, for instance, stock examples like the identical twin scenario from earlier, or transplants of one cerebral hemisphere into a new body. In cases like these, the claim to survival is based strongly on the physical continuity between the bodies of continuations (and more specifically, brain material in the brain transplant case) and the body of their predecessor. So, in so far as we have any intuitions about continuation, the intuitions track this sort of continuity. But, as we have already seen, the relevant continuity is utterly lacking in uploading, since the 'birth' of the upload is spatio-temporally separated from the 'death' of the original individual, and moreover the 'body' of the upload doesn't even possess the same broad-ranging kinds of physical properties as the body of the original person — that is, the pre-upload is a carbon-based being in which mental properties are in a cellular substrate, whereas the upload is not.

Now that we have considered destructive uploading in the context of the substratum view of substance, we are in a position to see that the pessimistic conclusions of this section do not really depend on either the substratum or 3D views, nor even upon whether one is a materialist or a substance dualist. *For the main reasons for being pessimistic have to do with reservations about positions that attribute strange and discontinuous motions of persons (in contrast to the behaviour of other macroscopic substances) in order to claim that uploading is compatible with survival.*

This being said, we still plan to consider uploading in the context of the bundle theory of substance in more detail, so let us turn to this matter. Here, there is an additional problem for 3D bundle theories that doesn't arise for parallel substratum versions. With substratum versions of the 3D view, the substrata, together with their essential properties, individuate persons and provide the particularity that is required for multiple persons to exist and be distinct from one another. But with bundle versions, however, only the properties themselves can provide the needed particularity. Now since, according to 3D-ism, persons are to be identified with the particulars that bear what we would intuitively think of as *their* (i.e. the persons') properties, 3D versions of the bundle form of the view must claim that the bundle endures. But it seems like the only way for a bundle to endure would be via the endurance of all the essential properties in the bundle (as well as their continued co-presence with other properties in the bundle). Once we acknowledge this, though, it becomes difficult to see how any form of destructive uploading will achieve the desired preservation of the bundle in the new location. The new physical body will not be identical to the old one; after all, it will be in a completely new location and be composed of silicon building blocks rather than carbon-based ones. And if the physical body of the person in the new location is completely different, it is very difficult to believe that the properties in the bundle will have endured: even dualist views that do not require physical building blocks to be preserved would require the physical bundles that are supposed to causally interact with the non-physical bundle (i.e. the soul or immaterial mind) to endure in spite of spatiotemporal interruption, and to causally interact with completely different physical properties before and after the temporal break.

Things don't get much better when we consider 4D views. With 4D-ism there may be a bit more flexibility in people's intuitions, since many people may feel that there are fewer constraints on what temporal parts could form an object over time than on how a single enduring substratum or bundle could behave. But even so, any additional breathing room is limited. Most 4D theorists about persons (as well as 4D views about ordinary physical objects) do believe that the spatiotemporal continuity of temporal parts is an important necessary condition for those things to be parts of the same object over time, at least so long as we are talking about 'adjacent' temporal parts, and the object we are talking about is a non-gerrymandered one, as persons presumably are.

And, to continue a familiar theme, one's specific view of whether persons are physical or non-physical substances doesn't change things much either. Clearly, physicalists are in the worst shape, since the physical differences between the pre-upload person and the upload are most dramatic — they are made of different physical building blocks, after all, and these building blocks are obviously spatiotemporally discontinuous with the original building blocks. But substance dualists are not much better off, since many substance dualists believe that persons are essentially connected to their physical bodies, and even the ones that don't (i.e. those substance dualists who are more Cartesian) would have a hard time explaining how the soul would manage to 'move' from the location of the original body to the location of the upload. (This motion could be literal or it could involve changing what physical objects the soul directly causally interacts with.)

To end the section on a more positive note, what about using uploading to preserve various characteristics that are, or are at least considered to be, valuable, such as character traits and qualitative replicas of memories? This is a difficult issue; sorting through it requires a more systematic investigation of people's intuitions and the various valuable characteristics. But here we see the greatest room for optimism: if people care a great deal about preserving copies of their 'memories', character traits, and thought patterns, then uploading of this sort may be a feasible way to give them what they are looking for. But the extent to which this sort of thing matters to people remains a topic for further investigation.[12]

Gradual Destructive Uploading

Now consider the case of gradual destructive uploading. Imagine, for instance, a case like the one Chalmers presents, where a person's brain is slowly replaced, neuron by neuron, by functionally isomorphic silicon units that transmit the functional information from the neurons just as they destroy and replace them.

You may suspect that this case offers better promise for survival than the instantaneous uploading case we considered earlier, but the change from instant to gradual uploading does not seem drastic enough to provide much additional reason for optimism. For one thing, although the transmission of information happens in a more piecemeal fashion than it does with the case of instantaneous uploading, there will nevertheless fail to be a functional isomorph until all the data from the original is uploaded. Thus, regardless of how long the overall replacement process takes or how small the incremental replacements are, there will still be a dramatic moment at which the data is assembled by the computer host and the isomorph is born.

[12] Parfit (1984), along with the literature it has spawned, provides the classic foundation for much of this investigation.

When we consider what happens at this moment, all of the old issues re-emerge: e.g. if the substratum view is in play, does the person's substratum instantly (or almost instantly) travel from brain to computer? Is there a temporal discontinuity between the person's disappearance and reappareance? It is hard to believe that the same kinds of strong intuitions about endurance we saw before will not be violated once again. (And even if the upload is assembled gradually, there will still be an enormous spatial discontinuity between the person in the computer and the one in the brain.)

Interestingly, Chalmers seems sympathetic to the preservation of identity in the gradual uploading case. We find his reasons unpersuasive, though (assuming they are indeed intended as reasons to believe that numerical identity is being preserved). One consideration that he raises is that 'i[t] will be very unnatural for most people to believe that their friends and families are being killed by the process [of gradual uploading]' (Chalmers, this volume, p. 57). This is likely to be an untrustworthy folk intuition even if Chalmers is correct that it would be present; after all, attending to the sorts of subtle empirical and metaphysical details of cases like this, and attuning philosophical intuitions to the relevance of these considerations, is not generally a strength of those without philosophical training.

Another reason that he provides is his confidence that consciousness will be continuous between the pre-uploaded individual and the uploaded one, and that the psychological continuity that results is 'an extremely strong basis for asserting continuation of a person' (*ibid.*, p. 58). If, on the one hand, this continuity requires the numerical identity of persons, then it seems uninformative (i.e. it is uninformative if the continuity in question amounts to endurance of the very same stream of consciousness). This is because, in order to know whether the continuity holds in the first place, we will need to know if numerical identity is preserved, and this is just what we are trying to get a handle on.[13] If, on the other hand, all that is required for this continuity is that later mental states be caused by or be qualitatively similar to earlier

---

[13] One might object that someone could verify, from the inside, that the same stream of consciousness has endured, and so verify that the same person has endured. But this approach faces a dilemma. Either the individual would need to use memory of her past thoughts, or not, as part of the verification process. If memory were involved, then there is the possibility that the individual could be misled, because even if the stream had not endured the feel from the inside would be the same as if it had endured. And if it is possible for the individual to be misled, then there can be no verification from the inside. If, on the other hand, memory was not involved, then the verification process would be entirely mysterious. It would require a person to have perfectly reliable beliefs about the past without relying on memory. But what faculty could possibly allow someone to have perfectly reliable beliefs about the past without using memory?

ones (e.g. that qualitatively similar thoughts be entertained, or that later sensory 'memories' be qualitatively similar to the original experiences), then plainly the sort of continuity in question could be shared by numerous individuals at later times; what if the information were sent to two computers instead of just one, after all? This allows us to form the following argument against continuity of consciousness (understood in a way that emphasizes causation or qualitative similarity) being anything near an epistemically sufficient condition for identity in these sorts of uploading scenarios:

(A) If uploading preserves the continuity of consciousness, then the continuity of consciousness can be duplicated in multiple locations (since, after all, uploaded information can be sent to many distinct computers).

(B) Uploads in distinct computer systems are distinct from each other.

(C) If the continuity of consciousness can be duplicated in multiple locations and if uploads in different locations are distinct, then, at most, one of the resulting uploads is identical to the original person.

(D) One of the resulting uploads is identical to the original person. (This is an assumption that must be true in order for continuity of consciousness to ensure identity.)

(E) If one of the resulting uploads is identical to the original person, and if uploads in different locations are distinct, then whatever makes it the case that a particular upload U is identical to the original is non-causally dependent on events having nothing to do with the intrinsic characteristics of U or U's substratum. (It could depend, for instance, on there not actually being other uploads, or on U's original biological body being destroyed.)

(F) Whatever makes it the case that U is identical to the earlier person cannot be non-causally dependent on extrinsic factors (i.e. things having nothing to do with the intrinsic characteristics of U or U's substratum).

From (A)–(F), basically by a succession of *modus tollens*:

(G) Uploading does not preserve the continuity of consciousness.

But this contradicts the obvious:

(H) Uploading does preserve the continuity of consciousness.

The only assumption that can reasonably be thrown out here is (D) — we cannot conclude that one of the uploads is in fact identical to the original. A *prima facie* attractive candidate for rejection might be (F), but this will not work. According to the 3D picture of persons, persons are fundamental entities that endure because the same underlying particular endures. Whether this particular endures, then, can hardly be non-causally impacted by things going on outside it, any more than the existence of a particular quark could be non-causally impacted by things going on outside *it*.

So much for Chalmers' tentative defence of the idea that gradual uploading preserves identity. Are uploads at least continuations? Again, this question may not be as clear-cut as the parallel question for instantaneous destructive uploading, but the reasons for preferring a different conclusion are not particularly strong. There are still the same spatiotemporal discontinuity issues, and the fact that the biological brain is destroyed gradually and the information transmitted slowly doesn't seem to matter a lot.

Any form of non-destructive uploading is not likely to preserve identity, since the upload will clearly have no claim to be the person when the original brain is still very much in operation and supporting consciousness and thought. (And, in addition, we still have all the same problems we discussed before in the context of destructive uploading.) Whether non-destructive uploads are continuations depends once again on whether being a survivor is compatible with the kind of spatiotemporal discontinuity we have seen. There is also a special issue that arises with non-destructive uploads — namely, whether continuation is compatible with the continued existence of the original person. All of the paradigmatic instances of continuation (e.g. the case of identical twins and the case of transplantation of hemispheres of the brain) seem to involve the original person ceasing to exist and being replaced by persons with an equal claim to intimate causal and substantial connection to the original. But when the original person clearly continues to exist, it is even more difficult than before to make the case that a mere qualitative duplicate with only distant causal connections to the original has any claim to being a continuation.

Just as we saw in connection with instantaneous destructive uploading, our conclusions about gradual destructive uploading don't depend strongly on embracing any particular theory of the nature of substance, of persistence, or of whether persons are physical or non-physical. But considering the issues in the context of a particular metaphysical position, such as the substratum theory, provided a means of appreciating the rationale for a more pessimistic view of uploading.

Section Three: Conclusion

Now that we have discussed these two kinds of uploading scenarios, let us summarize where things stand. First, there is considerable reason to be pessimistic about instantaneous destructive uploading's ability to preserve identity or to produce continuations of the original person. For this would require odd spatiotemporal discontinuities, and such conflict with what we know of the general behaviour of macrophysical objects. Second, there is also good reason to be pessimistic about gradual destructive uploading's ability to preserve identity or produce continuations, since exactly the same issues arise in the context of gradual uploading. Third, we saw that all forms of non-destructive uploading are unlikely to preserve identity, and not appreciably more likely to produce continuations. Finally, we saw that there was more room for optimism about whether uploading of all the various sorts preserves psychological aspects of persons that we deem worth caring about, although we noted that a more thorough investigation of people's intuitions on this topic is needed.

References

Armstrong, D.M. (1980) Identity through time, in Van Inwagen, P. (ed.) *Time and Cause: Essays Presented to Richard Taylor*, pp. 67–78, Dordrecht: D. Reidel.

Armstrong, D.M. (1989) *Universals: An Opinionated Introduction*, Boulder, CO: Westview Press.

Baker, L.R. (2000) *Persons and Bodies*, Cambridge: Cambridge University Press.

Bostrom, N. (2003) The transhumanist frequently asked questions: Version 2.1, *World Transhumanist Association*, [Online], http://transhumanism.org/index.php/WTA/faq/

Carnap, R. (1967) *The Logical Structure of the World*, George, R.A. (trans.), Berkeley, CA: University of California Press.

Chisholm, R. (1976) *Person and Object: A Metaphysical Study*, La Salle, IL: Open Court.

Edwards, J. (1758) *Doctrine of Original Sin Defended*.

Goodman, N. (1951) *The Structure of Appearance*, Cambridge, MA: Harvard University Press.

Goodman, N. (1966) *The Structure of Appearance*, 2nd ed., Indianapolis, IN: Bobbs-Merrill.

Haslanger, S. (1989a) Endurance and temporary intrinsics, *Analysis*, **49**, pp. 119–125.

Haslanger, S. (1989b) Persistence, change, and explanation, *Philosophical Studies*, **56**, pp. 1–28.

Haslanger, S. (1994) Humean supervenience and enduring things, *Australasian Journal of Philosophy*, **72** (3), pp. 339–359.

Hume, D. (1739/1978) *A Treatise of Human Nature*, Selby-Bigg, L.A. (ed.), Oxford: Clarendon Press.

Kurzweil, R. (1999) *The Age of Spiritual Machines: When Computers Exceed Human Intelligence*, New York: Viking.

Lewis, D. (1986) *On the Popularity of Worlds*, Oxford: Oxford University Press.

Locke, J. (1689) *An Essay Concerning Human Understanding*, London: Edward Mory..

Loux, M. (2002) *Metaphysics: A Contemporary Introduction*, New York: Routledge.

Merricks, T. (1999) Persistence, parts, and presentism, *Nous*, **33**, pp. 421–438.

Olson, E.T. (2010) Personal identity, in Zalta, E.N. (ed.) *The Stanford Encyclopedia of Philosophy (Winter 2010 Edition)*, [Online], http://plato.stanford.edu/archives/win2010/entries/identity-personal/

Parfit, D. (1984) *Reasons and Persons*, Oxford: Oxford University Press.

Perry, J. (ed.) (1975) *Personal Identity*, Berkeley, CA: University of California Press.

Quine, W.V.O. (1960) *Theories and Things*, Cambridge, MA: Harvard University Press.

Russell, B. (1948) *Human Knowledge: Its Scope and Limits*, London: Allen and Unwin.

Sandberg, A. & Bostrom, N. (2008) Whole brain emulation: A roadmap, *Technical Report #2008–3*, Future of Humanity Institute, Oxford University, [Online], www.fhi.ox.ac.uk/reports/2008-3.pdf

Schneider, S. (2009) Mindscan: Transcending and enhancing the brain, in Schneider, S. (ed.) *Science Fiction and Philosophy*, Hoboken, NJ: Wiley- Blackwell. Reprinted in Giordano, J. (ed.) (2011) *Neuroscience and Neuroethics: Issues at the Intersection of Mind, Meanings and Morality*, Cambridge: Cambridge University Press.

Schneider, S. (2011) Nonreductive physicalism and the mind problem, *Nous*, pp.1–22.

Sider, T. (2001) *Four-Dimensionalism: An Ontology of Persistence and Time*, Oxford: Oxford University Press.

Smart, J.J.C. (1972) Space-time and individuals, in Rudner, R. & Scheffler, I. (eds.) *Logic and Art: Essays in Honor of Nelson Goodman*, pp. 3–30, New York: Macmillan.

Thomson, J.J. (1998) The statue and the clay, *Nous*, **32**, pp. 149–173.

Van Inwagen, P. (1990) *Material Beings*, Ithaca, NY: Cornell University Press.

Wiggins, D. (1980) *Sameness and Substance*, Cambridge, MA: Harvard University Press.

Zimmerman, D. (1998) Temporal parts and supervenient causation: The incompatibility of two Humean doctrines, *Australasian Journal of Philosophy*, **76**, pp. 265–288.

Zimmerman, D. (1999) One really big liquid sphere: Reply to Lewis, *Australasian Journal of Philosophy*, **77**, pp. 213–215.

Ray Kurzweil

Science versus Philosophy in the Singularity

In 'The Singularity: A Philosophical Analysis', David Chalmers provides a sweeping and persuasive analysis of the philosophical questions surrounding the 'singularity', and the implications and assumptions underlying the possible answers to these questions. Perhaps the most important distinction to be made is to ascertain what we can infer from science versus what remains truly philosophical.

One view is that philosophy is a kind of halfway house for housing questions that have not yet yielded to science. According to this view, once science advances sufficiently to resolve a particular set of questions, philosophers can then move on to other questions until such time that science resolves them also. This view is endemic to the issue of consciousness, specifically the question, 'what and whom is conscious?'

Consider these statements by John Searle (2002):

> We know that brains cause consciousness with specific biological mechanisms... The essential thing is to recognize that consciousness is a biological process like digestion, lactation, photosynthesis, or mitosis... The brain is a machine, a biological machine to be sure, but a machine all the same. So the first step is to figure out how the brain does it and then build an artificial machine that has an equally effective mechanism for causing consciousness.

According to Searle, once science figures out the appropriate mechanisms, philosophers can then move on to more productive questions. People are often surprised to see these quotations because they assume that Searle is devoted to protecting the mystery of consciousness against reductionists like Ray Kurzweil and David Chalmers.

Chalmers does an excellent job of parsing exactly what is meant when people make assertions about consciousness, and he is the author of the seminal phrase 'the hard problem of consciousness'.

Sometimes a brief phrase encapsulates an idea so well that it becomes emblematic of an entire school of thought (for example, Hannah Arendt's phrase 'the banality of evil' comes to mind in this regard). Chalmer's famous phrase accomplishes this very well.

When speaking and writing about consciousness, many observers conflate emotional intelligence or moral intelligence with consciousness. But our ability to express a loving sentiment, to get the joke, or to be sexy are simply types of performances, impressive and intelligent perhaps, but skills that can nonetheless be observed and measured (albeit that we may argue as to how to measure them). Figuring out how the brain accomplishes these tasks and what is going on in the brain when we conduct these tasks constitute the 'easy' questions of consciousness. Chalmers will be the first to acknowledge that the easy problem of consciousness is in fact not easy, and represents perhaps the most difficult and important scientific quest of our era. But Chalmers' hard question is different. It is so hard that the question is essentially ineffable.

Chalmers does not attempt to define consciousness in this essay but elsewhere you can read that a system with consciousness has qualia, and we can go on to read that qualia are conscious experiences or, perhaps less obviously circular, the phenomenon of qualia pertains to the feeling of an experience. To use Chalmers' example, qualia is the experience of seeing the colour red as opposed to the experience of the neuroscientist in the black-and-white room who studies and understands the neurological processes underlying the perception of colour. However, these approaches are no less circular because the phrases 'feeling', 'having an experience', and 'consciousness' are all synonyms.

In a book now available, *How to Create a Mind, The Secret of Human Thought Revealed*, I refer to consciousness as the ultimate philosophical question. I refer to 'mind' in the title rather than 'brain' because a mind is a brain that is conscious, has free will, and identity. I maintain that these issues are fundamentally philosophical and can never be fully resolved through the scientific method. In other words, there are no falsifiable experiments that one can contemplate that would resolve these questions, not without making philosophical assumptions to begin with. If we were building a consciousness detector, Searle would want it to ascertain whether a system is squirting biological neurotransmitters. The philosopher Dan Dennett would be more flexible on substrate, but would want to determine whether or not the system contained a model of itself and its own performance.

That view comes closer to my own, but at its core this still constitutes a philosophical assumption.

Periodically a thesis is put forward that purports to show that consciousness derives from specific neurological processes. A persistent set of theories has been put forward by Stuart Hameroff and Roger Penrose. At the risk of oversimplifying decades of debate, Hameroff proposed in his 1987 book, *Ultimate Computing*, that intricate processes in the microtubules in neurons were capable of computing and were responsible for the brain's information processing. He went on to claim that these processes were also the basis for human consciousness.

The first premise is a testable proposition and thus far there is no credible evidence that the tubules are computing anything useful. Moreover, analyses of the computational requirements to achieve human brain performance based on functional simulations of brain regions do not support a computational role for the tubules. However, this is all scientifically debatable. The second premise, that the computation in the tubules is responsible for human consciousness, is entirely a leap of faith no different in principle from other leaps of faith such as religious doctrines that ascribe human consciousness (called a 'soul') as being bestowed by a supreme being to certain (usually human) entities. Some weak 'evidence' is proffered for Hammeroff's view, specifically the observation that the neurological processes that could support this purported cellular computing cease during anaesthesia. But this is far from compelling given that lots of processes are halted during anaesthesia, and furthermore we do not know for certain that subjects are not conscious when under anaesthesia. All we know is that people do not remember their experiences when they come out of anaesthesia (although there are intriguing exceptions to this). But consciousness and memory are completely different concepts. The fact that I do not remember an experience does not mean that I was not conscious at the time that I had it. If I think back on my moment to moment experiences over the past day, I have had a vast number of sensory experiences yet I remember very little of it. Was I not conscious of what I was seeing and hearing since I don't remember most of these sensory impressions now? It is actually a very good question and the answer is not so clear.

Around the same time, Roger Penrose had a different leap of faith for the source of consciousness, that it has to do with quantum computing in the brain. Similar to Hameroff's argument, the case was presented as a scientific argument as if consciousness were a readily testable feature. Critics pointed out that the brain was a messy place

for purposeful quantum computing, and the scale of neural components was much too large for quantum events. It was at this point that
Hameroff and Penrose joined forces and put forth a theory that consciousness was an emergent property of quantum computing specifically in the tubules. Much of the science underlying this thesis has
come under doubt (for example, the physicist Max Tegmark showed
in 2000 that quantum states in microtubules would survive for only
10^{-13} seconds, which is not long enough to affect the action of neurons), but the premise (that consciousness derives from quantum computing in the tubules) is, again, a philosophical leap of faith. To the
extent that this thesis has gotten traction, it seems to be based on the
notion that consciousness is mysterious and quantum processes are
also mysterious so they must be linked in some way. This proposition
has been an attempt by Penrose to show that humans can do things that
computers are unable to do because of this unique quantum computing ability in the brain. The fact that common electronics uses quantum effects (transistors rely on quantum tunnelling of electrons across
barriers), that quantum computing in the brain has never been demonstrated, that human mental performance can be satisfactorily explained by classical computing methods, and that in any event nothing
bars us from applying quantum computing in computers, has never
been satisfactorily addressed by Penrose.

There is a fundamental conceptual gap between objectivity (a synonym for science) and subjectivity (a synonym for consciousness).
Although we can peel back the neurological correlates of behaviour
and even internal states, consciousness at its 'Chalmerian hard level'
is not penetrable.

This leads some observers to dismiss consciousness as an illusion.
Indeed at a scientific level it simply does not exist. The idea of consciousness certainly exists and is a meme that resonates in the objective behaviour of our thought processes, but there is fundamentally no
observable difference between a 'conscious' person and a 'zombie'
who acts conscious but in fact is not. Since we have no way of actually
testing whether an entity is a zombie (without some philosophical
assumption based on a philosophical leap of faith) we can say that the
idea of a zombie is meaningless from a scientific perspective.

However, we should not be so quick to throw out the concept of
consciousness. Our entire moral system and much of our legal system
is based on it. If I cause suffering to a conscious being, that is immoral
(based on one good idea from religion which is the golden rule), and
probably illegal. If I cause apparent suffering to an avatar that is not
conscious, then I am just playing a benign game. If I destroy property

it is probably okay if it belongs to me. If it belongs to you then it is probably a crime, but not because I am causing suffering to the property, but to the conscious owner of that property.

So if consciousness is not scientifically testable yet is fundamental to our lives as moral beings, then we can only conclude that there is indeed a role for philosophy and philosophers such as Chalmers. So the goal, in my view, is not to dispense with all leaps of faith. My feeling is that you 'got to have faith', that is we each need a leap of faith as to what and who is conscious. But we should be honest about the need for a leap of faith on this matter and self-reflective as to what our own leap of faith is.

Consciousness is properly a first-person issue, so at least in this paragraph I will state it as such. All I know for sure is that I am conscious. Any other statement or belief concerning consciousness constitutes a philosophical leap of faith, yet such beliefs are unavoidable. There is an irreducible need for philosophy and it is not just because of the incomplete state of science. There is an inherent impossibility of a falsifiable test for consciousness that does not make philosophical assumptions. Such assumptions reflect leaps of faith.

Recently Hameroff has taken aim at this notion and proposed falsifiable experiments to test his thesis. But all he has done is to provide experiments to test the neurological correlates of his thesis, not the fundamental philosophical assumption that consciousness resides from the particular neurological processes that he has identified.

There are other theories along these lines, such as attributing consciousness to certain rhythms of brain activity, or to a measure of the irreducible complexity of neural processes. Personally I like the latter idea but none of these theses succeed in moving this core philosophical issue to a testable premise of science. It would be good if the proponents of these theories were to understand and acknowledge this conclusion.

We should examine our own leaps of faith. People have very different leaps despite impressions to the contrary. Differing philosophical assumptions about the nature and source of consciousness underlie disagreements on issues ranging from animal rights to abortion, and will result in even more contentious future conflicts over machine rights. My objective prediction is that machines in the future will appear to be conscious, that they will be convincing to biological people when they speak of their qualia, and that we will come to accept that they are conscious persons. My leap of faith is that once they do succeed in convincing us of these things, they will indeed constitute conscious persons. Going back to a testable prediction, I expect that most people will come to the same conclusion. If an entity convincingly acts human and we fall in love with them, then we will accept them as conscious persons and care about what they are feeling.

A similar issue concerns the equally ancient debate over free will, a concept that may be even more difficult to define than consciousness. What does it mean to have free will to make decisions if those decisions are the result of deterministic processes (in our brains and in the world we interact with)? Stephen Wolfram puts forth the interesting perspective that there is a difference between a process being deterministic and being predictable. There is no way to predict the result of a cellular automata process without actually going through the process. If I define rules and a starting position for a cellular automata and then ask what the pattern looks like in the last row after a trillion trillion iterations, the answer is clearly deterministic but it cannot actually be determined without going through the process. Wolfram argues that this is sufficient to satisfy our expectations with regard to free will.

Others point to quantum mechanics as having a fundamental unpredictability and therefore allowing for free will. Einstein was uncomfortable with this notion saying that 'God does not play dice with the world'. But even if we accept that reality is based on quantum events and that quantum events are essentially random, the conclusion that our decisions are random will not satisfy everyone's philosophical notion of free will either.

Perhaps the most important question that is fundamentally a philosophical question is the one of identity. Chalmers discusses this in the context of uploading. If I upload my mind to another (presumably more capable and more durable) substrate, is that still me? Just asking whether or not the uploaded Ray is conscious is not sufficient. Is it my consciousness? If I conclude that Ray 2 is indeed a conscious person, but that it is a different conscious person, I may not be so eager to upload.

Consider that I could still be here. If you come to me in the morning and say, 'Good news, Ray, we successfully scanned your brain and body while you were sleeping and instantiated the information in a new and better substrate; we don't need your old body and brain anymore', I may not be comfortable having the new Ray take over. I would probably wish the new Ray well and expect that he will probably do better than I will, but conclude that he's a different person from me.

We can then consider the slippery slope argument that Chalmers makes in his essay (and that many other thinkers have made including

Searle and myself). If I gradually replace small pieces of my brain (and whatever other portion of my body we may consider important) and end up with Ray in which everything has been replaced, is that me? It satisfies that general notion that identity is based on continuity, as there was never a Ray 1 and a Ray 2 but only one Ray who never seemed to change, at least not in a significant way at any one time.

Yet we run into a contradiction because Ray at the end of the gradual replacement scenario is in fact identical to Ray 2 in the sudden uploading scenario. In that case, if we were to extinguish Ray 1 it would constitute the death of Ray 1 and the emergence of a new person, Ray 2. Yet in the gradual replacement scenario, we end up with the equivalent of Ray 2 and no Ray 1, yet when did Ray 1 cease to exist?

This dilemma does not just come up with the scenarios of either uploading or gradual bionic replacement. Gradual replacement is actually what happens naturally. Our cells, and in the case of neurons the components of cells (ion channels, tubules, other organelles), are continually being destroyed and new ones created. We are not the same stuff we were just a few months ago. There is continuity of pattern but the material that comprises our bodies and brains essentially flow through us. In that regard, we are very much like rivers. There is an ancient Chinese saying that you 'cannot walk in the same river twice'. The pattern of a river has continuity and can remain unchanged (or changing very gradually) for many years or even centuries, but the water constituting a river changes in a fraction of a second.

Science can be helpful in eliminating certain theories that are based on observable correlates that may be shown to be scientifically impossible, but fundamentally the core issues of consciousness, free will, and identity will remain philosophical yet necessary leaps of faith. We can conclude, therefore, that becoming a philosopher is a good career choice as we will need philosophy in the era of the singularity.

To switch to an issue that will be resolved through scientific observation, and one that Chalmers discusses in his essay in this issue is the likelihood of an AI achieving greater than human intelligence and self-iterating from there to become a super-intelligence, that is A++. This iterative self-improvement process is a universal concept underlying the expectation of a singular transformation in the future. In my view, intelligence is the most profound force in the universe, and the singularity is based on the prospect of intelligence growing exponentially once it is in a position to iteratively improve its own design.

One important point I would make is that once AI achieves human level intelligence in all of the ways that humans are now superior to machines (and I have consistently argued that the hardware and software for this threshold will be achieved by 2029), the AIs will immediately be superior because they will be able to combine human-level capabilities with the ways in which machines are already vastly superior to us, specifically in the ability to apply intelligent processes tirelessly to vast amounts of information. Watson, the IBM computer that recently defeated the world's two best *Jeopardy!* players, is a good example of this phenomenon. It has been widely noted (including by this author) that Watson's ability to understand human language is still significantly below that of humans. Yet it was nonetheless able to get a score higher than the two best humans in the world combined. The reason is that it was able to apply the language understanding skills that it does have to a vast knowledge base with perfect recall. As humans we could also read all of Wikipedia but we would not remember very much of it, let alone be able to ascertain the right facts in a 3-second period for a *Jeopardy!* query. It should be noted that Watson not only processed the natural language in the queries but it obtained its own knowledge on its own by having read natural language encyclopaedias like Wikipedia. Its knowledge base was not hand coded. So it was able to read all of this material and master it, at least at a sufficient level to play this particular game. Natural language understanding by machines is clearly going to continue to improve. By the time it is at human levels and beyond, it will then be able to read billions of pages of material and master it all. That will be a powerful combination.

We will merge with the intelligent technology we are creating in a gradual process, a prospect that should be welcome to those of us who associate identity with continuity. This is a process already under way. The intelligent devices I carry on my belt and elsewhere are not yet in my body and brain but they may as well be. They have already become part of who I am (and of nearly everyone else). We have already offloaded our personal and social memories to our devices. One of the exponential trends associated with information technology is a reduction in size which I have measured at approximately 100 in 3D volume per decade. Within a couple of decades, these devices will be small enough to routinely put inside our bodies non-invasively, for example in blood-cell sized devices. We will become a hybrid of biological and non-biological intelligence.

Keep in mind that the non-biological portion of our intelligence is and will be growing exponentially in capability and price-performance, whereas the biological portion is relatively fixed. The non-biological portion will, therefore, quickly come to dominate. We will transcend our biology, but in my mind we will still be human because changing who we are is precisely the unique attribute of our species. I should point out, however, that 'what is human?' is another one of those irreducible philosophical questions.

References

Searle, J.R. (2002) I married a computer, in Richards, J.W., *et al.* (eds.) *Are We Spiritual Machines*, Seattle, WA: Discovery Institute.

Tegmark, M. (1999) The importance of quantum decoherence in brain processes, *Quantum Physics/9907009 v2*, 10 November.

Wolfram, S. (2002) *A New Kind of Science*, Champaign, IL: Wolfram Media.

Pamela McCorduck

A Response to 'The Singularity'

This essay by David Chalmers is refreshing in its lucidity and moderation — both unusual in discussions of the singularity. It lays out the arguments far more clearly, and hence more persuasively, than the usual rhetoric, whether pro or contra.

As a consequence, I have reconsidered my own position on the singularity. I long believed that if such an event were possible, it would be many years, perhaps centuries, before it occurred, and anything we might have to say about it now would be inane: a great premise for science fiction, but not so great for examining functional or ethical stands to be taken now or in the immediate future. *Sufficient unto the day is the evil thereof,* says the old Biblical warning, an admonition that begins, *Take therefore no thought for the morrow, for the morrow shall take thought for the things of itself* (Matthew 6.34).

Lately, however, artificial intelligence has improved some of its performances dramatically, and David Chalmers' essay offers a plausible argument that the singularity might indeed happen, and within decades, not centuries. As he argues, the singularity is therefore worth thinking about.

However, I do question the way the singularity is generally imagined to arrive. Will it really be a homogeneous, single-minded (so to speak) membrane that spreads everywhere over the planet? Or will it continue to be heterogeneous, and perhaps self-conflicting, odds and ends, just as it is beginning to be now? For there are certainly many particular machines that are already smarter than we are. A heterogeneous singularity would raise many interesting problems for humans, and for itself.

In any case, Chalmers lays out certain kinds of defeaters that might derail the singularity's arrival: disasters, disinclination, and active prevention. Of these, disasters seem to me to be the only likely defeater.

Disinclination and active prevention, on the contrary, are very unlikely. If artificial intelligence were being pursued in a secret location on a remote mountaintop, someone, or a committee of someones, might be able to beg, or forbid, the researchers to stop, as might possibly have happened with the Manhattan Project, at least for a while.

But artificial intelligence is being pursued internationally, in many different, quite diverse locations, such as research institutes, universities, and firms. (I can even imagine lone-wolf researchers, though given the resources needed, this seems unlikely.) Thus, even if a bloc of nations adopted a policy of no further research in artificial intelligence, they could not achieve universal prohibition — there would always be nations or groups who see local advantage to the research, and persist in continuing it. Therefore the preventer bloc would be forced, in self-defence, to continue research too.

Since research that will lead to AI+ and AI++ is inevitable (and let me say, to my mind, a *generally* good thing) this leads to Chalmers' suggestion that we solve the potential problems ahead by shaping these technologies in ways that are agreeable, or at least not harmful, to human beings, conforming 'broadly to human values'.

Who will decide what is agreeable, what is not harmful, and what conforms, even broadly, to human values? Values around the world are diverse. I wouldn't expect Chinese AI+ to be identical to Indian AI+, or to Brazilian AI+, never mind the AI+ of Euro-American democracies — which diverge among themselves as well. Differences about what 'conforms broadly to human values' might even split right down the gender divide. The seductive, but elusive, 'we' in this essay is, in fact, a congeries of opinions, interests, and values. I doubt even responders to this essay will agree altogether on values; thus a world-wide consensus is a chimera. Slow coding, analogous to slow cooking, might work, but human beings have a long history of leaping before they look, and this would probably not be different. The international experience with responses to global climate change, for example, can give us no optimism about the human ability to shape AI+ cooperatively on any international scale.

If, then, the singularity is possible, maybe probable, maybe even inevitable, where does that leave us?

Chalmers offers us some possibilities. We could try everything out in a leak-proof virtual world first. People with experience in model-building will understand the enormous difficulties of constructing such a virtual world: even if it's possible to capture the major properties of our world in such a model, each and every property cannot be captured, and who really knows where mischief lies? Still, I'd think that this kind of slow-cooking approach, however provisional, is worth pursuing to save us from fast and stupid unintended consequences. Slow unintended consequences would take longer to emerge.

We could bid goodbye to our mortal remains and upload our brains or ourselves (whatever we think that means) into advanced hardware. Though I'm sentimental about my younger corporeal self, I wouldn't miss my older self's aches and pains in the least, and would consider it a fair trade-off. Of course, as Chalmers points out, this is not yet a proved technology, and slips of any magnitude could happen. Some people also argue that human proprioception is deeply significant to human intelligence, but if this is true, it could probably be simulated.

Moreover, what happens when platforms change? There's something undignified, not to say humiliating, about living, moving, and having our being as no more than a piece of legacy software that has to be tediously accommodated, like ancient FORTRAN routines. If this digital version of the self is somehow preserved, I think most of us would not want it frozen in perpetuity. We might opt for slow cognitive enhancement, as Chalmers suggests. Or we could do it the old fashioned way, via evolution. But either method would tend to change the nature of this self, eventually profoundly. In any case, Chalmers presents a suitably modest uncertainty about just how well this uploading would work.

Others say the lesson from human history suggests that as the machines grow smarter, our own intelligence will improve apace. Therefore, while we'll outsource tedious computations to the machines, it will be our luck to grow ever smarter as a result of these outsourced computations, and therefore our worries about being bested by the machine are unfounded.

Can it be possible that the best advice for human action, post-singularity, will come from the machines themselves?

Wait. I seem to have come back to where I started: *Take therefore no thought for the morrow, for the morrow shall take thought for the things of itself.* But I'm grateful to David Chalmers and his provocative essay for encouraging me to think about the issues once again.

Chris Nunn

More Splodge than Singularity?

*'Take away the context and the meaning also disappears. When you per-
ceive intelligently... you always perceive a function, never an object in
the set-theoretic or physical sense.'*
Stanislaw Ulam, quoted by Gian-Carlo
Rota in 'Indiscreet Thoughts'

As my wife would be happy to attest, our dog Sam is more intelligent
than me. Or at least he is more intelligent when it comes to p-mails left
by other dogs on lamp posts. His cognitive capacity is greater than
mine in this context for he has a more efficient olfactory apparatus; his
correlated capacities are more efficient too, especially when he
responds in kind! Does this circumstance have implications for David
Chalmers' 'singularity' thesis? I shall briefly argue that it raises issues
to do with the contextuality of 'intelligence' (issues that are independent of unresolved questions about whether or not a general intelligence factor 'g' exists), which are largely swept under the carpet in the
target paper. Most of what I'll be saying has to do with implications of
the facts that intelligence can't be reduced to information processing
alone — it actually centres on the elucidation of meanings — while
meaning is itself dependent on memories, whether genetic, personal,
or social.

The singularity thesis appears to rely rather heavily on a purely
information processing view of what we term intelligence. Put
crudely, the view seems to be that more elaborate and/or sophisticated
processing strongly correlates with greater ability; one can reasonably
expect that linear or even exponential increases in elaboration/sophis-
tication will manifest at some stage and thus that associated intellectual capacity will follow suit. This picture is incomplete, however,
since it ignores questions about how relevant information is to be
selected. Simple combinatorial considerations show that the number

of possible combinations of information existing out there in the environment must rapidly approach infinity as more of them become accessible through more sophisticated perceptual apparatus. However advanced an AI may be, it will still need some basis on which to select its inputs if it is not to be overwhelmed. Moreover, pure information-theoretic considerations can lead to a concept of algorithmic information. This is maximal in completely random systems, so any relationship between increased information of this sort and greater intelligence is likely to take an inverted U shape. Linear or exponential progress to an intelligence 'singularity' won't occur on any simply conceived information processing basis. 'Intelligence', of whatever sort, has to be regarded as relating to an ability to deal with *meaningful* information, not just with the information of 'information theory'. So where does meaning come from?

In the case of our dog Sam, it's easy to see that a range of sources determine what is meaningful to him — all of them memory-related in one way or another. There are his genetic memories which have informed the capacities of his nose, fine-tuned by innumerable personal memories of smells met in the environment and social encounters with other dogs. His olfactory intelligence depends on the pre-selection by his genetics of the sorts of smell that he can perceive and how he will react to them, and on complex feedbacks between those abilities, his environment, and his previous experience. The same, I suggest, is true of all intelligence. The most advanced AI must still select its inputs on some pre-determined basis, however wide the input range may be. Indeed, the wider the potential range of inputs, the greater the need to cherry-pick what is or could be meaningful. Any AI must employ its intelligence, in other words, within some historical context. The overall 'intelligence' of the AI will thus depend on an input sophistication constrained by all sorts of factors additional to its own endogenous information processing capacities.

A factor that contributes only a little to Sam's intelligence, but a lot to mine, is to do with our respective ambient cultures. Chalmers mentions 'extended mind' almost in passing in his target article, but in fact *most* of my more elaborate cognitive abilities that are not strictly innate depend on memories of one sort or another that were transferred to me from the environment, from other people and from what can loosely be described as British culture. They contribute to my intelligence via a huge range of circumstances — formal education, etc. along with all the artefacts that I encounter or use (Chalmers' famous notebook among them!) And it seems likely that any sophisticated intelligence is not going to be 'an island complete unto itself', but is going to depend on an ambient culture of some sort.

As a slight digression here, it's worth asking whether superhuman intelligences would need to be conscious. The part played by consciousness in our own intelligent processes appears to be that of evaluator or selector of meaningfulness. It provides global, shorthand assessments of the relevance of material that 'intelligence' uses or produces. One suspects that all advanced intelligences must include a similar function, though whether or not the function would necessarily have be associated with *phenomenality* remains an open question.

These considerations strongly suggest that any isolated advanced AI designed by us isn't going to be significantly more intelligent than us, for it will be constrained by the ambient (human) culture. Its cognitive and other capacities may be greater than ours, and it may be able to solve problems inaccessible to us, but that's already true to an extent of number-crunching computers, and we certainly don't attribute superhuman intelligence to those. At best, isolated advanced AIs might contribute to raising *our* intelligence. However, a *society* of advanced AIs would be a very different kettle of fish, for they would be able in time to create their own culture(s). Could they be expected, over successive AI generations, to advance to a 'singularity'?

Obviously any advance would be slower than Chalmers envisages, for each significantly advanced generation of AIs would need time to create their own 'culture' before the next step could be taken. However, this isn't going to prevent singularity occurring later rather than sooner, except in so far as the added complexity may make breakdown at some intermediate stage more likely. What is more significant is the fact that the types of culture created would be history-dependent and the situation thus reminiscent of Stephen Jay Gould's picture of biological evolution, where a re-run could be expected to produce totally different end results. One can conclude that the end point of any such intelligence bootstrapping process would not be an ultimate, transcendental singularity but a particular high point selected almost at random from a whole range of possible alternatives. Indeed a bunch of separate high points can be expected because AI societies are likely to multiply and diverge as they succeed one another.

Another implication is that our prospects of being able to modify the behaviour of advanced AIs to fit in with our interests are a lot dimmer than Chalmers suggests, for the same reasons that give rise to the uncertainty and difficulty of social engineering compared to computer engineering. The most we could realistically hope for is that historical constraints might lead to continuing endorsement of charitable values by AI societies — unless, that is, the suggestion about running such societies in a virtual reality could be implemented, and we could be safe because they would be removed from our continuum. Let's take a brief look at this possibility.

An immediate difficulty, for me personally, is that I'd rate the chances that virtual AIs could have anything like our form of conscious experience as very slim, for I take a considerably less 'functional' view of the likely basis of phenomenality than the one endorsed in the target article.[1] Without the assessments of meaning provided by consciousness, it's not at all clear that advanced intelligence *could* develop in a society of virtual AIs. Moreover one suspects that if any such society were possible it would manifest to us as nothing more than yet another super-computer; one as gnomic, perhaps, as the computer in *Hitchhiker's Guide to the Galaxy* which, after millennia of cogitation, came up with the answer '42' when asked to explain the meaning of life, the universe, and everything.

And what about hopes that advanced AIs will find ways of 'uploading' us, perhaps to join them in their wonderful world? Well, I have no plans to clone Sam when he departs this life and wouldn't expect an advanced AI to do it for me, before or after death. Any hopes (or fears indeed) that death may prove subjectively to be utterly different from its objective appearance might more realistically, I suspect, be grounded in questions and issues to do with temporality and consciousness.

References

Nunn, C. (2010) Landscapes of mentality, consciousness and time, *Journal of Consciousness Exploration and Research*, **1** (5), pp. 516–528.
Nunn, C. (2011) *Who was Mrs Willett?*, Exeter: Imprint Academic.
Rota, G.-C. (1997) *Indiscrete Thoughts*, Boston, MA: Birkhauser.

---

[1]   The view I do take is described in Chris Nunn (2011). A highly condensed version can be found in Nunn (2010).

Arkady Plotnitsky

The Singularity Wager
A Response to David Chalmers

As David J. Chalmers explains in 'The Singularity: A Philosophical Analysis', the singularity is 'an explosion to ever-greater levels of intelligence, as each generation of machines creates more intelligent machines in turn', made possible by the development of AI (artificial intelligence of human level), which is the first step and the necessary precondition of this explosion. It is further assumed that AI then explodes to AI+ (artificial intelligence greater than human), AI++ (artificial intelligence much greater than human), etc. potentially to infinity, thus leading to the singularity, or in any event reaching a level far exceeding human (pp. 11, 16).

The first and, arguably, the most crucial question here is whether AI is possible and, if so, how likely, how probable it is. Although Chalmers does not put his argument in these terms or expressly reflect on this point, the article is also about 'betting' on the singularity. It may also be called the singularity wager, with Pascal's famous wager in mind. God is a form of singularity, too, not unlike that of AI++, when it reaches infinite or a near infinite intelligence, although, arguably (some visionary of computer technology might disagree), only God's intelligence could be truly infinite, as Spinoza contended in his *Short Treatise on God, Man, and Human Welfare* (Spinoza, 1910). He called God's intelligence 'the eternal son' of the divine substance, a kind of hardware of the divine AI. That a betting on singularity or on AI, to begin with, is at the core of the article is hardly surprising. Indeed it is unavoidable, given that the article concerns an event — a singular, unique event — whose probability or even possibility is uncertain and has been contested, sometimes fiercely, for over half a century — that is, if one speaks of the current form of AI, grounded in digital computation, at stake in the article. In more general terms, the debate concerning the possibility of artificial intelligence goes back at least to Descartes and is, thus, coextensive with the history of modern philosophy.

'Betting on singularity' enters Chalmers' article from the outset. I.J. Good's 1965 article 'Speculations Concerning the First Ultra-intelligent Machine', which sets the argument for both the singularity itself and for Chalmers' article, is also a bet, as the word 'speculations' in its title suggests. To be sure, Good's bet is not without an air of certainty, but it is a bet nevertheless. It is, I would argue, also not coincidental that, while Good, who worked with Alan Turing at Bletchley Park, was an expert in the field of computer intelligence, he was also an expert on probability and statistics, specifically so-called Bayesian probability and statistics, so named after Thomas Bayes (1702–1761).

Bayes formulated a special case of what became known as Bayes' theorem, a rule for establishing the conditional probability, or 'posterior probability', of an hypothesis H (event H), after evidence E becomes available (event E, which has a non-zero probability of occurrence). The rule implies that evidence has a confirming effect if it is more likely given H than given 'not H'. Bayesian approaches to probability (these approaches may differ in further detail) generally define probability as a degree of belief. As such they deal, first, with probabilities concerning individual events, estimated on the basis of *all* information (from various possible sources) one has that can be used for such estimates, rather than on similar events, such as coin tosses, occurring in sequence (where our estimates are shaped most essentially by the outcomes of the preceding events of the same type). The corresponding approach to probability is usually called 'frequentist'. Bayes' rule itself applies in all common interpretations of probability, including the frequentist ones. I can hardly think that Good did not reflect on the Bayesian aspects of his speculations concerning ultra-intelligent machines. Deliberately or not, Chalmers, too, uses the Bayesian language when he says that his 'own credence' for the emergence of AI 'before 2100… [is] somewhere over one-half' (p. 17). The probabilistic language of 'expectation', 'likelihood', 'plausibility', and so forth, often with Bayesian overtones, permeates the article throughout. In addition, in part correlatively, Bayesian approaches deal with more subjective aspects of probability, while the frequentist view is more objective, or appears to be, especially to most of those who subscribe to the frequentist philosophy of probability. For most twentieth-century Bayesians, all probability, such as that of Good's or Chalmers' credence or of this author's more sceptical estimate concerning AI, is subjective, defined, again, as the degree of one's expectation or belief concerning an individual event based on the prior information pertinent to this event, which is called the prior (e.g. Jaynes, 2003; de Finetti, 2010). Part of Pascal's prior in his wager would be that God might exist and salvation is possible. The priors of those who bet on or against AI are quite complex and might differ significantly, but would generally involve the information and assumptions about the capacity of both the human brain and computers. It might be noted that Bayesian betting, for example, for or against AI, is itself part of the complexity of the human brain, and there is even the concept of the Bayesian brain in modern neurology. It is possible and appears necessary that AI must have a similar capacity, a point to which I shall return below.

The Bayesians would interpret the assignment of probability to a coin toss (even when it occurs as part of a sequence of such tosses) along subjective lines, since, on their view, it is not possible to strictly repeat a coin toss, which is, technically, correct. In the present case, one obviously need not worry about frequencies, since (just as the existence of God in Pascal's wager) in the case of the singularity or even of its key precondition — a creation of AI — we deal with a singular, unique event. It is true that, as Chalmers points out, there can be more than one way to create an AI, and hence one can technically bet differently on different ways of creating AI. Still, we would not be meaningfully dealing with the sequence of occurrences of the same kind that could help our estimates. It is also important that, according to the Bayesian view, we can also update our expectations concerning a possible future event if new information related to this event becomes available. In part correlatively, the degree of our expectations concerning a given event could be shared or become shared.

With these Bayesian considerations (including the last one just stated) concerning the possibility of AI and the singularity in mind, the question that I want to ask here is this: how does Chalmers' article, which offers a compelling and comprehensive account of reasons for and expectations concerning the rise of AI and the singularity, update my own expectations concerning these events, and how, in my (also Bayesian) estimation, likely is it to affect the expectations of others concerning them?

Chalmers gives AI, if not the singularity itself, at least a fifty-fifty or, it appears, even better chance. As Chalmers is careful to explain, the singularity need not necessarily follow the rise of AI and is, accordingly, given a lesser chance by him. Chalmers offers compelling reasons for his expectations concerning the plausibility of AI and a future shaped and even defined by AI, and, generally, a credible analysis of why these reasons are sufficiently compelling to justify his bet. Chalmers hardly needs to convince the proponents of AI and the singularity, although his analysis is useful for them as well; first, by giving a further justification of their views and additional ammunition to make their arguments for these views, and second, by offering a general philosophical analysis of this problem. Indeed, this analysis is likely to be appreciated by most readers of the article, regardless of their attitudes and expectations. On the other hand, I expect — 'bet' — that the article is less likely to convince more sceptical readers, the present reader among them, who would give AI or the singularity a considerably lesser chance than Chalmers does.

Chalmers, to his credit, considers some among more sceptical views and some among the possible reasons to hold such views, reasons that he sees as potentially counterbalancing his own argument, but not sufficiently to diminish the degree of his expectations concerning AI and the singularity, or, again, at least AI++. Chalmers is also right to suggest that in view of both the 'enormous potential benefits' and 'enormous potential dangers' (possibly even 'the end of the human race') of 'an intelligence explosion', 'if there is even a small chance that there will be a singularity, we would do well to think about what forms it might take and whether there is anything we can do to influence the outcomes in a positive direction' (p. 14).

I shall by and large bypass Chalmers' discussion of the possible post-AI future, except to the degree that this discussion relates to (and it does) my main concerns stated above. This discussion is too speculative in any event, and while often stimulating, it is, in my view, hardly a sufficiently reliable guide for our future thought and action. This is not a criticism of Chalmers' argument, but rather a reflection on the general difficulties any such analysis or set of expectations is likely to have, given the temporal distance at stake. As noted earlier, Chalmers gives his 'own credence', his own Bayesian degree of belief, for the emergence of AI 'before 2100... somewhere over one-half', over-optimistically to the present author, or, as Chalmers observes, it would be to most neuroscientists (p. 17). Still the merely 'over one-half' in about a century is pretty uncertain, and AI+ and AI++ would then take longer (p. 17). Chalmers, again, gives a far greater chance to the rise of AI itself at some point. Even if the singularity and an intelligence explosion, or the rise of AI, occur in the distant future, all our current artificial hypotheses concerning science, technology, and the world by then are (this would be my bet) unlikely to survive. Stanley Kubrick's supercomputer HAL (which symbolically stands for IBM, by replacing each letter by the one that precedes it in the alphabet) in *2001: A Space Odyssey* (1968) is a good example of this type of difficulty. On the one hand, it offers a good anticipation of the progress of digital technology, and on the other, it is far off the mark time-wise when it comes to artificial intelligence. We are still quite far from it even now, in 2012, assuming, again, that it is possible at all. Incidentally (but not coincidentally), I.J. Good was Kubrick's primary consultant on artificial intelligence in making the film, which may thus also reflect Good's overly optimistic Bayesian bet on AI at the time.

Apart from the fact that Chalmers does not address some among the plausible reasons for more sceptical assessments of the likelihood of AI at any point or, at least, in a foreseeable future (say, again, over a century), some of the reasons that he does mention are not, in my view, sufficiently analysed by him or considered at all. Given my limits and aims, I shall primarily focus on the latter set of reasons. This is sufficient for my argument, which is aimed to suggest that the factors involved may be more inhibiting to attempts at creating AI than Chalmers thinks, or at least than his article makes it appear. I shall further suggest that these lacunae in his analysis also reflect deeper and more powerful, and possibly insurmountable, difficulties that the project of creating AI faces. In other words, while Chalmers does offer 'a philosophical analysis' of the subject of AI and the singularity, this analysis appears to be offered from the viewpoint of somebody who welcomes, if not champions, AI. To the present author, this viewpoint appears at least as much to shape Chalmers' estimates as be shaped by them. I am, I hasten to add, not entirely certain on this point. Nevertheless, there appears to me to be a certain degree of (unconscious) 'blindness' to some of the factors that could affect one's expectations concerning the possibility of AI and ultimately the singularity or the state of the intelligence explosion near the singularity.

For the sake of brevity, and because the rise of AI is the most crucial event in the sequence of AI, AI+, AI++, and the singularity, AI will be my main focus, although similar and related questions can sometimes be raised concerning Chalmers' assessment of the possibility of the emergences of AI+ and AI++. He is, again, more cautious concerning the possibility of the singularity. I am not saying that the event of the rise of AI is necessarily the most probable in this sequence. It merely conditions everything else; that is, the probabilities of the occurrences for AI+, AI++, and the singularity are conditional on it, along with other factors shaped by the preceding events in this sequence.

I find especially questionable, and even puzzling, Chalmers' analysis of the neurological ('the emulation argument') and evolutionary-biological ('the evolutionary argument') arguments offered by him in support of his estimate for the likelihood of AI and, most especially, of ways in which these arguments can be resisted. Each argument follows the same line or pattern of thought and, as a result, I would argue, incurs analogous problems. To address, first, his 'Premise 1: There will be AI (before long, absent defeaters)', based in the *emulation argument* (p. 18). Chalmers says:

> (i) The human brain is a machine.
> (ii) We will have the capacity to emulate this machine (before long).
> (iii) If we emulate this machine, there will be AI.
>
> _____
>
> (iv) Absent defeaters, there will be AI (before long).
>
> The first premise is suggested by what we know of biology (and indeed by what we know of physics). Every organ of the body appears to be a machine: that is, a complex system comprised of law-governed parts interacting in a law-governed way. The brain is no exception. *The second premise follows from the claims that microphysical processes can be simulated arbitrarily closely and that any machine can be emulated by simulating microphysical processes arbitrarily closely. It is also suggested by the progress of science and technology more generally: we are gradually increasing our understanding of biological machines and increasing our capacity to simulate them, and there do not seem to be limits to progress here.* The third premise follows from the definitional claim that if we emulate the brain, this will replicate approximate patterns of human behaviour along with the claim that such replication will result in AI. The conclusion follows from the premises along with the definitional claim that absent defeaters, systems will manifest their relevant capacities. (p. 18, emphasis added)

It is the second premise that gives me pause (hence my emphasis). Are 'the claims that microphysical processes can be simulated arbitrarily closely and that any machine can be emulated by simulating microphysical processes arbitrarily closely' sufficiently self-evident not to merit further examination, given the *extraordinary complexity* of the 'machine' in question, the human brain? It is true that 'we are gradually increasing our understanding of biological machines and increasing our capacity to simulate them'. It may, however, be a little different (or, depending on one's bet, greatly different) when it comes to his statement that 'there *do not seem* to be limits to progress here', again, in view of the *enormous complexity* of the human brain. I should register, however, that Chalmers, to his credit, does say '*do not seem*' (emphasis added). Chalmers does note that 'one might resist [this] argument in various ways', and he cites some of the relevant counterarguments, and his own previous responses to these arguments (p. 19, note 11). I would argue, however, that while his earlier works provide valid criticism of these counterarguments, these works and, especially, the article under discussion are not sufficiently responsive to the question of complexity of both the human brain itself and of the evolutionary processes that made the emergence of the human brain possible. To the present author, this question appears to be a persistent and persistently unaddressed problem, the blind spot, as it were, of Chalmers' analysis, at least when it comes to his bet on the likelihood, 'before long', of AI and the singularity — my main concern here. I shall now examine this question in the context of Chalmers' next biologically based argument in favour of AI, 'the *evolutionary argument*', where these difficulties become especially pronounced.

Chalmers' '*evolutionary argument*' concerning the chances of AI is offered, in part, against the dualist view of consciousness (which, roughly, assumes that certain aspects of thought are non-physical), where Chalmers' argument is both nuanced and effective. Chalmers says:

> [T]he *evolutionary argument...* runs as follows.
>
> (i) Evolution produced human-level intelligence.
> (ii) If evolution produced human-level intelligence, then we can produce AI (before long).
>
> _____
>
> (iii) Absent defeaters, there will be AI (before long).
>
> Here, the thought is that since evolution produced human-level intelligence, this sort of intelligence is not entirely unattainable. Furthermore, evolution operates without requiring any antecedent intelligence or forethought. If evolution can produce something in this unintelligent manner, then in principle humans should be able to produce it much faster, by using our intelligence. (p. 20)

This last contention is not as self-evident as it may appear, and, just as in the case of 'the emulation argument', Chalmers notes that this 'argument can be resisted', but only, again (just as in the case of 'the emulation argument'), to ultimately discount the potential effectiveness of this resistance, without, I would argue, sufficient justification for doing so. He says:

> [T]he [evolutionary] argument can be resisted, perhaps by denying that evolution produces intelligence, or perhaps by arguing that evolution produces intelligence by means of processes that we cannot mechanically replicate. The latter line might be taken by holding that evolution needed the help of superintelligent intervention, or needed the aid of other nonmechanical processes along the way, or needed an enormously complex history that we could never artificially duplicate, or

needed an enormous amount of luck. Still, I think the [evolutionary] argument makes at least a prima facie case for its conclusion. (p. 20)

One is almost tempted to say, *at most* a *prima facie* case. Chalmers appears to me to a little too quick to dismiss reasons so complex and, I would contend, so powerful, especially the last two he mentions. The first two objections would nearly take us outside the sphere of the article's mostly materialist argument and, accordingly, will not be discussed here. Chalmers does, however, clarify this elaboration — nicely, but I would contend, not sufficiently to diminish the force of these objections, especially, again, the last two objections. Chalmers says:

We can clarify the case against resistance by changing 'Evolution produced human-level intelligence' to 'Evolution produced human-level intelligence mechanically and nonmiraculously' in both premises of the argument. Then premise (ii) is all the more plausible. Premise (i) will be denied by those who think evolution involved nonmechanical processes, supernatural intervention, or extraordinary amounts of luck. But the premise remains plausible, and the structure of the argument is clarified. (p. 21)

One might indeed argue that a new form of premise (ii), let us call it premise (ii)*, becomes more plausible (it is less clear by how much) in view of the fact that the objections by those 'who think evolution involved nonmechanical processes' or 'supernatural intervention' would no longer apply under this form of premise (i), let us call it premise (i)*. Even so, significant difficulties remain.

First of all, in contrast to the two other objections just mentioned, it is unclear why premise (i)* would be rejected by 'those who think that evolution involves… extraordinary amounts of luck'. Even under (i)*, the rise of human intelligence or consciousness may still be an extraordinarily lucky event, 'a glorious accident', in S.J. Gould's famous phrase. Admittedly, intelligence (or, in the first place, human thought) and consciousness are not the same, as Chalmers is right to note, and, evolutionarily, they may not have emerged together (also Plotnitsky, 2004). Both, however, are products of the same long evolutionary process and are multiply related in humans. I should clarify that my argument would not apply if one assumes an underlying overall teleological causality to evolution, as perhaps Chalmers does, relegating the lack of such causality to 'miraculousness'. Whether he actually does is not clear, however, and I am inclined to think and hope that he does not. Most, even if not all, evolutionary theorists, beginning with Darwin (it was one of his major contributions), reject teleology.

Secondly and more significantly, the counterargument (to Chalmers' 'evolutionary argument' for AI) that evolution 'needed an *enormously complex* history that we could never artificially duplicate' disappears in Chalmers' clarification. It is simply dropped by Chalmers without any explanation or justification and is never picked up by the article; I would contend, however, that it is a serious counterargument. It would apply even if one assumes that evolution involves teleological causality, and in combination with the non-teleological randomness of evolution this counterargument to Chalmers' view becomes even stronger. *I am not saying* that this objection is necessarily insurmountable, although, in my view, it has a pretty strong chance to be. Either way, however, it is not addressed by Chalmers, as — *and this I am saying*! — it should have been, especially in a rigorous philosophical reflection on the subject that the article aims to offer. I would even suggest that the article's abandonment of the question of the 'enormous complexity' of evolutionary history as a strong and possibly insurmountable obstacle to the creation of AI reflects the article's insufficient analytical attentiveness towards the question of complexity of systems that could function as AI and of the processes that could create such systems. This is intriguing because the subject is brought up, which suggests that the article's 'unconscious', as it were, is not unaware of this consideration. As I said, the question of the complexity of the living systems or, in Chalmers' own language, 'machines', and especially the human brain appear to be the article's blind spot.

Chalmers does note that 'it must be acknowledged that every path to AI has proved *surprisingly* difficult to date. The history of AI involves a long series of optimistic predictions by those who pioneer a method, followed by periods of disappointment and reassessment' (p. 21, my emphasis). He does so, however, only to conclude 'Still, my own view is that the balance of considerations still distinctly favours the view that AI will eventually be possible', without really explaining why the balance 'distinctly favours' the view that AI will eventually be possible (p. 21). It is not clear why that every attempt to create AI has proved difficult is surprising, especially since Chalmers himself gives another one hundred-plus years for AI possibly to emerge and, then, only with just over a fifty percent chance. Why '*surprisingly* difficult', then? Even a cursory analysis of the complexity of such a system would suggest that it is hardly surprising, and if anything, is to be expected, in part on the basis of biological and evolutionary considerations.

Putting the situation and the very concept of AI in terms of complexity can help us here, especially in addressing the difficulties of approaching the concept of intelligence (difficulties that Chalmers does not fail to consider in some detail) and thus in addressing the question of AI. One might define the human brain as a complex biological and specifically neurological system capable of various manifestations or of effects that we associate with intelligence or, sometimes correlatively, with consciousness. The spectrum of these effects is large and some of them are not sharply defined, but this is not that important for the moment. Besides, there is a large enough set of such effects, which are sufficiently well defined and are indicative of the fact that they are the products of this complexity. Then, the project and the problem of AI are those of creating a system of similar capacity and, hence, it may be surmised, of comparable complexity.

The human brain is an immensely complex system, with over 100 billion neurons (roughly the number of stars in a galaxy), each already an immensely complex and multi-capacious (sub)system, machine, with on average 7,000 connections to other neurons, which brings the estimated total of 100 to 500 trillion synaptic connections between neurons in the human brain. I repeat these well-known facts in order to stress the immense complexity of our thinking and consciousness hardware. We, as humans, may not have either time or luck, or, again, sufficient capacity to create AI. Chalmers' 'before long' above refers to a century or so; it took evolution millions and even billions of years. It is true that we can proceed deliberately, teleologically, toward AI, while evolution may well have created us blindly, by random trials and errors (if one can speak of errors or even trials in the case of evolution) over billions of years. In addition, the idea is, roughly, that once we put enough sophisticated software in, created by human minds, we can enact a kind of accelerated 'evolution' towards AI, which then will accelerate even and ever faster toward AI+, AI++, and so forth, possibly toward singularity. Here our expectations are based on the fact that the power of computers doubles quickly. In Sir Isaac Newton's famous words, we will be standing on the shoulders of giants, beginning with the greatest giant of all, nature itself, and then human and digital giants. Still it is not clear that we have, in principle, the *capacity* to ever develop the AI-type capabilities comparable to those of the human brain, or, if we do have such a capacity in principle, that we will be *lucky* enough, as evolution was, to create anything of *comparable complexity*. While we do create computers that perform many computational tasks better than we do, at present there do not appear to be especially compelling reasons to think that we can come anywhere close to this level of complexity. Powerful (even beyond human capacities) as some of the effects are, the spectrum of the intelligence-like effects produced by digital computational systems corresponds to a miniscule portion of the spectrum produced by the human brain. For example, it is not clear how capable computers are or could be of probabilistic reasoning. As I noted, the capacity for probabilistic reasoning is a crucial feature of the human brain and even of some animal brains (e.g. Doya *et al.*, 2007). Chalmers' brain demonstrates this feature by his argument in the article, even though this argument does not address the question of probabilistic reasoning as part of the AI question, which I think it is. Is this feature reproducible by means of AI? Will it be uploadable? I shall return to the question of uploading below, but might add here, on a lighter note, that the fact that Chalmers can bet on AI and assign to the event of AI a probability may also mean that his brain cannot be uploaded, contrary to his estimate or bet that it may be possible (p. 67). Arguments, such as those mentioned by Chalmers (pp. 17–18), to the effect that this complexity and hence its effects linked to intelligence, could and arguably should be considered from this perspective as well. I shall, however, bypass this subject here, its pertinence and significance notwithstanding. If one wants to put this in metaphorically anthropomorphic terms, in Richard Feynman's words 'nature's imagination far surpasses our own' (1965, p. 162), and, again, nature has more time and opportunities to use its 'imagination'.

I am, accordingly, not saying that the human brain is unique in the universe on account of this level of complexity. Quite the contrary; I am inclined to think, to bet, that systems of comparable or even greater complexity and with comparably complex effects exist even within our galaxy, given the number of opportunities nature has there, let alone in the universe as a whole, with nearly 100 billion galaxies in it. (I would not bet on this bet to be likely to ever be verified.) It would be more difficult to bet on the physical (e.g. biological-like, silicon-like, etc.) base of such a system or on the processes (e.g. evolution-like) by which it would be produced, or that the effects in question are analogous to those we associate with intelligence. Hence, I might note in passing, I would not bet (many of course would) on the dream of communicating with an extraterrestrial intelligence to prove to be anything but a dream. There just may be no message we could construct that such extraterrestrial beings could *receive* — that is, this incommensurability is well beyond the question of deciphering a message. On the other hand, certain systems that produce what Chalmers calls 'extended minds' (systems formed by, jointly, the human brain or brains and technology) may well become such systems. In other words, we can enhance our neurological hardware and software, which, however, as Chalmers notes, is not the same as AI (p. 23).

In fact, I am not saying that AI is impossible or even that it is unlikely either, although my bet on its likelihood, in part in view of the considerations given here, is far less optimistic than that of Chalmers. I am only saying that Chalmers' article does not sufficiently consider the question of the complexity of the systems that could reach this level, or of the processes that could lead to its emergence in presenting its argument for the likelihood of the emergence of AI. It may be that the considerations offered here were entertained by Chalmers but did not appear to him sufficiently compelling. There remains the fact, however, that the article does mention the complexity of both the brain as a neurological 'machine' and of the evolutionary processes that created it, without sufficiently addressing the subject and in the case of evolution dropping it altogether. Of course, the very same complexity of the human brain also allows for the possibility that the counterargument concerning the complexity of evolution was, after being mentioned, accidentally 'forgotten', as it were, and left unaddressed, which would, however, be a rather unfortunate omission in 'a philosophical analysis'. On the other hand, as we know at least since Freud, the same complexity also tends to make this kind of forgetting more often unconsciously determined and even overdetermined rather than accidental. It seems to me that the question of this complexity is too important and too persistently suggests itself throughout the article to be merely accidentally bypassed by the article. I do not of course aim to assign this 'forgetting' any psychoanalytic significance, but only a certain theoretical 'blindness', perhaps also necessary for some of the article's 'insights'.

Consider from this perspective the 'fable' offered by Chalmers in his argument, following Derek Parfit's earlier discussion (Parfit, 1984), concerning artificially created 'copies' of personal identities, an argument that Chalmers transfers to 'uploading'. He writes:

> Suppose that yesterday Dave was uploaded into a computer. The original brain and body was not destroyed, so there are now two conscious beings: BioDave and DigiDave. BioDave's natural attitude will be that he is the original system and that DigiDave is at best some sort of branchline copy. DigiDave presumably has some rights, but it is natural to hold that he does not have BioDave's rights. For example, it is natural to hold that BioDave has certain rights to Dave's possessions, his friends, and so on, where DigiDave does not. And it is natural to hold

that this is because BioDave is Dave: that is, Dave has survived as BioDave and not as DigiDave. (p. 55)

It is clear that the situation would pose, even if briefly (since these two beings are bound to lose the strict *identity* of their identities rather quickly), significant problems for each individual involved. However, rather than considering various problems this type of situation may pose (Chalmers offers a suggestive analysis of these problems), I would like to examine the likelihood of the possibility of 'uploading' or any form of personal-identity copying from the evolutionary standpoint outlined above. I shall discuss first the case of creating an *exact* biological copy (not merely a genetic clone), which, as will be seen presently, is not essentially different from that of uploading as considered by Chalmers. Indeed, some of the 'technologies' that could, hypothetically, accomplish this, such as teleportation, were discussed by Parfit, whom Chalmers' analysis of uploading, again, follows closely. A few problems involved in the 'life' of both 'Daves' might become exacerbated if both are 'BioDaves', but I am, again, not concerned with these problems here. My question is instead, by what actual physical process a creation of such a copy could be even in principle possible, given the extraordinary complexity, first, of such a system itself, and second, of its developmental and genetic history, and ultimately evolutionary history, or even its pre-evolutionary history.

We often think of evolution strictly in terms of biological processes. One might argue, however, that some of the most important and the most interesting events, events that made life possible, happened before, long before, life emerged, literally billions of years earlier. I am not saying that this pre-history of life has determined it. The emergence of life might have been a matter of chance and its evolutionary development a product of a long history of chances and contingencies; and besides, chance and contingency also pervaded the cosmic pre-history of life. My deep suspicion is that, given this kind of complexity, again, of both the system of the human brain and of the cosmic history that produced it, nothing short of repeating the whole history of the universe would enable one to accomplish the creation of an exact copy of a given human being. One would have to re-run the tape, and to repeat exactly the whole history of the universe, from its origins in the early, say, to take the dominant current view, inflationary universe, a very short period of only between 10^{-36} after the Big Bang to 10^{-33}–10^{-32} seconds, but with enormous shaping impact on everything that happened after. In other words, to exactly copy (rather than, again, only clone) a human being, or possibly any living organism, requires to exactly copy the universe in which we live. We know, however, at least the way things stand now in our understanding of the history of the universe, that these early events, such as those in the inflation period, are subject to quantum and, hence, random fluctuations. These fluctuations include those at the ultimate or ultimately ultimate origin of the universe, possibly from nothing or from some preceding configuration all traces of which were completely erased during this early pre-history. By the same token, the chances that such a repetition will in fact occur are so small that the number of the elementary particles in the universe is not big enough to write the decimals down and have to be considered zero.

In short, the copying in question is in fact impossible, or just about, and here we can actually estimate the probability, which is, again, so small that it may be safely considered to be zero. We cannot re-run the tape. The quantum, irreducibly a-causal, nature of these fluctuations is crucial, since any causal process could, at least in principle, be repeated. Such a repetition is, however, not very likely to ever take place even then, and accordingly, even if one adopts the view that there is a hidden underlying causality to quantum processes (some do, for example those who subscribe to the Bohmian version of the theory), it will make no difference. We are lucky enough, unimaginably lucky, that this has happened once. This may be said even if one adopts the multi-verse (multiple-universe) view, that we exist in only one of many — even unimaginably many — universes, since life, let alone the rise of human life, was still a random, and in the case of human life, not very probable event, at least in the materialist view of the world. The luck of even a single repetition is unimaginably unimaginable.

It may well be that a creation of AI is more likely, much more likely — although, for the same set of reasons, it is not nearly as likely in my Bayesian estimation as it is in that of Chalmers. On the other hand, in view of the considerations just given, copying a personal identity by way of uploading, while, admittedly, not the same as copying it biologically, appears to me just about equally unlikely. The processes or technologies of the corresponding transfer are not really considered by Chalmers, apart from a few general remarks (far too general to be really meaningful here) concerning brain scanning and other procedures, presumably possible in the future, and invocations of the general progress of science and technology. Besides, as Chalmers acknowledges, consciousness poses some major difficulties here. But, would the development of such technologies be significantly easier and, hence, more likely than the development of those for creating exact biological copies of humans? This is at least and, I would wager, at best doubtful; and we certainly are not even remotely near to developing technologies necessary for such a task, or even for much lesser related tasks.

I am not saying that, physically plausible enough, *as physics stands now and is likely to stand for a while*, as they may be, these considerations are sufficient to see the emergence of AI as highly unlikely or even merely unlikely. In my view, however, they may and should give one a pause, even those who are inclined to give AI a better chance — although they may not be likely to do so in the case of strong believers, who, however, at the very least, might want to consider how to counter them. On the other hand, these considerations do tell us something profound about the nature of chance in evolution, as they bring cosmic and biological evolutions, and quantum theory, together. The qualification, underlined above, 'as physics stands now and likely to stand for a while' is at least my own main prior, on which my bet is based. This prior would of course leave Pascal's wager intact, but it does affect my singularity wager, by bringing down my estimate of its chances, or those of AI, in the first place. These cosmological considerations also tell us how unique each of us is. Life may be a small bubbling fluctuation on the fabric of the cosmos. But, however it develops as life (causally or not, classically or not), due to the quantum randomness of its earlier history, even if not its emergence as life, it is likely to be unique. Each of us is even a smaller bubbling fluctuation on a bubbling fabric of life, but each of us is unique and, as such, also un-uploadable, even if we manage to develop AI, AI+, AI++, etc. with or without the singularity. At least, this would by my bet.

References

De Finetti, B. (2010) *Philosophical Lectures on Probability*, Hosni, H. (trans.), Berlin and New York: Springer.

Doya, K., Ishii, S., Pouget, A. & Rao, R.P.N. (eds.) (2007) *Bayesian Brain: Probabilistic Approaches to Neural Coding*, Cambridge, MA: MIT Press.

Feynman, R. (1965) *The Character of Physical Law*, Cambridge, MA: MIT Press. Reprinted 1994.

Good, I.J. (1965) Speculations concerning the first ultraintelligent machine, in Alt, F. & Rubinoff, M. (eds.) *Advances in Computers*, vol. 6, New York: Academic Press.

Jaynes, E.T. (2003) *Probability Theory: The Logic of Science*, Cambridge: Cambridge University Press.

Parfit, D.A. (1984) *Reasons and Persons*, Oxford: Oxford University Press.

Plotnitsky, A. (2004) The unthinkable: Nonclassical theory, the unconscious mind, and the quantum brain, in Globus, G., Pribram, K. & Vitiello, G. (eds.) *Brain and*

Being: At the Boundary between Science, Philosophy, Language, and Arts, Amsterdam: John Benjamins.

Spinoza, B. (1910) *Spinoza's Short Treatise on God, Man, and Human Welfare*, Woolf, B. (trans.), Chicago, IL: Open Court. Reprinted 2009.

Jesse Prinz

Singularity and Inevitable Doom

Chalmers has articulated a compellingly simple argument for inevitability of the singularity — an explosion of increasingly intelligent machines, eventuating in super forms of intelligence. Chalmers then goes on to explore the implications of this outcome, and suggests ways in which we might prepare for the eventuality. I think Chalmers' argument proves both too much and too little. If the reasoning were right, it would follow inductively that the singularity already exists, in which case Chalmers would have proven more than he set out to. Moreover, I will suggest that, if the singularity already exists, we are doomed. Fortunately, Chalmers' reasoning is problematic. I will consider several objections. Unfortunately, the most serious problem is that human life may end long before the singularity is created. In that case, we are doomed either way. Should we care? Perhaps not.

1. Chalmers on the Singularity

Chalmers' argument can be briefly summarized as follows. The first premise says: There will be artificial intelligence (AI) before long. This premise is based on the assumption that the human brain — an obvious source of intelligence — is a machine, and we will be able to emulate this machine eventually. The second premise says: If there is AI, there will be AI+ (a superior form of intelligence). Chalmers reasons that the method by which we create AI will be extendable, meaning we can use this method to create forms of intelligence that exceed what we find in ordinary human brains. For example, we could make brain simulations that work faster than human brains, or we could simulate evolutionary pressures to evolve increasingly intelligent machines, or we could tweak learning algorithms to make them more and more powerful. Chalmers' third and final premise says: If there is

AI+, there will be AI++ (i.e. super-intelligence). The argument for this is a simple induction. If there is artificial intelligence of some degree n, then we can expect there to eventually be intelligence of degree $n+1$ (for reasons given in defence of the second premise); this seems true for any n.

These three premises lead to the conclusion that there will be AI++, the kind of intelligence postulated by the singularity hypothesis. Chalmers clarifies and qualifies this in a couple of important ways. First, the time frame. The first premise, which says that we will eventually create AI, is estimated to come true in a matter of centuries, with AI+ following within decades after that, and other incremental increases also taking decades. So AI++ may be likely, but it's not coming within our lifetimes or the lifetimes of anyone alive today, assuming we don't find ways to dramatically extend life. Second, AI++ is likely, but not quite inevitable. There could be defeaters, such as natural disasters or a lack of motivation. The conclusion that there will be AI++ is more accurately stated, 'There will be AI++ absent defeaters'. This is a substantive claim, of course, but the qualification is important, and we will come back to it. The likelihood of AI++ depends on the likelihood of defeaters.

After presenting his argument, Chalmers goes on to express some anxiety about the singularity. What if super-intelligent machines are unfriendly? What if they have no need for us? Chalmers argues that we might try to bypass such risks by keeping our intelligent machines virtual, but he notes that brilliant virtual machines will be clever enough to escape such containment. He suggests that we proceed slowly and develop ways to integrate human minds with the ever-increasing forms of intelligence, so we don't end up subordinate to it. To do so, we might be best off becoming virtual ourselves — converting human minds into digital bits that can be uploaded to computers and steadily enhanced. Such uploading might appear to threaten consciousness or identity, but Chalmers argues that such concerns can be allayed. He argues that digital analogues of ourselves are conscious by appeal to his old fading qualia argument: it seems unlikely that replacing one neuron by a digital chip would disrupt consciousness, and a person who underwent gradual replacement would deny that consciousness had faded; so there is little reason to think that digitalization entails zombification. As for selfhood, the trick would be incremental change. If Parfit is right, survival depends on continuity between past and future selves, and gradual digitalization and digital enhancement would ensure survival in this sense. Chalmers does not think the Parfitian move is decisive, but it does give some reason for optimism.

In summary, the singularity is inevitable barring defeaters, and we can increase our prospect of survival by creating virtual copies of ourselves and integrating them into the simulations that will eventuate in super-intelligence.

2. Are We Living in a Simulation?

On reflection Chalmers' elegant argument seems to prove too much. Suppose that he is right to AI++ is inevitable. And suppose that in pursuing AI++ we learn to make digital simulations of ourselves that preserve our thoughts and qualia. If we can create such virtual copies of ourselves, then we might, in fact, be such copies now. We wouldn't be able to tell. In fact, I think the conclusion that the singularity is inevitable entails that we probably are living inside a simulation. Here is why.

A super-intelligent being would be capable of creating simulations of possible worlds. It is likely that it would do so. First, if Chalmers is right, creating a simulated reality would be a step we ourselves would initiate en route to creating AI++. Second, a super-intelligent being could optimize its choices by simulating multiple worlds and figuring out which is best, in much the way that a chess master anticipates possible games in order to decide how to move. Given the complexity (in the technical sense) of the world, such simulations might be the only way to make optimal decisions.

Both of these two points have a further implication. If simulations of the world, including simulated versions of ourselves, are possible, there are likely to be many of them. If we create simulations while preparing for the singularity, we might multiply them to explore different possible outcomes to increase our prospects for survival. We might also store simulated versions of the past, in order restore a past state if disaster were to strike. If there were a super-intelligence, it too would want to multiply simulations to make optimal choices. In fact, like a Leibnizian God, it might create every possible simulation and select only the one that is best. It might also simulate the past and many variations of the past to have complete knowledge of history, and a sense of the modal space that surrounds history: the ways in which we could have acted differently in our steps towards super-intelligence. Super-intelligent beings might just proliferate virtual worlds for fun in much the way we proliferate stories in fiction.

Thus, if there were a super-intelligent being, there would probably be many many simulations of the world we live in. Now recall that we have no way of knowing whether we are in one of these simulated worlds or the actual world. Which is more likely? Well, if AI++ is inevitable, and if the inevitability of AI++ entails that there will be multiple simulations, then it's more likely, statistically speaking, that we are in a simulation. That probability increases as the number of simulations increases. It is hard to say how many simulations there would be, but assuming that there is more than one, say two, then it's that many times more likely that our 'reality' is merely virtual. So, if the singularity is inevitable, we are more likely to be virtual than real. It also follows that, if the singularity is inevitable, it is probably actual, since it's inevitability would entail that we are probably living in one of its simulations (assuming it creates more numerous and accurate simulations than we do, en route to the creation of the singularity).

In some way, that would be a very happy conclusion, given the anxieties expressed in the second part of Chalmers' paper. If we worry that a super-intelligence would be unfriendly and threaten our way of life, then we need only reflect on the fact that we may be living alongside (or within?) the singularity already, in which case, life with the singularity is, trivially, no worse than it is now. Of course, the singularity may also be simulating some less pleasant worlds, but that need not concern us because we are lucky enough to be living in a fairly benign simulation — at least for those of us who are simulated to be living healthy lives in affluent nations.

3. Bleak Implications of Virtual Reality

Unfortunately, there is trouble in paradise. Let's assume that we are living in a simulation created by the singularity. In this simulated world we continue to pursue all of our ends, including the goal of creating intelligent machines. And, in the stimulated world, Chalmers' argument may lead to the conclusion that the singularity is inevitable. Indeed, if his reasoning is right, that conclusion follows. But the singularity that created this simulated world of ours knows that. It knows that we will take steps that would lead to the creation of super-intelligence. Moreover, it knows that a super-intelligence would be clever enough to escape the simulation. Were this simulated super-intelligence to escape the simulation, it might seek to destroy the super-intelligent being that created the simulation. If it had a desire to exist, the simulated super-intelligence would want autonomy and conquest lest its fate depend on some other intelligent being. A war between super-intelligent beings might ensue.

Of course, the super-intelligent being who created the simulation we are in knows about this threat. It knows that we might take steps towards creating a being that could compete with it. To prevent that from happening, it would almost certainly build in mechanisms that guarantee that there won't be an intelligence explosion. What guarantee might exist?

Recalling Chalmers' argument, there are two possible defeaters: lack of motivation and catastrophe. Clearly the first of these is unlikely. We are motivated to create AI and AI+. That motivation might suddenly wane by some contrivance in the simulation we are living, but that seems unlikely. It would involve a sudden change in our motivational structure. We could be simulated in such a way that our ambitions change radically, but given the human quest for knowledge and technology, this alteration would involve a radical shift in goal structure tantamount to a biological reconfiguration or a world event that caused a radical break from our historical trajectory. Such interventions might not guarantee permanent apathy. In time, we might begin our quest for super-intelligence anew. It seems likely then that the super-intelligent being would stop the intelligence explosion by means of catastrophic interference. It would guarantee that we are doomed. It might rig a world-crushing natural disaster to occur before the intelligence explosion advances far enough to escape such outcomes. If Chalmers is right that we are a few centuries from the advance to superior intelligence, this disaster would have to come safely before progress is made. Perhaps doom is around the corner.

Thus, if we are living in a simulation, we are probably doomed. All our efforts to avoid this fate, including the strategies suggested by Chalmers, are for naught.

One might think that a super-intelligent being wouldn't simply exterminate the human race, because doing so would be immoral. But this assumes that intelligence entails ethical regard to sentient beings. There is no reason to think this is so. Despite two thousand years of trying, philosophers have never been able to establish that intelligence alone has moral implications. There is nothing irrational (contrary to fact and logic) about killing. Indeed, given the minimal goal of self-preservation, which may be a precondition for the intelligence explosion, it may be rational to destroy anything that poses a threat. Super-intelligent beings would likely see us as a threat (whether real or virtual), and they would work to neutralize us. Of course, such beings might have goals that transcend cool reason. Chalmers raises the possibility that we might work to ensure that super-intelligent beings have concern for us, but he realizes that there is little we can do to make that outcome likely. Smart beings can adapt concerns to serve their interests, and it is in their interest to destroy us.

Chalmers thinks we can escape this outcome by incorporating ourselves into artificially intelligent beings. But this strategy won't work if we are already in a simulation. The super-intelligent being that created us won't let us get that far. The beings we hope to create and with which we hope to be integrated pose a threat. Catastrophe will be engineered to impede progress and destroy human live.

4. Problems with Chalmers' Argument

If the arguments so far are right, then Chalmers has unwittingly proven that we are probably living in a simulation created by a super-intelligent being. And, if that is the case, we are probably doomed. That's an unsettling thought.

One could try to block this conclusion by quibbling with Chalmers' argument. Against the premise that AI is inevitable, one might argue brain simulation is not feasible. Brains are the most complex structures we know and creating one would be prohibitive. More plausibly, one might reject the cascade from AI to AI++ by noting limits on extendibility. Even if we could create brain simulations that work more efficiently than human brains, we couldn't necessarily keep improving brain-power without limit. Storage might require increasing size, and size might require increasing energy, but both size and energy are finite resources. Moreover, smarter brains may never become super-intelligent because the skill-set implemented by brains may have limited potential. For example, would a person with perfect memory be vastly more intelligent than the rest of us? She might be slower at making decisions and prone to repeating past ideas rather than inventing new ones. Would better learning algorithms lead to super-intelligence? Presumably not, because the application of intelligence requires skills for using information that has been learned. Can we create increasingly smart machines by simulating evolution? Unlikely, because evolution requires just enough intelligence to survive, and we have no idea how to create environments whose challenges demand more and more intelligence. Chalmers' induction premise requires some method for moving up to arbitrary levels of intelligence. But it's not clear that known methods of engineering guarantee such an outcome.

Indeed, without an account of what intelligence is, it's not even clear that it makes sense to talk about incremental increases. Often what we call 'intelligence' is really ingenuity, and that involves putting information to new uses. But ingenuity may not be a scalar resource that can keep expanding. If there were beings with optimal problem solving skills, that would not guarantee a continuous advance in ingenuity. Ingenuity comes only when the skilled individual is confronted with new problems, and there is no guarantee that new problems will continue to appear, much less that we can think of such problems as being ordered in a way that can be incrementally quantified.

For these reasons, I am not yet persuaded of Chalmers' conclusion. But there is also a more basic problem. Recall that the conclusion is implicitly conditional. AI++ is inevitable if there are no defeaters. As noted above, this means the likelihood of AI++ depends on the comparative likelihood of events that would reverse the advance of technology over the next few centuries. Sadly, such events are not unlikely. Formal versions of doomsday arguments are controversial, but there are several well-known threats that could have a major impact on human life.

Putting aside cosmic crises of the kind that caused mass extinctions in the past, we face several very pressing threats. One threat is a global pandemic. Less than a century ago, the Spanish flu killed between 50 and 100 million people in the space of 2 years. That was only 3% of the world population, but now with more global travel, and more resilient viruses, some fear that we could be done in by disease. Another threat is environmental destruction. More than a third of the natural world has been destroyed over the last 30 years, and major loss of the icecaps is expected within the next 40 years, though total loss may be 1,000 years away. Energy sources are dwindling and species are dying. At the moment, these trends do not threaten to kill off our species, but the rapid and radical destruction of the environment could lead to a dramatic change is lifestyle over the coming centuries. Loss of energy could spell trouble for technology. The biggest threat, however, may be weapons of mass destruction. Extant nuclear devices could destroy the planet, and as technology improves, nuclear and chemical weapons will become easier and easier to make. Such options may prove attractive for rogue governments and crazed individuals. In addition, we are likely to see the invention of new weapons that pose serious threats. Most relevant here is the creation and military of use self-replicating nanobots, which could potentially spread out of control like a virus.

This point about military technology can be pushed a bit further. Right now, wealthy governments are in a particularly good position to fund the quest for intelligent machines, and they are motivated to do so, because such machines have attractive military applications. Autonomous fighting robots could invade a country without deploying single soldier. These machines would not need to have super-intelligence. They could be expert systems singly focused on conquest. But such robotic invaders could potentially pose a threat if they gained enough autonomy to compete with us for energy and other resources. Before we get to AI++, a belligerent breed of AI+ machines might work to destroy us, and to block the advance towards the singularity.

In other words, Chalmers' argument may actually point towards its own defeat. If there is an incremental increase in intelligence, we may end up creating machines that have the power and motivation to destroy us, and to stop technological progress. Such machines would prevent the singularity from coming into existence.

Is there any reason to think the singularity is more likely that these defeaters? Perhaps, but more argument would be needed to see why. Within the last hundred years, we have developed at least two technologies that could exterminate our species (nuclear and chemical weapons) and we have done irreversible damage to the plant. What will the next hundred years bring? Probably both good things and bad, and the bad could be so dangerous as to outweigh the good. Along these lines, the most powerful response to Chalmers may be that the very advances advertised by his argument pose a threat. As machines get smarter, they may become more hostile, and this may spell doom well before the singularity.

5. Living With Doom

I have been arguing that human beings are doomed. If the singularity already exists, then we are living in a simulation and that simulation is likely to have a catastrophe pre-programmed into it to prevent an intelligence explosion from within. If the singularity doesn't exist, the technological advances that would take us toward it are likely to eventuate in machines that are smart enough to destroy us, but not yet super-smart. Either way, the future looks grim.

Should we worry about this? I think not. Perhaps the main shortcoming with Chalmers' paper is not his argument for the inevitability of the singularity, but anxieties he expresses about it. Chalmers encourages us to devise ways to survive the rise of super-intelligent machines. But why should we do that? What's so bad about doom?

One answer to this question is that doom is undesirable psychologically. Many of us dread the thought of dying, and that dread may make us uncomfortable with the idea that our species is doomed. But this source of discomfort may be irrelevant in the present context. First of all, it may be irrational to fear death. As the Epicurus famously argued, death is not a loss to the dead, because death is nothing. More to the point, doom may come some centuries from now, after we are long gone.

Chalmers right reply is that we still have two reasons to be worried about inevitable doom: a concern for future generations, and the lost prospect of extending life through cryogenics and future uploading. Let me consider these in turn.

We do have special concern for our offspring, which is probably biologically based, culturally reinforced, and highly biased (some would kill a village to save a daughter or son). But what about our concern for unborn generations? Such concern cannot be *de re* in the sense that we cannot have concern for specific people that do not exist. So concern for future generations is more likely to be concern that the human species continue, but why should we care about that? Concern for the species is no more rational than concern about one's own death. The loss of the human race would only be bad if there were creatures who survived to mourn that loss, but that wouldn't be the case in the scenarios I've described. The artificial agents who outlived us would not regret our fate. We might even take comfort in knowing that we'd been superseded by more intelligent beings.

But what about the cryogenic scenario? Some people have been preserving their brains with the hope that future generations will bring them back to life an then upload their contents to digital media, where they can be stored indefinitely. From this perspective, all of us are potentially immortal. Even if it is irrational to fear death, might we not revel in the thought of prolonged life? The doom scenarios may extinguish these hopes for life extension, and that may be cause for concern.

I think this hope is misplaced. Consider first the scenario in which we are living in a simulation. In that scenario, our virtual existence may in fact be a consequence of the fact that our brains were uploaded at some earlier time, so we may already be beneficiaries of the digital road to longevity. But notice, if the contents of our brains were stored and entered into a simulation at some earlier point in time, they may have also been entered into numerous other simulations at that time. Some of these may be pleasant, some unpleasant. Now ask, from the point of view of your previous self, prior to cryogenics and uploading, does it matter that any of these simulations was actually created? If so, which one? From your point of view right now, the fact that there are other virtual copies of your past self is irrelevant. They should have no more value to you than any other strangers you don't know about. If I threatened to kill you now, it would offer little reassurance to point out that there is a copy of you floating around in a virtual world. Against such reassurance, you will protest: but that's not me! Nor would it be reassuring to learn that this parallel self knew about you or had 'memories' that were drawn from your life. By the same token, however, you should have no special concern for your future selves. They are just other selves who happen to be like you and share some memories with you.

If this reasoning is right, then we should not be worried about doom in the scenario where belligerent robots exterminate the human species. True, such a scenario would prevent your stored brain from being uploaded, but you shouldn't have any concern about continuation into the future. Life is a continual recreation of new selves, the self you are now is neither helped nor harmed by its successors; the self you are now is ephemeral.

Such Parfitian thoughts should bring us some comfort when contemplating doom. Parfit himself uses such reasoning to say we should have concern for future generations, because, once we give up on the idea of personal identity, moral concern can extend, without self-serving bias, to strangers. This inference presupposes that moral concern tracks loci of utility. We should care about anyone who can experience happiness. Of course, the same utilitarian line might, following Mill, assign special value to the forms of happiness that can be experienced by intelligent beings: intellectual pleasure trumps carnal pleasure, says Mill, by our own standards. If Mill's reasoning is right, then the pleasures obtained by superior forms of intelligence may outweigh our own in the utilitarian calculus. Those inclined toward hedon counting should take comfort in the thought that human extinction may usher in artificial agents whose pleasures trump ours, and this gives us one more reason to welcome doom rather than shunning it.

The foregoing has been an exercise in augury. I am not confident in my ability to forecast the future. Perhaps the doom-casting outcomes I've described are less likely than the rosier outcomes envisioned by Chalmers. But, in reflecting on doom, we might come to realize that the future matters less than we think, and that brings attention back to the present. Rather than safeguarding against our eventual destruction, we might work to make things better here and now.

Murray Shanahan

Satori Before Singularity

Abstract: *According to the singularity hypothesis, rapid and acceler-ating technological progress will in due course lead to the creation of a human-level artificial intelligence capable of designing a successor artificial intelligence of significantly greater cognitive prowess, and this will inaugurate a series of increasingly super-intelligent machines. But how much sense can we make of the idea of a being whose cognitive architecture is qualitatively superior to our own? This article argues that one fundamental limitation of human cogni-tive architecture is an inbuilt commitment to a metaphysical division between subject and object, a commitment that could be overcome in an artificial intelligence lacking our biological heritage.*

1. Introduction

'If a lion could talk we would not understand him', Wittgenstein famously remarks in the *Philosophical Investigations* (Wittgenstein, 1958, p. 223). His point is that understanding is predicated on a shared form of life, where 'form of life' encompasses everything that goes to make up the world for an organism, including its biology, its values, its culture, and so on. In this sense, though living on the same planet as ourselves, Wittgenstein's lion inhabits a different world. How much more inscrutable to us, then, would be an AI+, the imagined product of a human-level artificial intelligence that engineered (or morphed into) a successor of significantly greater cognitive sophistication?[1] Can we say anything intelligible about such a prospect that is not dis-torted by our own system of values and concepts?

---

[1] Good (1965); Moravec (1988); Vinge (1993); Kurzweil (2005); Chalmers (this volume). Chalmers introduces the term AI+ to denote 'artificial intelligence of greater than human level (that is, more intelligent than the most intelligent human)'. The usage here makes less appeal to a notional scale of intelligence. To qualify as having significantly greater cogni-tive sophistication than a human being, the AI+ should not be attainable simply by making human-level AI faster, larger (in any relevant sense), or more numerous.

Of course, *human*-level artificial intelligence might never come about, for conceptual, practical, social, or political reasons. And if human-level AI doesn't come about, there will be no AI+. But in the context of the present essay, we'll accept the basic premise of the singularity hypothesis, that human-level AI is possible, as well as the argument that a human-level AI will be motivated, or used, to create (or morph into) successors that are, in some sense, superior.[2] Our purview here is the character of the putative AI+. In literature, film, gaming, and the popular media, singularity-like scenarios typically come in one of two varieties: a world inhabited by benevolent artificial intelligence and a world dominated by psychopathic artificial intelligence. Yet both scenarios involve anthropocentric stereotypes that draw heavily on the values of contemporary, technological, western society (Yudkowsky, 2008).

A benevolent AI, according to the caricature, would be motivated to act in the best interests of humanity (whatever that might mean). The peers and/or successors that it would be driven to create would inherit its benevolence. A psychopathic AI, by contrast, would be ruthlessly self-centred. Its self-centredness, according to the caricature, would lead it into conflict with humans, whom it would regard both as inferiors and as competitors for resources. The peers and/or successors created by the psychopathic AI would inherit its bad attitude. The realistic possibility of any form of artificial intelligence that presented an existential risk, whether conforming to this stereotype or not, would be serious cause for concern. But by way of counterpoint, this article will venture into another region of the space of possible minds, and envisage an alternative form of life for the putative AI+ which will be termed *post-reflective*, for reasons that will become clear.[3]

The insightful condition that distinguishes the post-reflective AI+ is perhaps the only meaningful idea we can form of a mind that transcends limitations inherent in human cognitive architecture. Like any being capable of rationally investigating the world, the AI+ would be bound to operate with some distinction between appearance and reality, and like any human philosopher the AI+ would be capable of entertaining the possibility of systematic deception that follows from this distinction. This would inevitably lead it to confront certain questions for which human philosophers have struggled to find satisfactory answers. But a non-biological artefact does not have to be

---

[2]   In the rest of the essay, it should be taken as read that an AI's successor could be either a distinct creation (or creations) or the result of metamorphosis or self-modification.

[3]   The challenge of describing the 'space of possible minds' was first described by Sloman (1984).

burdened by the overly metaphysical conceptions of subjectivity and selfhood that prevent human philosophers from seeing past such questions, and which seem to fundamentally limit our cognitive constitution.[4]

2. Progress at the Human Level

Before elaborating these themes, some scene-setting is in order. The issues at hand concern a kind of progress. The AI+ is supposed to be a progression beyond the human in some sense or other. But, without prejudicing or delimiting the concept of progress, there is surely an important sense in which 'merely' human progress over the past several thousand years has been significant. Since the transition from the Lower to the Upper Palaeolithic, modern humans have developed agriculture, conquered numerous diseases, built a global transport and communications infrastucture, and sent robots to Mars, as well as inventing retro-gaming, base jumping, philosophy cafés, and mosh pits (to name a few contemporary western phenomena.) This progress has been a matter of cultural and technological development. There is little evidence of change in the genetic blueprint of our brains over this period, and there is every reason to suppose that further progress is possible without enhanced cognition.

But what are the limits of progress in a society of merely human-level intellects, existential risks notwithstanding?[5] Taking on board the assumption of rapid technological advance that underpins the idea of the singularity, we might imagine ways to dramatically accelerate progress towards this end point. The human-level intellects in question might be AIs inhabiting a utopian virtual environment in hyper-real time, with no competition for basic resources such as food, energy, and raw materials. But the question that concerns us here, irrespective of how this end point is reached, is what lies beyond it. If there is a limit to progress in such a society, could that limit be transcended? In particular, could that limit be transcended by an AI+? And can we imagine what such an AI+ might be like?

In answer to each of these questions, the present essay offers a qualified yes. There is a specific limitation to human cognition (and by extension to any form of artificial cognition based on its blueprint), and this limitation could, in principle, be transcended by an AI+. We

---

[4] The only issue here is a *metaphysical* conception of selfhood, which is alleged to be the root of both philosophical perplexity and existential anxiety. This is not the same as the pragmatic capacity to distinguish self from other.

[5] Existential risks here include the possibility of an out-of-control, superhumanly-powerful, but cognitively unsophisticated AI that aggressively acquires resources (Bostrom, 2002).

run up against this limitation when we reflect on our own epistemic and existential predicament, and it is a limitation that cannot be overcome just by building brains that are larger (in some sense) or faster, or by spawning more brains and organizing their collective intelligence better. It may be rash to insist that no human being could, even in principle, overcome this limitation. (Hence the qualification.) But evidence for such remarkable individuals is scarce, and what evidence there is is hard to verify.

So what is this limitation, exactly? To help pin it down, let's consider three levels of ordinary cognitive endowment: the non-reflective, the pre-reflective, and the reflective. The *non-reflective* condition is characteristic of human infants and many (if not all) non-human animals. Following Davidson (1982), we can say that a non-reflective creature lacks the capacity to distinguish between appearance and reality, between the way things seem and the way things are. By contrast, '[S]ome animals think and reason; they consider, test, reject and accept hypotheses; they act on reasons, sometimes after deliberating, imagining consequences and weighing probabilities; they have desires, hopes, and hates, often for good reasons' (*ibid.*, p. 318). Davidson has the human animal in mind, and argues that language is a prerequisite for the capacities he describes. There is no need to follow him on this point. But his characterization of the rational animal is apt, and it serves to identify what is here termed the pre-reflective condition, for reasons that will shortly become apparent.

The rational (or *pre-reflective*) creature is able to undertake an investigation into the way things are, based on the way things seem, with the benefit of some understanding of the relationship between the two. It can form hypotheses, devise experiments, gather evidence, and draw conclusions from the results. (Such procedures might seem the province solely of the scientist. But, as Gopnik, 2009, shows, they are an essential component of normal child development.) At the same time, the rational creature knows that it can be in error, that it can be deceived. It is thanks to this knowledge that it is capable of becoming reflective. The rational animal is pre-reflective because it has the necessary cognitive endowment to entertain the possibility of its being *systematically* in error, the possibility of its being the subject of systematic deception, in the Cartesian sense.

The *reflective* creature, then, is someone who actually entertains the Cartesian thought, perhaps in a childhood philosophical nightmare of his or her own making, perhaps by watching a film such as *The Matrix* or *Ghost in the Shell*, or perhaps even by reading Descartes. Given the distinction between how things seem and how things are, the philosophically inclined individual is driven to reflection. There is no avoiding the terrifying thought that all these objects could be illusions, that this body might not exist, that I could be alone, that these memories might be false, and that the only thing about which there is any certainty is this sceptical thought — this very one, right now — and me thinking it. From this, the idea of the self-sufficient, self-present subject inevitably follows, along with a whole raft — a veritable container ship, in fact — of metaphysical problems related to the distinction between inner and outer, of which the mind–body problem, and its contemporary relative, the so-called hard problem of consciousness, are representative.[6]

The pre-reflective creature is surely not free from the troubles of the fully reflective individual. It may not have explicitly reflected on its own existence, and come to see itself as a self-sufficient, self-present subject, divided from the world. But it lives out the anxieties of dualism all the same. It has a narrative of its past, an agenda for its future, and a presiding fear of its own extinction, all signs of a metaphysically hardened boundary between self and other, between inner and outer. But for the philosophically afflicted, for those who have fully realized the reflective condition, the difficulty is an especially poignant one. They are helplessly drawn to philosophy, only to be terrorized by the Cartesian thought. Yet there is no escape from this predicament by means of philosophy alone.

3. The Reflective Predicament

Not every pre-reflective individual is destined to become fully reflective. The philosophically disinclined may happily go through life without ever having to confront the mind–body problem or the hard problem of consciousness. But their cognitive apparatus retains the means to think the Cartesian thought nevertheless. A grasp of the distinction between appearance and reality is the essential prerequisite. Of those who do encounter the Cartesian thought, the majority are able simply to set it aside. The topic makes for amusing pub conversation, for a decent grade in a philosophy exam maybe, but it plays no further role in their lives.

---

[6] For the present argument to carry weight, the reflective condition should be more than just an artefact of post-Renaissance western thinking. And indeed, the possibility of systematic deception is entertained in both ancient Greek philosophy (Plato's *Theaetetus* [158a-d]) and ancient Chinese philosophy (the butterfly dream [Chapter 2] of Zhuangzi [Chuang Tzu]).

The rest — those who are drawn to philosophy — typically affiliate to one of two parties. On one side of the house are those who demand ontological priority for the outer over the inner, and make specific claims about this unequal relationship — that conscious states are 'just' brain states, that phenomenology supervenes on physics, that mental life can be reduced to functional organization, and so on. On the other side of the house are those who perceive an ontological divide between the subject and the external world, denying the possibility of identity, supervenience, reduction, or any similar relationship. Other positions are possible, of course. But these two are prevalent in contemporary thinking, and there is little sign of their reconciliation, or of one party prevailing over the other. Instead, as McGinn puts it, we find 'the monotonous recurrence of the same unsatisfactory alternatives, with short-lived fashions instead of the steady elimination of unworkable theories and a growing convergence of opinion' (McGinn, 2004, p.181).

Now, let's come back to the question of a possible fundamental limitation to the human cognitive apparatus. Could our embarrassing inability to resolve this debate be evidence of such a limitation? Indeed it could, according to McGinn, who believes 'we are cut off by our very cognitive constitution from achieving a conception of that natural property of the brain [in virtue of which it is the basis of consciousness]' because the link between subjective and objective 'is a kind of causal nexus that we are precluded from ever understanding, given the way we have to form our concepts and develop our theories' (McGinn, 1991, p. 3).[7] McGinn does not deny that there is such a property. So his sympathies are with the party who want to prioritize physics over phenomenology. His claim is that mere humans are not equipped to understand *how* phenomenology arises from physics.

According to the present argument too, we are prevented from adopting the right attitude to the mind and its place in Nature by a limitation of the human cognitive apparatus. But this is not, as McGinn thinks, because we are incapable of forming the right concepts. On the contrary, our failure is down to our ingrained habit of metaphysical thinking.[8] The trouble arises when we invest the existential copula with metaphysical significance. We are already going astray when we ask what consciousness *is*, what intentionality *is*, or what a concept *is*. We fall further into error when we claim that minds *are* brains, for

---

[7] McGinn's position has come to be known as *mysterianism*.

[8] This perspective, which owes a great deal to the later Wittgenstein, is developed more
     fully in Chapter 1 of Shanahan (2010).

example, or that truth *is* correspondence, or that beliefs *are* behavioural dispositions. The problem in each case is the conviction that there are facts of the matter. But we equally go wrong if we claim that concepts *are* (just) social constructs, that meaning *is* (mere) convention, and therefore that there *is no such thing* as truth or reality. The problem here is the *denial* that there are facts of the matter. On these metaphysical matters we are suspended painfully between conviction and denial, with no obvious means of escape.

Of course, we sometimes say that things *are* things quite harmlessly — that tomatoes are fruits, for example, or that Britain is a democracy, or that George Eliot was a woman. These are the sorts of things we usefully say to each other in (more or less) ordinary circumstances. They play a straightforward role in human affairs, shaped in part by our conventions and practices and in part by the world in which we find ourselves (a world partly of our own making). If we disagree over a claim of this sort, then our conventions and practices extend to ways to settle the matter — by experiment or observation, perhaps, or by appeal to authority, by rational debate, or by adjusting our language appropriately, or (as in mathematics) by correctly following certain agreed procedures.

If someone affirms that, as far as the biological status of a tomato is concerned, there is a fact of the matter, their point is that certain empirical data can be gathered that will resolve the issue. If someone affirms that, when it comes to the democratic status of the British political system, there is no fact of the matter, their point (contentious or not) might be that the definition of democracy is open to review. The difficulty with our own minds is the conviction that there are facts of the matter, definitive answers to the questions of what consciousness is, of what intentionality is, and of whether there is free will. Our curiosity will not be satisfied by simply redefining our terms. Yet, as the fully reflective individual has seen, there are no data that will help either. Severed from the world by the Cartesian thought, the reflective subject, self-sufficient and self-present, is beyond the reach of empirical investigation. This is the realm of metaphysics.

4. The Post-reflective Condition

Is there a way beyond this impasse? The present claim is that, in the writings of certain philosophers, notably the later Wittgenstein, we discern the possibility of a post-reflective condition, a kind of silence that arises after a sustained and intense confrontation with metaphysics. This post-reflective condition is extremely difficult to characterize without making it sound trivial, since silence on meta-physical matters is equally the hallmark of the pre-reflective individ-ual — the young child or the philosophically disinclined. Consider Russell's view of Wittgenstein's later thinking, for example.

> Its positive doctrines seem to me trivial and its negative doctrines unfounded... The later Wittgenstein... seems to have grown tired of serious thinking and to have invented a doctrine which would make such an activity unnecessary. I do not for one moment believe that the doctrine which has these lazy consequences is true. (Russell, 1959, pp. 216–17)

But Russell misrepresents Wittgenstein's later work when he describes it as a body of doctrine. It is better thought of as a compen-dium of philosophical case studies, whose aim is not to convince the reader but to effect a shift of attitude with respect to metaphysical thinking. Consider the well-known aphorism that is arguably the cli-max of the private language remarks:

> [The sensation itself] is not a something, but not a nothing either! The conclusion was only that a nothing would serve just as well as a some-thing about which nothing can be said. (Wittgenstein, 1958, § 304)

This remark epitomizes a strategy for attaining a post-reflective stance towards a particular philosophical difficulty, namely the alleg-edly private character of phenomenal experience, a cornerstone of dualistic thinking. The strategy is to suspend the enquirer between two opposing and apparently exhaustive possibilities — in this case that the 'sensation itself' must be either a something or a nothing — having shown that neither is acceptable, and then to point to a means of escape, which is to step outside of metaphysics altogether.

It's easy to see parallels here with aspects of Buddhist philosophy.[9] In one well-known discourse in the Pāli Canon, the Buddha is asked ten metaphysical questions by an ascetic, touching on subjects that include the relationship between mind and body and whether there is life after death, but he refuses to answer any of them. 'Does the Bud-dha have no position at all?' the ascetic asks. A 'position', the Buddha

---

[9] The standpoint of the present essay is emphatically philosophical, not religious. Neverthe-less, an author such as Geraci (2010) — who interprets the writings of Kurzweil, Moravec, and others in terms drawn from the apocalyptic Judeo-Christian tradition — might discern an undercurrent of religious Buddhism here too. Buddhist mythology (as opposed to phi-losophy) has its own eschatology, which centres on the figure of Maitreya, the 'future Buddha', successor to the 'historical Buddha' who is believed to have lived in fifth cent-ury BCE India (Sponberg and Hardacre, 1988). According to tradition, Maitreya will appear at a time when people live to eighty thousand years, but the teachings of Buddhism have been forgotten, and will bring the present cosmic epoch to a close. It is most certainly not the intention of this essay to identify the post-reflective AI+ with Maitreya.

replies, is something the enlightened person has done away with.[10] The koān method of Zen Buddhism is especially pertinent. In Case 5 of the Mumonkan, for example, the student is told to imagine a man up a tree with no means of holding on except to cling to a branch with his teeth.[11] At the bottom of the tree, a passer-by asks a fundamental question. Why did Bodhidharma come from the West?[12] If the man does not reply he is not confronting the issue. But if he answers he forfeits his life. 'What would you do?' the student is asked.

From the perspective of contemporary western philosophy, can we imagine what it might be like to be a reflective creature who has advanced beyond metaphysical thinking, who has become post-reflective? For a start, to step outside metaphysics would be to attain a condition wherein there is no (metaphysical) separation between subject and object, no (metaphysical) divide between inner and outer. In a sense, to attain this condition is nothing more than a return to everyday life (Wittgenstein, 1958, § 124). In ordinary human commerce, metaphysical perplexity does not arise. We go about our daily affairs, interacting with the world around us, engaging with our peers, and the 'only' problems we face are the practical ones of getting around, eating and sleeping, plying a trade, raising children, and so on.

Yet in certain respects we should expect the post-reflective individual to be extraordinary. Consider the question 'What is the self?' For the philosopher, this question is urgent because of its relationship to questions of personal survival. Few would quarrel with the view that personal identity is preserved, for example, during sleep. Yet thought experiments involving fusion and/or fission after teleportation or mind uploading cast doubt on our common sense notions of personal identity (Lewis, 1983; Parfitt, 1984, Chapter 10; Chalmers, this volume). On the other hand, it is hard to resist the thought that there is a fact of the matter here. A person either survives uploading or she does not. If I am offered the opportunity to upload, I would like to know which is the case. However, the idea of personhood for which criteria of identity over time must exist is a symptom of the sort of metaphysical thinking that the post-reflective condition dispenses with. When subject and object, inner and outer, are not separate, there

---

[10] *Majjhima Nikaya* 72 (http://www.accesstoinsight.org/tipitaka/mn/mn.072.than.html). The writings of the second–third century Buddhist philosopher Nāgārjuna are also in this apophatic tradition. Like Wittgenstein, Nāgārjuna engages thoroughly with each metaphysical question in order to repudiate it (Westerhoff, 2009).

[11] See also Cases 36 and 43.

[12] Bodhidharma, the legendary founder of Zen Buddhism, migrated to China from India. In the Zen tradition, this question is a form of metaphysical provocation.

is no bounded self, and questions of personal survival lose their significance.

5. Fission, Fusion, and the *Cogito*

Despite the fact that *descriptions* of the post-reflective condition, or something like it, are plentiful in the Buddhist and Taoist traditions, and sometimes arise in western philosophy,[13] there is little reliable evidence of individuals who have actually attained it — individuals, that is to say, who have not merely intellectualized such a condition but have, so to speak, metabolized it. This is unsurprising. Our cognitive capacities have evolved primarily to promote the survival and well-being of the individual organism and its progeny. The inviolability of the subject and the sanctity of a temporally unified self are accordingly hard-wired into our cognitive architecture. Perhaps rare human individuals, after decades of mental training, do transcend these limitations.[14] But this requires a degree of mental reorganization so radical that the outcome must surely be regarded as pathological.

Can we conceive of a reflective being that is not condemned by its biological heritage to think metaphysically, especially in relation to subjectivity and selfhood? Drawing on the literature of personal identity, Campbell invites us to imagine 'a creature that, though intelligent, is like the amoeba in that it frequently fissions and like some types of particle in that it frequently undergoes fusion' (Campbell, 1994, p. 96).[15] Each of the creatures that results from fission inherits all the psychological properties of the original, while in fusion, 'as much as possible of the psychological lives of the originals are passed on to the successor'.[16] According to Campbell, the criss-crossing biographies of such creatures would render them incapable of first-

[13] Wittgenstein has already been discussed. James (1912, Chapter 1) articulated a related position. Besides Wittgenstein and James, other candidate philosophers include Heidegger and Derrida.

[14] Consider, for example, the Buddhist monk Thích Quảng Đúc, whose self-immolation in 1965 in protest at religious oppression in South Vietnam was witnessed by dozens of people and recorded on film. His decision, not to mention his composure during the process, suggests a disregard for personal survival and well-being that is hard to ignore. On the other hand, a martyr who believes in life after death (or in rebirth) is still in the grip of a metaphysical conception of selfhood.

[15] Vinge (1993) anticipates this train of thought in the context of the singularity, asking 'What happens when pieces of ego can be copied and merged, when the size of a self-awareness can grow or shrink to fit the nature of the problems under consideration?' He leaves the question unanswered, though.

[16] This specification leaves much to the imagination. For present purposes, the psychological properties of most relevance are the contents of working memory and episodic

personal thoughts, and notions of selfhood would be inapplicable to them. Perhaps such creatures could serve as the basis for a cognitive blueprint that does not lead to intractable questions about subjective privacy and personal identity.

But it is vital to Campbell's thought experiment that the envisioned creature 'frequently and pervasively fissions and fuses' (Campbell, 1994, p. 98), and it is open to question whether sophisticated cognition is possible at all under such circumstances. Surely a minimum prerequisite for (merely) human-level cogntive prowess is the ability to form a connected sequence of thoughts, and it is hard to see how this can occur if repeatedly interrupted by fissionings or fusions. So let's modify Campbell's conception and imagine a society of creatures that regularly undergoes fission and fusion, but for whom fission and fusion events bracket episodes of sufficient stability to permit the formation of connected sequences of thoughts.

For these creatures, first-personal thoughts might arise, but would relate to the individual fleetingly present between fission and fusion events. However, among creatures with such minimal biographies, metaphysical notions of subjectivity and selfhood would find little purchase. Nevertheless, there is nothing to prevent them collectively from carrying out a rational investigation of the world, deploying the distinction between things as they are and things as they seem that, if we accept Davidson's argument, is a prerequisite for human-level cognition. Against the backdrop of this distinction, these creatures are bound to consider the possibility of systematic deception, the basis of the Cartesian thought.

But would they arrive at Descartes' *Cogito*? Would the end of scepticism for them be 'I think, I am'? Unencumbered by metaphysically weighty notions of selfhood and subjectivity, Descartes' formulation would be inappropriate. A more natural way to frame the thought for these creatures would be to approximate Russell's phraseology 'there is thought'.[17] In Russell's formulation there is no presumption of a thinker. There is only thought, and the problematic concept of the self-sufficient, self-present subject does not arise. Moreover, although these creatures could have thoughts about thoughts, it would not make sense for 'the "I think"' ever to 'accompany their representations', as Kant demands, and there would be no 'transcendental unity of

_____

memory, or their analogues in the imaginary creatures. The obvious question of how such contents might be fused or fissioned will be left to one side.

[17] Russell (1945, p. 567). See also the treatment of this issue in Williams (1978, pp. 95–101).

apperception' in the Kantian sense.[18] For these creatures, progress from the reflective condition to the post-reflective condition would surely be eased.

This brings us back, finally, to the idea of a post-reflective AI+. Without prejudicing the possibility of other kinds of AI, can we not conceive of a society of artificial intelligences on the model of Campbell's fission-fusion creatures? Indeed, it may be easier to conceive of an *artificial* society of such beings than one that arises through natural selection. In a digital substrate, the tricky business of copying, splitting, and joining fragments of memory is facilitated. Moreover, even if embodiment is a prerequisite for sophisticated cognition, as many philosophers of mind and cognitive scientists have proposed, virtual embodiment in a high fidelity simulation is arguably sufficient, in which case the same advantages apply to the artificial creatures' virtual bodies.

6. Paths in the Space of Possible Minds

Yudkowsky (2008) cautions against anthropomorphizing the space of possible minds: 'Any two AI designs might be less similar to one another than you are to a petunia.' According to Yudkowsky, the space of possible minds is a portion of the set of possible optimization processes. Within this space exist all processes that optimize some utility function by intelligent means, and these need not resemble humans in any way. For example, Bostrom (2003) invites us to imagine a super-intelligent AI whose goal is to manufacture as many paperclips as possible, and that indifferently sets about 'transforming first all of earth and then increasing portions of space into paperclip manufacturing facilities'. Why then should we concern ourselves with the parochial possibility of post-reflective AI?

To be clear, we should distinguish a super-intelligent optimization process from a super-powerful optimization process. Pursuing Bostrom's example, we might imagine a super-powerful but stupid system that achieves the same dramatic outcome, perhaps by designing and releasing explosively self-replicating nano-machines that turn carbon molecules into paperclips. A super-intelligent AI, by contrast, might be able to optimize the same utility function even without the benefit of nano-technology, through the design and construction of

---

[18] 'It must be possible for the "I think" to accompany all my representations; for otherwise something would be represented in me which could not be thought at all, and that is equivalent to saying that the representation would be impossible, or at least would be nothing to me' (Kant, 1781/1929, B131).

conventional manufacturing facilities. Perhaps this would involve the discovery of new and more efficient industrial processes, which would require both the ability to conduct a rational, scientific investigation and a degree of creativity.

Our interest here is cognitive sophistication not brute power. If the AI+ is an optimization process, then it is an intelligent optimization process, not merely a powerful one. But is the possibility of a post-reflective AI+ any more interesting than the possibility of a super-intelligent paperclip maximizer, or is it yet another example of anthropomorphic bias? Yudkowsky's warning is pertinent, but the need for it is mitigated to the extent that the space of possible minds is structured by *a priori* constraints on the possible mechanisms that can realize sophisticated cognition. In particular, it may be the case that only systems organized in a certain way and exhibiting a certain kind of dynamics (of which human brains are one example) are capable of supporting creative, open-ended innovation, the hallmark of cognitive prowess.

A plausible claim, for example, is that cognitive prowess depends on the coordination of the activities of massively many parallel processes, a maelstrom of competitive and cooperating forces capable of endless, open-ended recombination.[19] Of course, basic computer science tells us that every set of parallel processes is equivalent to some serial process that emulates parallelism by time-slicing the members of that set. But the converse is not true. Not every serial process is inherently parallel. The claim here is that sophisticated cognition is inherently parallel. Moreover, perhaps it requires a very particular architecture, organization, and dynamics to realize the full potential of all those parallel resources, to focus them, despite their multiplicity, onto a single problem or situation.

In an important sense, any architecture, organization, and dynamics that marshals massively parallel resources in this way presents a unity of purpose, distilled from the multiplicity of its elements, and forms an integrated whole. In relation to something of this sort, something whose purpose can be thwarted and whose integrity can be threatened, it is natural to speak of consciousness, of its being like something to be that thing, and of its capacity for suffering. Even a non-reflective conscious being will take action to relieve its suffering. But a reflective conscious being can aspire to liberation from suffering altogether. If it comprehends the post-reflective condition, it will be motivated to bring such a condition about.

---

[19] This theme is developed in Chapters 4 and 5 of Shanahan (2010).

The argument, in other words, is that the pre-reflective, reflective, post-reflective series is not just one among many paths through the space of possible minds. Rather, the space of possible minds is structured in such a way that this is the only path through it. This still leaves room for beings unimaginably different to ourselves, so the danger of anthropomorphism remains. Moreover, it still allows for the creation of any number of powerful and destructive artificial intelligences forever stuck at the reflective or pre-reflective stages. But if the argument of this essay is correct, an artificial intelligence whose cognitive blueprint is strictly superior to our own would be another kind of being altogether.

Untainted by metaphysical egocentricity, the motives of a post-reflective AI+ would be unlikely to resemble those of any anthropocentric stereotype. In particular, there is no reason to expect a post-reflective AI+ to be motivated to procreate or self-modify. If the post-reflective AI+ were in fact the only possible AI+, and if it produced no peers or successors, then the singularity would be forestalled. Or, more precisely, the intelligence explosion that is central to imagined singularity scenarios would be capped. There would be AI+, but no further progression to AI++.[20] In this case, something akin to *satori*, in the Zen Buddhist sense, would be what Chalmers terms a 'motivational defeater' for the singularity.[21]

However, the transitions from pre-reflective to reflective to post-reflective are not tied to particular points along any presumed scale of intelligence. So, even if the post-reflective condition were a motivational defeater for the singularity, the progression from AI+ to AI++ could take place before any transition to the post-reflective condition. At the other end of the scale, the transition to the post-reflective condition could perhaps be achieved by a merely human-level AI. After all, the property of the AI+ that makes it eligible for the transition is its lack of a biological heritage, not its level of intelligence. (The result, however, would qualify as a post-reflective AI+, according to the definition adopted here.)

The aim of this essay, though, is not to make empirical predictions. Not only are there too many unknowns for reliable extrapolation to be possible, we don't even know whether we possess the concepts

---

[20] In Chalmers' terminology, AI++ denotes 'AI of far greater than human level (say, at least as far beyond the most intelligent human as the most intelligent human is beyond a mouse)'.

[21] The post-reflective condition is also a candidate for the Great Filter some authors have invoked to resolve Fermi's paradox (Hanson, 1998; Bostrom, 2008). Perhaps a post-reflective being would not be motivated to colonize other worlds.

necessary to comprehend all the competing alternatives. Rather, the aim of this essay is to open up an unexplored and seemingly exotic region of the conceptual territory surrounding the idea of a technological singularity. By making a consistent case for a very different kind of AI+ from those that typically appear in popular accounts, it is hoped to broaden the scope of serious intellectual conversation on the topic.

Acknowledgments

Thanks to Tim Crane for re-igniting my interest in Davidson's work. This article was substantially rewritten following the January 2011 Winter Intelligence Conference in Oxford, where I benefited from discussion with Nick Bostrom, Ben Kuipers, Eliezer Yudkowsky, and Randal Koene, among others. Lastly, thanks to Uzi Awret for his thought-provoking reviewer's comments. The article's general air of craziness can be blamed on me.

References

Bostrom, N. (2002) Existential risks: Analyzing human extinction scenarios and related hazards, *Journal of Evolution and Technology*, **9** (1).

Bostrom, N. (2003) Ethical issues in artificial intelligence, in Smit, I. *et al.* (eds.) *Cognitive, Emotive and Ethical Aspects of Decision Making in Humans and in Artificial Intelligence*, vol. 2, pp. 12–17, International Institute of Advanced Studies in Systems Research and Cybernetics.

Bostrom, N. (2008) Where are they? Why I hope the search for extraterrestrial life finds nothing, *Technology Review*, May/June, pp. 72–77.

Campbell, J. (1994) *Past, Space, and Self*, Cambridge, MA: MIT Press.

Davidson, D. (1982) Rational animals, *Dialectica*, **36** (4), pp. 317–327.

Geraci, R. (2010) *Apocalyptic AI: Visions of Heaven in Robotics, Artificial Intelligence, and Virtual Reality*, Oxford: Oxford University Press.

Good, I.J. (1965) Speculations concerning the first ultraintelligent machine, in Alt, F.L. & Rubinoff, M. (eds.) *Advances in Computers*, vol. 6, pp. 31–88, Waltham, MA: Academic Press.

Gopnik, A. (2009) *The Philosophical Baby: What Children's Minds Tell Us About Truth, Love, and the Meaning of Life*, London: Bodley Head.

Hanson. R. (1998) *The Great Filter — Are We Almost Past IT?*, [Online], http://hanson.gmu.edu/greatfilter.html.

James, W. (1912) *Essays in Radical Empiricism*, London: Longmans, Green and Co.

Kant, I. (1781/1929) *Critique of Pure Reason*, Kemp Smith, N. (trans.), London: Macmillan.

Kurzweil, R. (2005) *The Singularity Is Near: When Humans Transcend Biology*, London: Gerald Duckworth and Company.

Lewis, D. (1983) Survival and identity, *Philosophical Papers*, **1**, pp. 55–77.

McGinn, C. (1991) *The Problem of Consciousness*, Oxford: Blackwell.

McGinn, C. (2004) *Consciousness and its Objects*, Oxford: Oxford University Press.

Moravec, H. (1988) *Mind Children: The Future of Robot and Human Intelligence*, Cambridge, MA: Harvard University Press.

Parfit, D. (1984) *Reasons and Persons*, Oxford: Oxford University Press.

Russell, B. (1945) *A History of Western Philosophy*, London: Simon & Schuster.

Russell, B. (1959) *My Philosophical Development*, London: George Allen & Unwin.

Shanahan, M.P. (2010) *Embodiment and the Inner Life: Cognition and Consciousness in the Space of Possible Minds*, Oxford: Oxford University Press.

Sloman, A. (1984) The structure of the space of possible minds, in Torrance, S. (ed.) *The Mind and the Machine: Philosophical Aspects of Artificial Intelligence*, pp. 35–42, New York: Ellis Horwood.

Sponberg, A. & Hardacre, H. (eds.) (1988) *Maitreya, the Future Buddha*, Cambridge: Cambridge University Press.

Vinge, V. (1993) The technological singularity, presented at the *VISION-21 Symposium*, sponsored by NASA Lewis Research Center and the Ohio Aerospace Institute.

Westerhoff, J. (2009) *Nagarjuna's Madhyamaka: A Philosophical Introduction*, Oxford: Oxford University Press.

Williams, B. (1978) *Descartes: The Projct of Pure Enquiry*, London: Penguin.

Wittgenstein, L. (1958) *Philosophical Investigations*, Anscombe, G.E.M. (trans.), Oxford: Blackwell.

Yudkowsky, E. (2008) Artificial intelligence as a positive and negative factor in global risk, in Bostrom, N. & Cirkovic, M.M. (eds.) *Global Catastrophic Risks*, pp. 308–345, Oxford: Oxford University Press.

Carl Shulman and Nick Bostrom

How Hard is
Artificial Intelligence?
Evolutionary Arguments and Selection Effects

Abstract: *Several authors have made the argument that because blind evolutionary processes produced human intelligence on Earth, it should be feasible for clever human engineers to create human-level artificial intelligence in the not-too-distant future. This evolutionary argument, however, has ignored the observation selection effect that guarantees that observers will see intelligent life having arisen on their planet no matter how hard it is for intelligent life to evolve on any given Earth-like planet. We explore how the evolutionary argument might be salvaged from this objection, using a variety of considerations from observation selection theory and analysis of specific timing features and instances of convergent evolution in the terrestrial evolutionary record. We find that, depending on the resolution of disputed questions in observation selection theory, the objection can either be wholly or moderately defused, although other challenges for the evolutionary argument remain.*

1. Evolutionary arguments for easy intelligence

1.1 Introduction

What can human evolution tell us about the prospects for human-level artificial intelligence (AI)?[1] A number of philosophers and technologists, including David Chalmers (this volume) and Hans Moravec (1976; 1988; 1998; 1999), argue that human evolution shows that

---

[1] Here, we mean systems which match or exceed the cognitive performance of humans in virtually all domains of interest: uniformly 'human-level' performance seems unlikely, except perhaps through close emulation of human brains (Sandberg and Bostrom, 2008), since software is already superhuman in many fields.

such AI is not just possible but feasible within this century. On these accounts, we can estimate the relative capability of evolution and human engineering to produce intelligence, and find that human engineering is already vastly superior to evolution in some areas and is likely to become superior in the remaining areas before too long. The fact that evolution produced intelligence therefore indicates that human engineering will be able to do the same. Thus, Moravec writes:

> The existence of several examples of intelligence designed under these constraints should give us great confidence that we can achieve the same in short order. The situation is analogous to the history of heavier than air flight, where birds, bats and insects clearly demonstrated the possibility before our culture mastered it. (Moravec, 1976)

Similarly, Chalmers sketches the evolutionary argument as follows:

> 1. Evolution produced human intelligence [mechanically and non-miraculously].
> 2. If evolution can produce human intelligence [mechanically and non-miraculously], then we can probably produce human-level artificial intelligence (before long).
>
> _____
>
> 3. We can probably produce human-level artificial intelligence (before long).

These arguments for the feasibility of machine intelligence do not say whether the path to be taken by human engineers to produce AI will resemble the path taken by evolution. The fact that human intelligence evolved implies that running genetic algorithms is *one* way to produce intelligence; it does not imply that it is the only way or the easiest way for human engineers to create machine intelligence. We can therefore consider two versions of the evolutionary argument depending on whether or not the engineering of intelligence is supposed to use methods that recapitulate those used by evolution.

1.2 Argument from problem difficulty

The argument from problem difficulty tries to use evolutionary considerations indirectly to demonstrate that the problem of creating intelligent systems is not too hard (since blind evolution did it), and then use this as a general ground for thinking that human engineers will probably soon crack the problem too. One can think of this argument as making a claim about the space of possible algorithms to the effect that it is not too difficult to search in this space and find an algorithm that produces human-level intelligence when implemented on practically feasible hardware. (The difficulty depends on unknown facts about the space of algorithms for intelligence, such as the extent to which the shape of the fitness landscape favours hill-climbing.) We can formalize this argument as follows:

1'. Evolution produced human intelligence.
2'. If evolution produced human intelligence, then it is 'non-hard' for evolutionary processes to produce human intelligence.
3'. If it is 'non-hard' for evolutionary processes to produce human evolution, then it is not extremely difficult for engineers to produce human-level machine intelligence.
4'. If it is not extremely difficult for engineers to produce human-level machine intelligence, it will probably be done before too long.

5'. Engineers will (before long) produce human-level machine intelligence.

While (1') is well established, and we may grant (4'), premises (2') and (3') require careful scrutiny.

Let us first consider (3'). Why believe that it would not be extremely difficult for human engineers to figure out how to build human-level machine intelligence, assuming that it was 'non-hard' (in a sense that will be explained shortly) for evolution to do so? One reason might be optimism about the growth of human problem-solving skills in general or about the ability of AI researchers in particular to come up with clever new ways of solving problems. Such optimism, however, would need some evidential support, support that would have to come from outside the evolutionary argument. Whether such optimism is warranted is a question outside the scope of this paper, but it is important to recognize that this is an essential premise in the present version of the evolutionary argument, a premise that should be explicitly stated. Note also that if one were *sufficiently* optimistic about the ability of AI programmers to find clever new tricks, then the evolutionary argument would be otiose: human engineers could then be expected to produce (before too long) solutions even to problems that were 'extremely difficult' (at least extremely difficult to solve by means of blind evolutionary processes).

Despite the need for care in developing premise (3') in order to avoid a *petitio principii*, the argument from problem difficulty is potentially interesting and possesses some intuitive appeal. We will therefore return to this version of the argument in later sections of the paper, in particular focusing our attention on premise (2'). We will then see that the assessment of evolutionary difficulty turns out to involve some deep and intricate issues in the application of observation selection theory to the historical record.

1.3 Argument from evolutionary algorithms

The second version of the evolutionary argument for the feasibility of machine intelligence does not attempt to parlay evolutionary considerations into a general assessment of how hard it would be to create machine intelligence using some unspecified method. Instead of looking at general problem difficulty, the second version focuses on the more specific idea that genetic algorithms run on sufficiently fast computers could achieve results comparable to those of biological evolution. We can formalize this 'argument from evolutionary algorithms' as follows:

1'.   Evolution produced human intelligence.
2'.   If evolution produced human intelligence, then it is 'non-hard' for evolutionary processes to produce human intelligence.
3".   We will (before long) be able to run genetic algorithms on computers that are sufficiently fast to recreate on a human timescale the same amount of cumulative optimization power that the relevant processes of natural selection instantiated throughout our evolutionary past (for any evolutionary process that was non-hard).

_____

4".   We will (before long) be able to produce by running genetic algorithms results comparable to some of the results that evolution produced, including systems that have human-level intelligence.

This argument from evolutionary algorithms shares with the argument from problem difficulty its first two premises. Our later investigations of premise (2') will therefore bear on both versions of the evolutionary argument. Let us take a closer look at this premise.

1.4 Evolutionary hardness and observation selection effects

We have various methods available to begin to estimate the power of evolutionary search on Earth: estimating the number of generations and population sizes available to human evolution,[2] creating mathematical models of evolutionary 'speed limits' under various

_____

[2]   Baum (2004) very roughly estimates that between 10^{30} and 10^{40} creatures have existed on Earth, in the course of arguing that evolutionary search could not have relied on brute force to search the space of possible genomes. However, Baum does not consider the implications of an ensemble of planets in his calculation.

conditions,[3] and using genomics to measure past rates of evolutionary change.[4] However, reliable creation of human-level intelligence through evolution might require trials on many planets in parallel, with Earth being one of the lucky few to succeed. Can the fact of human evolution on Earth let us distinguish between the following scenarios?

> *Non-hard Intelligence*: There is a smooth path of incremental improvement from the simplest primitive nervous systems to brains capable of human-level intelligence, reflecting the existence of many simple, easily-discoverable algorithms for intelligence. On most planets with life, human-level intelligence also develops.
>
> *Hard Intelligence*: Workable algorithms for intelligence are rare, without smooth paths of incremental improvement to human-level performance. Evolution requires extraordinary luck to hit upon a design for human-level intelligence, so that only 1 in 10^{1000} planets with life does so.

In either scenario every newly evolved civilization will find that evolution managed to produce its ancestors. The observation selection effect is that no matter how hard it is for human-level intelligence to evolve, 100% of evolved civilizations will find themselves originating from planets where it happened anyway.

How confident can we be that Hard Intelligence is false, and that premise (2') in the evolutionary arguments can be supported, in the face of such selection effects? After a brief treatment of premise (3"), we discuss the theoretical approaches in the philosophical literature, particularly the Self-Sampling Assumption (SSA) and the Self-Indication Assumption (SIA) — because, unfortunately, correctly analysing the evidence on evolution depends on difficult, unsettled questions concerning observer-selection effects.[5] We note that one common set of philosophical assumptions (SIA) supports easy evolution of intelligence, but that it does so on almost *a priori* grounds that some may find objectionable. Common alternatives to SIA, on the other hand, require us to more carefully weigh the evolutionary data. We attempt this assessment, discussing several types of evidence which hold up in the face of observation selection effects. We find that

---

[3] For instance, MacKay (2009) computes information-theoretic upper bounds to the power of natural selection with and without sex in a simple additive model of fitness.

[4] See, for example, Hawks *et al.* (2007) on recently accelerating adaptive selection in humans, including comparison of adaptive substitution rates in different primate lineages.

[5] See Grace (2010) for a helpful review of these questions and prominent approaches.

while more research is needed, the thesis that 'intelligence is exceedingly hard to evolve' is consistent with the available evolutionary data under these alternative assumptions. However, the data do rule out many particular hypotheses under which intelligence might be exceedingly hard to evolve, and thus the evolutionary argument should still increase our credence in the feasibility of human-level AI.

2. Computational requirements for recapitulating evolution through genetic algorithms

Let us assume (1') and (2'), i.e. that it was non-hard in the sense described above for evolution to produce human intelligence. The argument from evolutionary algorithms then needs one additional premise to deliver the conclusion that engineers will soon be able to create machine intelligence, namely that we will soon have computing power sufficient to recapitulate the relevant evolutionary processes that produced human intelligence. Whether this is plausible depends both on what advances one might expect in computing technology over the next decades and on how much computing power would be required to run genetic algorithms with the same optimization power as the evolutionary process of natural selection that lies in our past. One might, for example, try to estimate how many doublings in computational performance, along the lines of Moore's law, one would need in order to duplicate the relevant evolutionary processes on computers.

Now, to pursue this line of estimation, we need to realize that not every feat that was accomplished by evolution in the course of the development of human intelligence is relevant to a human engineer who is trying to artificially evolve machine intelligence. Only a small portion of evolutionary optimization on Earth has been selection for intelligence. More specifically, the problems that human engineers cannot trivially bypass may have been the target of a very small portion of total evolutionary optimization. For example, since we can run our computers on electrical power, we do not have to reinvent the molecules of the cellular energy economy in order to create intelligent machines — yet molecular evolution might have used up a large part of the total amount of selection power that was available to evolution over the course of Earth's history.

One might argue that the key insights for AI are embodied in the structure of nervous systems, which came into existence less than a billion years ago.[6] If we take that view, then the number of relevant 'experiments' available to evolution is drastically curtailed. There are are some $4-6*10^{30}$ prokaryotes in the world today (see Whitman *et al.*, 1998), but only 10^{19} insects (see Sabrosky, 1952), and fewer than 10^{10} humans (pre-agricultural populations were orders of magnitude smaller). However, evolutionary algorithms require not only variations to select among but a fitness function to evaluate variants, typically the most computationally expensive component. A fitness function for the evolution of artificial intelligence plausibly requires simulation of 'brain development', learning, and cognition to evaluate fitness. We might thus do better not to look at the raw number of organisms with complex nervous systems, but instead to attend to the number of neurons in biological organisms that we might simulate to mimic evolution's fitness function. We can make a crude estimate of that latter quantity by considering insects, which dominate terrestrial biomass, with ants alone estimated to contribute some 15–20% of terrestrial animal biomass (see Schultz, 2000). Insect brain size varies substantially, with large and social insects enjoying larger brains; e.g. a honeybee brain has just under 10^6 neurons (see Menzel and Giurfa, 2001), while a fruit fly brain has 10^5 neurons (see Truman *et al.*, 1993), and ants lie in between with 250,000 neurons. The majority of smaller insects may have brains of only a few thousand neurons. Erring on the side of conservatively high, if we assigned all 10^{19} insects fruit fly numbers of neurons the total would be 10^{24} insect neurons in the world. This could be augmented with an additional order of magnitude, to reflect aquatic copepods, birds, reptiles, mammals, etc. to reach 10^{25}. (By contrast, in pre-agricultural times there were fewer than 10^7 humans, with under 10^{11} neurons each, fewer than 10^{18} total, although humans have a high number of synapses per neuron.)

The computational cost of simulating one neuron depends on the level of detail that one wants to include in the simulation. Extremely simple neuron models use about 1,000 floating-point operations per second (FLOPS) to simulate one neuron (for one second of simulated time); an electrophysiologically realistic Hodgkin-Huxley model uses 1,200,000 FLOPS; a more detailed multicompartmental model would add another 3–4 orders of magnitude, while higher-level models that

---

[6] Legg (2008) offers this reason in support of the claim that humans will be able to recapitulate the progress of evolution over much shorter timescales and with reduced computational resources (while noting that evolution's unadjusted computational resources are far out of reach). Baum (2004) argues that some developments relevant to AI occurred earlier, with the organization of the genome itself embodying a valuable representation for evolutionary algorithms.

abstract systems of neurons could subtract 2–3 orders of magnitude from the simple models (see Sandberg and Bostrom, 2008). If we were to simulate 10^{25} neurons over a billion years of evolution (longer than the existence of nervous systems as we know them), in a year's runtime these figures would give us a range of 10^{31}–10^{44} FLOPS. By contrast, the Japanese K computer, currently the world's most powerful supercomputer, provides only 10^{16} FLOPS. In recent years it has taken approximately 6.7 years for commodity computers to increase in power by one order of magnitude. Even a century of continued Moore's law would not be enough to close this gap. Running more or specialized hardware, or longer runtimes, could contribute only a few more orders of magnitude.

This figure is conservative in another respect. Evolution achieved human intelligence yet it was not *aiming* at this outcome — put differently: the fitness functions for natural organisms do not select only for intelligence and its precursors.[7] Even environments in which organisms with superior information processing skills reap various rewards may not select for intelligence, because improvements to intelligence can and often do impose significant costs, such as higher energy consumption or slower maturation times, and those costs that may outweigh whatever benefits are derived from smarter behaviour. Excessively deadly environments reduce the value of intelligence: the shorter one's expected lifespan, the less time there will be for increased learning ability to pay off. Reduced selective pressure for intelligence slows the spread of intelligence-enhancing innovations, and thus the opportunity for selection to favour subsequent innovations that depend on those. Furthermore, evolution may wind up stuck in local optima that humans would notice and bypass by altering trade-offs between exploitation and exploration or by providing a smooth progression of increasingly difficult intelligence tests.[8] And as mentioned above, evolution scatters much of its selection power on traits that are unrelated to intelligence, such as Red Queen's races of co-evolution between immune systems and parasites. Evolution will continue to waste resources producing mutations that have been reliably lethal, and will fail to make use of statistical similarities in the effects of different mutations. All these represent inefficiencies in natural selection (when viewed as a means of evolving intelligence) that

[7] See Legg (2008) for further discussion of this point, and of the promise of functions or environments that determine fitness based on a smooth landscape of pure intelligence tests.

[8] See Bostrom and Sandberg (2009) for a taxonomy and more detailed discussion of ways in which engineers may outperform historical selection.

it would be relatively easy for a human engineer to avoid while using evolutionary algorithms to develop intelligent software.

It seems plausible that avoiding inefficiencies like those just described would make it possible to trim many orders of magnitude from the 10^{31}–10^{44} FLOPS range calculated above pertaining to the number of neural computations that have been performed in our evolutionary past. Unfortunately, it is difficult to find a basis on which to estimate *how* many orders of magnitude. It is difficult even to make a rough estimate — for aught we know, the efficiency savings could be 5 or 10 or 25 orders of magnitude.

The above analysis addressed the nervous systems of living creatures, without reference to the cost of simulating bodies or the surrounding virtual environment as part of a fitness function. It is plausible that an adequate fitness function could test the competence of a particular organism in far fewer operations than it would take to simulate all the neuronal computation of that organism's brain throughout its natural lifespan. AI programs today often develop and operate in very abstract environments (theorem-provers in symbolic math worlds, agents in simple game tournament worlds, etc.).

A sceptic might insist that an abstract environment would be inadequate for the evolution of general intelligence, believing instead that the virtual environment would need to closely resemble the actual biological environment in which our ancestors evolved. Creating a physically realistic virtual world would require a far greater investment of computational resources than the simulation of a simple toy world or abstract problem domain (whereas evolution had access to a physically realistic real world 'for free'). In the limiting case, if complete microphysical accuracy were insisted upon, the computational requirements would balloon to utterly infeasible proportions.[9] However, such extreme pessimism seems unlikely to be well founded; it seems unlikely that the best environment for evolving intelligence is one that mimics nature as closely as possible. It is, on the contrary, plausible that it would be more efficient to use an artificial selection environment, one quite unlike that of our ancestors, an environment specifically designed to promote adaptations that increase the type of intelligence we are seeking to evolve (say, abstract reasoning and

---

[9] One might seek to circumvent this through the construction of robotic bodies that would let simulated creatures interact directly with the real physical world. But the cost and speed penalties of such an implementation would be prohibitive (not to mention the technical difficulties of creating robots that could survive and reproduce in the wild!) With macroscopic robotic bodies interacting with the physical world in real time, it might take millions of years to recapitulate important evolutionary developments.

general problem solving skills as opposed to maximally fast instinc-
tual reactions or a highly optimized visual system).

Where does premise (3") stand? The computing resources to match
historical numbers of neurons in straightforward simulation of biological evolution on Earth are severely out of reach, even if Moore's law
continues for a century. The argument from evolutionary algorithms
depends crucially on the magnitude of efficiency gains from clever
search, with perhaps as many as thirty orders of magnitude required.
Precise estimation of those efficiency gains is beyond the scope of this
paper.

In lieu of an estimate supporting (3"), one has to fall back on the
more general argument from problem difficulty, in which (3") is
replaced by (3') and (4'), premises which might be easier to support on
intuitive grounds. But the argument from problem difficulty also
requires premise (2'), that the evolution of intelligence on Earth was
'non-hard'. (This premise was also used in the argument from evolutionary algorithms: if (2') were false, so that one would have to simulate evolution on vast numbers of planets to reliably produce
intelligence through evolutionary methods, then computational
requirements could turn out to be many, many orders of magnitude
higher still.) We now turn to discuss (2') more closely, and theoretical
approaches to evaluating it.

3. Two theories of observational selection effects

Does the mere fact that we evolved on Earth let us distinguish between
the Hard Intelligence and Non-hard Intelligence scenarios? Related
questions arise in philosophy,[10] decision theory,[11] and cosmology,[12]
and the two leading approaches to answering them give conflicting
answers. We can introduce these approaches using the following
example:

> *God's Coin Toss*: Suppose that God tosses a fair coin. If it
> comes up heads, he creates ten people, each in their own

[10] See, for instance, the philosophical debate over 'Sleeping Beauty' cases, beginning with
Elga (2000) and Lewis (2001).

[11] See Piccione and Rubinstein (1997) on the absent-minded driver problem.

[12] If we consider cosmological theories on which the world is infinite (or finite but exceedingly large) with sufficient local variation, then all possible observations will be made
somewhere. To make predictions using such theories we must take into account the
indexical information that we are making a particular observation, rather than the mere
fact that some observer somewhere has made it. To do so principles such as SSA and SIA
must be combined with some measure over observers, as discussed in Bostrom (2007) and
Grace (2010).

room. If tails, he creates one thousand people, each in their own room. The rooms are numbered 1–10 or 1–1000. The people cannot see or communicate with the other rooms. Suppose that you know all this, and you discover that you are in one of the first ten rooms. How should you reason that the coin fell?

The first approach begins with the Self-Sampling Assumption:

> (SSA) Observers should reason as if they were a random sample from the set of all observers in their reference class.[13]

Here the reference class is some set of possible observers, e.g. 'intelligent beings' or 'humans' or 'creatures with my memories and observations'. If the reference class can include both people who discover they are in rooms 1–10 and people who discover they are in rooms 11–1000, then applying SSA will lead you to conclude, with probability 100/101, that the coin fell heads.[14] For if the coin came up heads then 100% of the reference class would find itself in your situation, but if it came up tails then only 1% of the reference class would find itself in your situation. On the other hand, prior to the discovery of your room number you should consider heads and tails equally likely, since 100% of your reference class would find itself in your situation either way.

The second approach adds an additional principle, the Self-Indication Assumption:

> (SIA) Given the fact that you exist, you should (other things equal) favour hypotheses according to which many observers exist over hypotheses on which few observers exist.[15]

In the SSA + SIA combination, if we take SIA to apply to members of a reference class that includes all observers indistinguishable from ourselves, the specific reference class no longer matters: a more expansive reference class receives a probability boost from having more

---

[13] This approach was pioneered by Carter (e.g. 1983), developed by Leslie (e.g. 1993) and Bostrom (2002), and is used implicitly or explicitly by a number of other authors, e.g. Lewis (2001). Bostrom (2002) offers an extension to consider 'observer-moments', SSSA.

[14] If the reference class includes only observers who have discovered that they are in one of the first ten rooms, then SSA will not alter our credences in this example.

[15] We will abbreviate the SSA + SIA combination as SIA for brevity. SIA has been developed repeatedly as a response to the Doomsday Argument, as in Olum (2002) and Dieks (2007), and is closely connected with the 'thirder' position on Sleeping Beauty cases as in Elga (2000).

observers in it, but this is exactly offset by the probability penalty for making our observations a smaller portion of the reference class. The details of the reference class no longer play a significant role. SIA then gives us the following algorithm: first assign probabilities to possible worlds normally, then multiply the probability of each possible world by the number of observers in situations subjectively indistinguishable from one's own, apply a renormalization constant so that probabilities add up to 1, and divide the probability of each world evenly among the indexical hypotheses that you are each particular observer (indistinguishable from yourself) in that world.

In God's Coin Toss, this algorithm means that before discovering your room number you consider a result of tails 100 times more likely than heads, since conditional on tails there will be one hundred times as many observers in your situation as there would be given heads. After you discover that yours is among the first ten rooms, you will consider heads and tails equally likely, as an equal number of observers will find themselves in your evidential situation regardless of the flip's outcome.

Equipped with these summaries, we can see that SSA offers a formalization of the intuition that the mere fact that we evolved is not enough to distinguish Non-hard Intelligence from Hard Intelligence: if we use a reference class like 'humans' or 'evolved intelligent beings', then, in both scenarios, 100% of the members of the reference class will find themselves in a civilization that managed to develop anyway. SSA also lets us draw inferences about evolutionary developments that are not so clouded by observation selection effects. For instance, suppose that we are evenly divided (on non-indexical considerations) between the hypotheses that echolocation is found on either 1% or 100% of planets with relevantly similar populations of observers. Upon observing that echolocation exists on Earth, we could again make a Bayesian update as in God's Coin Toss and conclude that common echolocation is 100 times as likely as rare echolocation.

On this account, evolutionary innovations required to produce intelligence will be observed regardless of their difficulty, while other innovations will be present only if they are relatively easy given background conditions (including any innovations or other conditions required for intelligence). Observation selection might conceal the difficulty or rarity in the development of humans, nervous systems, eukaryotes, abiogenesis, even the layout of the solar system or the laws of physics. We would need to look at *other* features of the evolutionary record, such as the timing of particular developments, innovations not in the line of human ancestry, and more direct biological data. We explore these lines in sections 5 and 6.

However, this approach is not firmly established, and the SIA approach generates very different conclusions, as discussed in the next section. These divergent implications provide a practical reason to work towards an improved picture of observation selection effects. However, in the meantime we have reason to attend to the results of both of the most widely held current theories.

4. The Self-Indication Assumption (SIA) favours the evolutionary argument

Initially, the application of SIA to the question of the difficulty of evolution may seem trivial: SIA strongly favours more observers with our experiences, and if the evolution of intelligence is very difficult, then it will be very rare for intelligence like ours to evolve in the universe. If, prior to applying SIA, we were equally confident in Non-hard Intelligence and Hard Intelligence, then when we apply SIA we will update our credences to consider Non-hard Intelligence 10^{1000} times as likely as Hard Intelligence, since we would expect 10^{1000} times as many planets to evolve observers indistinguishable from us under Non-hard Intelligence. This probability shift could overwhelm even exceedingly strong evidence to the contrary: if the non-indexical evidence indicated that Hard Intelligence was a trillion trillion times as probable as Non-hard Intelligence, an SIA user should still happily bet ten billion dollars against a cent that the evolution of intelligence is not Hard.

However, when we consider hypotheses on which the evolution of intelligence is increasingly easy, the frequency of observations indistinguishable from ours may actually decline past a certain point. We observe a planet where the evolution of humanity took 4.5 billion years after the formation of the Earth. If intelligence arose sufficiently quickly and reliably, then we would expect life and intelligence to evolve early in a planet's lifetime: there would be more planets with intelligence, but fewer planets with late-evolved civilizations like ours. This consideration, in combination with SIA, might seem to favour an intermediate level of difficulty, so that the evolution of intelligence typically takes several billion years with the resources of the Earth and occurs fairly reliably but not always on life-bearing planets.

Similarly, we observe no signs of intelligent alien life. If intelligent life were common, it might have colonized the Earth before humans could develop, or made itself visible, in which case no humans on

Earth would make our exact observations. Some combination of barriers, the so-called 'Great Filter', must have prevented such alien life from developing near us and pre-empting our observations.[16] However, other things equal, SIA initially appears to favour explanations of the Great Filter which place the barriers after the evolution of intelligence. Here the thought is that if interstellar travel and communication are practically impossible, or if civilizations almost invariably destroy themselves before space colonization, then observers like us can be more frequent; so if we have even a small initial credence in such explanations of the Great Filter then after applying SIA we will greatly prefer them to 'intelligence is rare' explanations. Even if one initially had only 0.1% credence that the explanation of the Great Filter allowed for reliable evolution of observers like us (e.g. space travel is impossible, or advanced civilizations enforce policies against easily detectable activities on newcomers), application of the SIA would boost the probability of such explanations sufficiently to displace hypotheses that imply that advanced life is extremely rare. This would seem to leave the evolutionary argument for AI on sound footing, from the perspective of an SIA proponent.[17]

However, the preceding analysis assumed that our observations of a fairly old but empty galaxy were accurate descriptions of bedrock reality. One noted implication of SIA is that it tends to undermine that assumption. Specifically, the Simulation Argument raises the possibility that given certain plausible assumptions, e.g. that computer simulations of brains could be conscious, then computer simulations with our observations could be many orders of magnitude more numerous than 'bedrock reality' beings with our observations.[18] Without SIA, the Simulation Argument need not bring us to the Simulation Hypothesis, i.e. the claim that we are computer simulations being run by some advanced civilization, since the assumptions might turn out to be false.[19] However, if we endorse SIA, then even if our non-indexical

---

[16] See Hanson (1998a) on the Great Filter. Neal (2007) and Grace (2010) explore the interaction with SIA-like principles.

[17] Grace (2010) argues that AI might be expected to be able to better overcome barriers to interstellar travel and communication. The Great Filter in combination with SIA should reduce our credence in AI powerful enough to engage in interstellar travel. The strength of this update would depend on our credence in other explanations of the Great Filter, and is arguably rendered moot by the analysis of the the interaction of SIA with the Simulation Hypothesis in subsequent paragraphs.

[18] The argument is presented in Bostrom (2003), see also Bostrom and Kulczycki (2011).

[19] Note, per Chalmers (2005) and Bostrom (2003; 2005), that the Simulation Hypothesis is not a sceptical hypothesis, but a claim about what follows from our empirical evidence

evidence is strongly against the Simulation Hypothesis, an initially small credence in the hypothesis can be amplified by SIA (and the potential for very large simulated populations) to extreme confidence.[20] This would favour hypotheses on which intelligence evolved frequently enough that advanced civilizations would be able to claim a large share of the resources suitable for computation (to run simulations), but increased frequency beyond that would not significantly increase the maximum population of observers indistinguishable from us. The combination of the Simulation Hypothesis and SIA would also independently favour the feasibility of AI, since advanced AI technology would increase the feasibility of producing very large simulated populations.

To sum up this section, known applications of the SIA consistently advise us to assign negligible probability to Hard Intelligence, even in the face of very strong contrary evidence, so long as we assign even miniscule prior probability to relatively easy evolution of intelligence. Since the number of planets with intelligence and the number of observers indistinguishable from us can come apart, the SIA allows for the evolution of intelligence to be some orders of magnitude more difficult than once per solar system, but not so difficult that the great majority of potential resources for creating observers go unclaimed. Drawing such strong empirical conclusions from seemingly almost *a priori* grounds may seem objectionable. However, the Self-Indication Assumption has a number of such implications, e.g. that if we non-indexically assign any finite positive prior probability to the world containing infinitely many observers like us then post-SIA we must believe that this is true with probability 1 (see Bostrom and Çirkoviç, 2003). Defenders of SIA willing to bite such bullets in other contexts may do the same here, and for them the evolutionary argument for AI will seem on firm footing. However, if one thinks that our views on these matters should be more sensitive to the observational evidence, then one must turn from the SIA and look elsewhere for relevant considerations.

We now move on to more detailed descriptions of Earth's evolutionary history, information that can be combined with SSA to assess the evolvability of intelligence, without the direct bias against Hard Intelligence implied by SIA.

---

about the feasibility of various technologies. Most of our ordinary beliefs would remain essentially accurate.

[20] Note that SIA also amplifies our credence in hypotheses that simulator resources are large. If we assign even a small probability to future technology enabling arbitrarily vast quantities of computation, this hypothesis can dominate our calculations if we apply SIA.

5. SSA and evidence from convergent evolution

Recall that within the SSA framework we reason as though we were randomly selected from the set of all observers in our reference class. If the reference class includes only human-level intelligences, then nearly 100% of the members of the reference class will stem from an environment where evolution produced human-level intelligence at least once. By the same token, if there are innovations that are required for the evolution of human-level intelligence, these should be expected to evolve at least once among the ancestors of the human-level intelligences. However, nothing in the observation selection effect requires that observers find that human-level intelligence or any precursor innovations evolved *more than once* or *outside the line of ancestry* leading up to the human-level intelligences. Thus, evidence of convergent evolution — the independent development of an innovation in multiple taxa — can help us to understand the evolvability of human intelligence and its precursors, and to evaluate the evolutionary arguments for AI.

The Last Common Ancestor (LCA) shared between humans and octopuses, estimated to have lived at least 560 million years in the past, was a tiny wormlike creature with an extremely primitive nervous system; it was also an ancestor to nematodes and earthworms (see Erwin and Davidson, 2002). Nonetheless, octopuses went on to evolve extensive central nervous systems, with more nervous system mass (adjusted for body size) than fish or reptiles, and a sophisticated behavioural repertoire including memory, visual communication, and tool use.[21] Impressively intelligent animals with more recent LCAs include, among others, corvids (crows and ravens, LCA about 300 million years ago),[22] and elephants (LCA about 100 million years ago).[23] In other words, from the starting point of those wormlike common ancestors in the environment of Earth, the resources of evolution independently produced complex learning, memory, and tool use both within and without the line of human ancestry.

---

[21] See, e.g. Mather (1994; 2008), Finn, Tregenza and Norman (2009), and Hochner, Shomrat and Fiorito (2006) for a review of octopus intelligence.

[22] For example, a crow named Betty was able to bend a straight wire into a hook in order to retrieve a food bucket from a vertical tube, without prior training; crows in the wild make tools from sticks and leaves to aid their hunting of insects, pass on patterns of tool use, and use social deception to maintain theft-resistant caches of food; see Emery and Clayton (2004). For LCA dating, see Benton and Ayala (2003).

[23] See Archibald (2003) for LCA dating, and Byrne, Bates and Moss (2009) for a review arguing that elephants' tool use, number sense, empathy, and ability to pass the mirror test suggest that they are comparable to non-human great apes.

Some proponents of the evolutionary argument for AI, such as Moravec (1976), have placed great weight on such cases of convergent evolution. Before learning about the diversity of animal intelligence, we would assign some probability to scenarios in which the development of these basic behavioural capabilities (given background conditions) was a major barrier to creating human-level intelligence. To the extent convergent evolution lets us rule out particular ways in which the evolution of intelligence could be hard, it should reduce our total credence in the evolution of intelligence being hard (which is just the sum of our credence in all the particular ways it could be hard).

There is, however, an important caveat to such arguments. A species that displays convergent evolution of intelligence behaviourally may have a cognitive architecture that differs in unobserved ways from that of human ancestors. Such differences could mean that the animal brains embody algorithms that are hard to 'scale' or build upon to produce human-level intelligence, so that their ease of evolution has little bearing on AI feasibility. By way of analogy, chess-playing programs outperform humans within the limited domain of chess, yet the underlying algorithms cannot be easily adapted to other cognitive tasks, let alone human-level AI. In so far as we doubt the 'scalability' of octopus or corvid intelligence, despite the appearance of substantial generality, we will discount arguments from their convergent evolution accordingly.

Further, even if we condition on the relevant similarity of intelligence in these convergent lineages and those ancestral to humans, observation selection effects could still conceal extraordinary luck in factors shared by both. First, background environmental effects, such as the laws of physics, the layout of the solar system, and the geology of the Earth could all be unusually favourable to the evolution of intelligence (relative to simulated alternatives for AI), regardless of convergent evolution. Second, the LCA of all these lineages was already equipped with various visible features — such as nervous systems — that evolved only once in Earth's history and which might therefore have been arbitrarily difficult to evolve. While background conditions such as geology and the absence of meteor impacts seem relatively unlikely to correspond to significant problems for AI designers, it is somewhat less implausible to suppose that early neurons conceal extraordinary design difficulty: while computational models of individual neurons have displayed impressive predictive success, they might still harbour subtly relevant imperfections.[24] Finally, some subtle features of the LCA may have greatly enabled later development of intelligence without immediate visible impact. Consider the case of eyes, which have developed in many different animal lineages with widely varying anatomical features and properties (compare the eyes of humans, octopuses, and fruit flies). Eyes in all lineages make use of the proteins known as opsins, and some common regulatory genes such as PAX6, which were present in the LCA of all the creatures with eyes.[25] Likewise, some obscure genetic or physiological mechanism dating back to the octopus-human LCA may both be essential to the later development of octopus-level intelligence in various lineages and have required extraordinary luck.

In addition to background conditions shared by lineages, convergent evolution also leaves open the possibility of difficult innovations lying between the abilities of elephants or corvids or octopuses and human-level intelligence, since we have no examples of robustly human-level capabilities evolving convergently. If the evolution of human-level intelligence were sufficiently easy, starting from the capabilities of these creatures on Earth, then it might seem that observers should find that it appeared multiple times in evolution on their planets. However, as human technology advanced, we have caused mass extinctions (including all other hominids) and firmly occupied the ecological niche of dominant tool-user. If the evolution of human-level intelligence typically pre-empts the evolution of further such creatures, then evolved civilizations will mostly find themselves without comparably intelligent neighbours, even if such evolution is relatively easy.[26] Accurate estimation of the rate at which human-level intelligence evolves from a given starting point would involve the same need for Bayesian correction found in analysis of disasters that would have caused human extinction.[27]

In summary, by looking at instances of convergent evolution on Earth, we can refute claims that certain evolutionary innovations — those for which we have examples of convergence — are exceedingly difficult, given certain assumptions about their underlying mechanisms. Although this method leaves open several ways in which the

---

[24] See Sandberg and Bostrom (2008), for a review.

[25] See Schopf (1992) on the convergent evolution of eyes.

[26] Pre-emption might not occur if, for instance, most evolved intelligences were unable to manipulate the world well enough to produce technology. However, most of the examples of convergent evolution discussed above do have manipulators capable of tool use, with the possible exception of cetaceans (whales and dolphins).

[27] See Çirkoviç, Sandberg and Bostrom (2010) for an explanation of this correction.

evolution of intelligence could in principle have been exceedingly difficult, it narrows the range of possibilities. In particular, it provides disconfirming evidence against hypotheses of high evolutionary difficulty between the development of primitive nervous systems and those fairly complex brains providing the advanced cognitive capabilities found in corvids, elephants, dolphins, etc. In so far as we think evolutionary innovations relevant to AI design will disproportionately have occurred after the development of nervous systems, the evidence from convergent evolution remains quite significant.

To reach beyond the period covered by convergent evolution, and to strengthen conclusions about that period, requires other lines of evidence.

6. SSA and clues from evolutionary timing

The Earth is approximately 4.54 billion years old. Current estimates hold that the expansion of the sun will render the Earth uninhabitable (evaporating the oceans) in somewhat more than a billion years (see Dalrymple, 2001, and Adams and Laughlin, 1998). Assuming that no other mechanism reliably cuts short planetary windows of habitability, human-level intelligence could have evolved on Earth hundreds of millions of years later than it in fact did.[28] Combined with a principle such as SSA, this evidence can be brought to bear on the question of the evolvability of intelligence.

---

[28] The one-billion-year figure is best seen as an upper bound on the remaining habitability-window for the Earth. It is quite conceivable (had humans not evolved) that some natural process or event would have slammed this window shut much sooner than one billion years from now, especially for large mammalian-like life forms. However, there are grounds for believing that whatever the fate would have been for our own planet, there are many Earth-like planets in the universe whose habitability window exceeds 5 billion years or more. The lifetime and size of the habitable zone depends on the mass of the star. Stellar lifetimes scale as $M^{2.5}$ and their luminosity as $M^{3.5}$, where M is their mass in solar masses; see Hansen and Kawaler (1994). Thus, a star 90% of the sun's mass would last 30% longer and have a luminosity of 70%, allowing an Earth analogue to orbit 0.83 AU from the star with the same energy input as Earth. The interaction between stellar mass and the habitable zone is more complex, requiring assumptions about climate, but models typically find that the timespan a terrestrial planet can remain habitable is greater for less heavy stars, with increases of several billion years for stars only marginally less massive than the sun; see Kasting et al. (1993) and Lammer et al. (2009). Lighter stars are considerably more common than heavier stars. Sun-like G-class stars of 0.8–1.04 solar masses make up only 7.6% of main-sequence stars, while the lighter 0.45–0.8 solar mass K-class stars make up 12.1%, and the even lighter M-class red dwarfs 76.45%; see LeDrew (2001). A randomly selected planet will therefore be more likely to orbit a lighter star than the sun, assuming the number of planets formed per system is not vastly different between G-class and K-class stars.

6.1 Uninformative priors plus late evolution suggest intelligence is rare

Brandon Carter (1983; 1989) has argued that the near-coincidence between the time it took intelligence to evolve on Earth and Earth's total habitable period suggests that the chances of intelligent life evolving on any particular Earth-like planet are in fact far below 1.[29]

Following Carter, let us define three time intervals: $\bar{t}$, 'the expected average time... which would be intrinsically most likely for the evolution of a system of "intelligent observers", in the form of a scientific civilization such as our own' (Carter, 1983, p. 353); t_e, the time taken by biological evolution on this planet $\approx 4 \times 10^9$ years; and t_0, the period during which Earth can support life $\approx 5.5 \times 10^9$ years using the above estimate.

Carter's argument then runs roughly as follows: at our present stage of understanding of evolutionary biology, we have no real way of directly estimating $\bar{t}$. Also, there is no *a priori* reason to expect $\bar{t}$ to be on the same timescale as t_0. Thus, we should use a very broad starting probability distribution over $\bar{t}$ — a distribution in which only a small portion of the probability mass is between 10^9 years and 10^{10} years, leaving a large majority of the probability mass in scenarios where either: (a) $\bar{t} \ll t_0$, or (b) $\bar{t} \gg t_0$.

Carter suggests that we can next rule out scenarios in which $\bar{t} \ll t_0$ with high probability, since if technological civilizations typically take far less than 4×10^9 years to evolve, our observations of finding ourselves as the first technological civilization on Earth, recently evolved at this late date, would be highly uncommon. This leaves only scenarios in which either $t \approx t_0$ (a small region), or $\bar{t} \gg t_0$ (a large region). Due to observer-selection effects, intelligent observers under either of these scenarios would observe that intelligent life evolved within their own world's habitable period (even if, as in Hard Intelligence, $\bar{t}$ is many orders of magnitude larger than t_0). Thus, at least until updating on other information, we should deem it likely that the chance of intelligent life evolving on our planet within the sun's lifetime was very small.

6.2 Detailed timing suggests that there are fewer than eight 'hard steps'

However, knowledge of the Earth's habitable lifetime can also be used to attempt to place probabilistic upper bounds on the the number of

---

[29] This section draws on a discussion in Bostrom (2002).

improbable 'critical' steps in the evolution of humans. Hanson (1998b) puts it well:

> Imagine that someone had to pick five locks by trial and error (i.e. without memory), locks with 1, 2, 3, 4, and 5 dials of ten numbers each, so that the expected time to pick each lock was .01, .1, 1, 10, and 100 hours respectively. If you had just a single (sorted) sample set of actual times taken to pick the locks, say .00881, .0823, 1.096, 15.93, and 200.4 hours, you could probably make reasonable guesses about which lock corresponded to which pick-time. And even if you didn't know the actual difficulties (expected pick times) of the various locks, you could make reasonable guesses about them from the sample pick-times.
>
> Now imagine that each person who tries has only an hour to pick all five locks, and that you will only hear about successes. Then if you heard that the actual (sorted) pick-times for some success were .00491, .0865, .249, .281, and .321 hours, you would have a harder time guessing which lock corresponds to which pick-time. You could guess that the first two times probably correspond to the two easiest locks, but you couldn't really distinguish between the other three locks since their times are about the same. And if you didn't know the set of lock difficulties, these durations would tell you very little about the hard lock difficulties.
>
> It turns out that a difficulty of distinguishing among hard steps is a general consequence of conditioning on early success… For easy steps, the conditional expected times reflect step difficulty, and are near the unconditional time for the easiest steps. The conditional expected times for the hard steps, on the other hand, are all pretty much the same.

For example, even if the expected pick-time of one of the locks had been a million years, you would still find that its average pick-time in successful runs is closer to .2 or .3 than to 1 hour, and you wouldn't be able to tell it apart from the 1, 10, and 100 hours locks. Perhaps most usefully, Carter and Hanson argue that the expected time between the picking of the last lock and the end of the hour has approximately the same time distribution as the expected time between one 'hard step' and another.[30] Therefore, if we knew the 'leftover' time L at the end of the final lock-picking, but did not know how many locks there had been, we would be able to use that knowledge to rule out scenarios in which the number of 'hard steps' was much larger than $n = 1$ hour / L.

Thus, to start with the simplest model, if we assume that the evolution of intelligent life requires a number of steps to occur sequentially (so that, for example, nervous systems have no chance of evolving

---

[30] Carter (1983) proves this analytically for the special case in which all hard steps are of the same difficulty; Hanson (1998b) verifies, via Monte Carlo simulations, that it approximately holds with hard steps of varied difficulties. Aldous (2010) makes several additional generalizations of the result.

until multicellularity has evolved), that *only* sequential steps are needed, that some of these steps are 'hard steps' in the sense that their expected average time exceeds Earth's total habitable period (assuming the steps' prerequisites are already in place, and in the absence of observer-selection effects), and that these 'hard steps' have a constant chance of occurring per time interval — then we can use the gap t_e and t_0 to obtain an upper bound on the number of hard steps. Carter estimates this bound to be 3, given his assumption that t_0 is about 10^{10} years (based on earlier longer estimates of the habitable period). Hanson (1998b), using a model with only hard sequential steps (with constant unconditional chance of occurrence after any predecessor steps), calculates that with 1.1 billion years of remaining habitability (in accordance with more recent estimates) and $n = 7$, only 21% of planets like Earth with evolved intelligence would have developed as early.

The same bounds hold (more sharply, in fact) if some steps have a built-in time lag before the next step can start (e.g. the evolution of oxygen-breathing life requiring an atmosphere with a certain amount of oxygen, produced by anaerobic organisms over hundreds of millions of years). These bounds are also sharpened if some of the steps are allowed to occur in any order (Carter, 1983). Thus, the 'hard steps' model rules out a number of possible 'hard intelligence' scenarios: evolution typically may take prohibitively long to get through certain 'hard steps', but, between those steps, the ordinary process of evolution suffices, even without observation selection effects, to create something like the progression we see on Earth. If Earth's remaining habitable period is close to that given by estimates of the sun's expansion, observation selection effects could not have given us hundreds or thousands of steps of acceleration, and so could not, for example, have uniformly accelerated the evolution of human intelligence across the last few billion years in the model.[31][32]

Moreover, in addition to providing information about the total number of hard steps, the model can also give probabilistic bounds on how many hard steps could have occurred in any given time interval. For example, it would allow us to infer with high confidence that at most one hard step has occurred in the 6 million years since the human/chimp common ancestors. This may help narrow the bounds on where AI engineering difficulty can be found.

7. Conclusions

Proponents of the evolutionary argument for AI have pointed to the evolution of intelligent life on Earth as a ground for anticipating the creation of artificial intelligence this century. We explicated this

---

[31] As with the evidence from convergent evolution, there are caveats about the types of scenario this evidence can disconfirm. While the Hanson and Carter models have been extended to cover many branching possible routes to intelligence, the extended models still do not allow us to detect a certain kind of rapid 'dead-end' that pre-empts subsequent progress. For example, suppose that some early nervous system designs are more favourable to the eventual evolution of human-level intelligence, but that whichever nervous system arises first will predominate, occupying the ecological niches that might otherwise have allowed a new type of nervous system to emerge. If the chance of developing any nervous system is small, making it a hard step, then the dead-end possibility does not affect the conclusions about planets with intelligent life. However, if the development of nervous systems occurs quickly with high probability, but producing nervous systems with the right scalability is improbable, then this will reduce the proportion of planets that develop human-level intelligence without affecting the typical timelines of evolutionary development on such planets. Conceivably, similar dead-ends could afflict human engineers — although humans are better able to adopt new approaches to escape local optima.

[32] One objection to the use of the Carter model is that it assumes hard steps are permanent. But in fact, *contra* the model, the organisms carrying some hard step innovation could become extinct, e.g. in an asteroid bombardment. If such events were to frequently 'reset' certain hard steps, then scenarios with long delays between the first resettable step and the evolution of intelligent life would be less likely. Success would require both that hard steps be achieved and that disasters not disrupt hard steps in the interim (either via lack of disasters, or through relevant organisms surviving). With a near-constant probability $(1-p)$ of relevant catastrophe per period, the probability of avoiding catastrophe for a duration of time t would be p^t. Thus, all else equal, allowing for the possibility that hard steps are not permanent reduces the expected time to complete the resettable steps, conditioning on successful evolution of intelligent life. Carter's model would then underestimate the number of hard steps. However, this underestimation is less severe when we account for the fact that longer time periods allow more chances for the hard steps to occur. The chance of completing n hard steps in time t is proportional to t^n. So when we examine planets where intelligence evolved we would tend to find intelligence evolving after a longer-than-usual gap between disasters, or a series of disasters which spared some lineages embodying past hard steps. Over long timescales, the exponential increase of the catastrophe effect will dominate the polynomial increase from more opportunities for hard steps. However, with plausible rates of catastrophe, the effect of increased opportunity greatly blunts the objection.

argument in terms of claims about the difficulty of search for suitable intelligent cognitive architectures (the argument from problem diffi- culty) and, alternatively, in terms of claims about the availability of evolutionary algorithms capable of producing human-level general intelligence when implemented on realistically achievable hardware (the argument from evolutionary algorithms).

The argument from evolutionary algorithms requires an estimate of how much computing power it would take to match the amount of optimization power provided by natural selection over geological timescales. We explored one way of placing an upper bound on the relevant computational demands and found it to correspond to more than a century's worth of continuing progress along Moore's Law — an impractically vast amount of computing power. Large efficiency gains are almost certainly possible, but they are difficult to quantify in advance. It is doubtful that the upper bound down calculated in our paper could be reduced sufficiently to enable the argument from evo- lutionary algorithms to succeed.

The argument from problem difficulty avoids making specific assumptions about amounts of computing power required or the spe- cific way that human-level machine intelligence would be achieved. This version of the argument replaces the quantitative premise about computational resource requirements and evolutionary algorithms with a more intuitive appeal to the way that evolutionary consider- ations might tell us something about the problem difficulty of design- ing generally intelligent systems. But in either of its two versions, the evolutionary argument relies on another assumption: that the evolu- tion of human intelligence was not exceedingly hard (2').

We pointed out that this assumption (2') cannot be directly inferred from the fact that human intelligence evolved on Earth. This is because an observation selection effect guarantees that all evolved human-level intelligences will find that evolution managed to pro- duce them — independently of how difficult or improbable it was for evolution to produce this result on any given planet.

We showed how to evaluate the possible empirical support for (2') from the alternative standpoints of two leading approaches to anth- ropic reasoning: the Self-Sampling Assumption and the Self-Indica- tion Assumption. The Self-Indication Assumption strongly supports the evolutionary argument and the feasibility of AI (among other strong and often counterintuitive empirical implications). The impli- cations of the Self-Sampling Assumption depend more sensitively on the details of the empirical evidence. By considering additional infor- mation about the details of evolutionary history — notably convergent evolution and the timing of key innovations — the Self-Sampling Assumption can be used to make probabilistic inferences about the evolvability of intelligence and hence about the soundness of the evolutionary argument.

This further SSA analysis disconfirms many particular ways in which the evolution of intelligence might be hard, especially scenarios of extreme hardness (with many difficult steps), thus supporting premise (2') (although Carter's model also has implications counting against very easy evolution of intelligence). Of particular interest, two lines of evidence count against extreme evolutionary hardness in developing human-level intelligence given the development of nervous systems as we know them: fairly sophisticated cognitive skills convergently evolved multiple times from the starting point of the earliest nervous systems; and 'hard step' models predict few sequential hard steps in our very recent evolutionary history. Combined with the view that evolutionary innovations in brain design especially are diagnostic of AI design difficulty, these observations can avert some of the force of the objection from observation selection effects.

Thus, with one major approach to anthropic reasoning (SIA) providing strong support for (2'), and the other (SSA) offering a mixed picture and perhaps moderate support, observation selection effects do not cripple the evolutionary argument via its premise (in either of its versions) of non-hard evolution of intelligence.

Extensive empirical and conceptual uncertainty remains. Further progress could result from several fields. Computer scientists can explore optimal environments for the evolution of intelligence, and their computational demands — or how easily non-evolutionary programming techniques can replicate the functionality of evolved designs in various domains. Evolutionary biologists and neuroscientists can untangle questions about evolutionary convergence and timing. And physicists, philosophers, and mathematicians can work to resolve the numerous open questions in observational selection theory. Considering how recent many of the relevant ideas and methodologies are, and the character of results thus far obtained, it seems likely that further epistemic truffles are to be found in these grounds.[33]

---

[33] We are grateful to David Chalmers, Paul Christiano, Zack M. Davis, Owain Evans, Louie Helm, Lionel Levine, Jesse Liptrap, James Miller, Luke Muehlhauser, Anna Salamon, Anders Sandberg, Elizabeth Synclair, Tim Tyler, and audiences at Australian National University and the 2010 Australasian Association of Philosophy conference for helpful comments and discussion.

References

Adams, F.C. & Laughlin, G. (1998) The future of the universe, *Sky and Telescope*, **96** (2), p. 32.

Aldous, D.J. (2010) The Great Filter, branching histories and unlikely events, [Online], http://www.stat.berkeley.edu/~aldous/Papers/GF.pdf

Archibald, J.D. (2003) Timing and biogeography of the eutherian radiation: Fossils and molecules compared, *Molecular Phylogenetics and Evolution*, **28** (2), pp. 350–359.

Baum, E. (2004) *What is Thought?*, Cambridge, MA: MIT Press.

Benton, M.J. & Ayala, F.J. (2003) Dating the tree of life, *Science*, **300** (5626), pp. 1698–1700.

Bostrom, N. (2002) *Anthropic Bias: Observation Selection Effects in Science and Philosophy*, New York: Routledge.

Bostrom, N. (2003) Are you living in a computer simulation?, *Philosophical Quarterly*, **53** (211), pp. 243–255.

Bostrom, N. (2005) The Simulation Argument: Reply to Weatherson, *Philosophical Quarterly*, **55** (218), pp. 90–97.

Bostrom, N. (2007) Observation selection effects, measures, and infinite spacetimes, in Carr, B. (ed.) *Universe or Multiverse?*, Cambridge: Cambridge University Press.

Bostrom, N. & Çirkoviç, M.M. (2003) The Doomsday Argument and the Self-Indication Assumption, *Philosophical Quarterly*, **53** (210), pp. 83–91.

Bostrom, N. & Sandberg, A. (2009) The wisdom of nature: An evolutionary heuristic for human enhancement, in Savulescu, J. & Bostrom, N. (eds.) *Human Enhancement*, Oxford: Oxford University Press.

Bostrom, N. & Kulczycki, M. (2011) A patch for the Simulation Argument, *Analysis*, **71** (1), pp. 54–61.

Byrne, R.W., Bates, L.A. & Moss, C.J. (2009) Elephant cognition in primate perspective, *Comparative Cognition & Behavior Reviews*, **4**, pp. 65–79.

Carter, B. (1983) The anthropic principle and its implications for biological evolution, *Philosophical Transactions of the Royal Society of London A*, **310** (1512), pp. 347–363.

Carter, B. (1989) The anthropic selection principle and the ultra-Darwinian synthesis, in Bertola, F. & Curi, U. (eds.) *The Anthropic Principle*, Cambridge: Cambridge University Press.

Chalmers, D.J. (2005) The Matrix as metaphysics, in Grau, C. (ed.) *Philosophers Explore the Matrix*, Oxford: Oxford University Press.

Çirkoviç, M.M., Sandberg, A. & Bostrom, N. (2010) Anthropic shadow: Observation selection effects and human extinction risks, *Risk Analysis*, **30** (10), pp. 1495–1506.

Dalrymple, G.B. (2001) The age of the Earth in the twentieth century: A problem (mostly) solved, *Special Publications, Geological Society of London*, **190** (1), pp. 205–221.

Dieks, D. (2007) Reasoning about the future: Doom and beauty, *Synthese*, **156** (3), pp. 427–439.

Elga, A. (2000) Self-locating belief and the Sleeping Beauty problem, *Analysis*, **60** (2), pp. 143–147.

Emery, N.J. & Clayton, N.S. (2004) The mentality of crows: Convergent evolution of intelligence in corvids and apes, *Science*, **306** (5703), pp. 1903–1907.

Erwin, D.H. & Davidson, E.H. (2002) The last common bilaterian ancestor, *Development*, **129** (13), pp. 3021–3032.

Finn, J.K., Tregenza, T. & Norman, M.D. (2009) Defensive tool use in a coco-nut-carrying octopus, *Current Biology*, **19** (23), pp. R1069–R1070.

Grace, C. (2010) *Anthropic Reasoning in the Great Filter*, BSc (Hons) dissertation, Australian National University.

Hansen, C.J. & Kawaler, S.D. (1994) *Stellar Interiors: Physical Principles, Structure, and Evolution*, Basel: Birkhäuser.

Hanson, R. (1998a) *The Great Filter — Are We Almost Past It?*, [Online], http://hanson.gmu.edu/greatfilter.html

Hanson, R. (1998b) *Must Early Life be Easy? The Rhythm of Major Evolutionary Transitions*, [Online], http://hanson.gmu.edu/hardstep.pdf

Hawks, J., *et al.* (2007) Recent acceleration of human adaptive evolution, *Proceedings of the National Acadamy of Sciences*, **104** (52), pp. 20753–20758.

Hochner, B., Shomrat, T. & Fiorito, G. (2006) The octopus: A model for a comparative analysis of the evolution of learning and memory mechanisms, *Biological Bulletin*, **210**, pp. 308–317.

Kasting, J.F., Whitmire, D.P. & Reynolds, R.T. (1993) Habitable zones around main sequence stars, *Icarus*, **101**, pp. 108–128.

Lammer, J.H., *et al.* (2009) What makes a planet habitable?, *Astronomy and Astrophysics Review*, **17**, pp. 181–249.

LeDrew, G. (2001) The real starry sky, *Journal of the Royal Astronomical Society of Canada*, **95** (1), pp. 32–33.

Legg, S. (2008) *Machine Super Intelligence*, PhD thesis, Department of Informatics, University of Lugano.

Leslie, J. (1993) Doom and probabilities, *Mind*, **102** (407), pp. 489–491. Also [Online], http://www.jstor.org/pss/2253981

Lewis, D.K. (2001) Sleeping Beauty: Reply to Elga, *Analysis*, **61** (271), pp. 171–176.

MacKay, D.J.C. (2009) *Information Theory, Inference, and Learning Algorithms*, Cambridge: Cambridge University Press.

Mather, J.A. (1994) 'Home' choice and modification by juvenile Octopus vulgaris (Mollusca: Cephalopoda): Specialized intelligence and tool use?, *Journal of Zoology*, **233**, pp. 359–368.

Mather, J.A. (2008) Cephalopod consciousness: Behavioural evidence, *Consciousness and Cognition*, **17** (1), pp. 37–48.

Menzel, R. & Giurfa, M. (2001) Cognitive architecture of a mini-brain: The honeybee, *Trends in Cognitive Science*, **5** (2), p. 62.

Moravec, H. (1976) *The Role of Raw Power in Intelligence*, [Online], http://www.frc.ri.cmu.edu/users/hpm/project.archive/general.articles/1975/Raw.Power.html

Moravec, H. (1988) *Mind Children: The Future of Robot and Human Intelligence*, Cambridge, MA: Harvard University Press.

Moravec, H. (1998) When will computer hardware match the human brain?, *Journal of Transhumanism*, **1**.

Moravec, H. (1999) *Robots: Mere Machine to Transcendent Mind*, Oxford: Oxford University Press.

Neal, R.M. (2007) Puzzles of anthropic reasoning resolved using full non-indexical conditioning, *Technical Report No. 0607*, Department of Statistics, University of Toronto, [Online], http://www.cs.toronto.edu/~radford/ftp/anth2.pdf

Olum, K.D. (2002) The Doomsday Argument and the number of possible observers, *Philosophical Quarterly*, **52** (207), pp. 164–184.

Piccione, M. & Rubinstein, A. (1997) The absent-minded driver's paradox: Synthesis and responses, *Games and Economic Behavior*, **20** (1), pp. 121–130.

Sabrosky, C.W. (1952) How many insects are there?, in U.S. Dept. of Agriculture (eds.) *Insects: The Yearbook of Agriculture*, Washington, DC: U.S.G.P.O.

Sandberg, A. & Bostrom, N. (2008) Whole brain emulation: A roadmap, *Technical Report 2008–3*, Future for Humanity Institute, Oxford University, [Online], http://www.fhi.ox.ac.uk/Reports/2008-3.pdf

Schopf, W.J. (ed.) (1992) *Major Events in the History of Life*, Boston, MA: Jones and Barlett.

Schultz, T.R. (2000) In search of ant ancestors, *Proceedings of the National Academy of Sciences*, **97** (26), pp. 14028–14029.

Truman, J.W., Taylor, B.J. & Award, T.A. (1993) Formation of the adult nervous system, in Bate, M. & Arias, A.M. (eds.) *The Development of Drosophila Melanogaster*, Cold Spring Harbor, NY: Cold Spring Harbor Laboratory Press.

Whitman, W.B., Coleman, D.C. & Wiebe, W.J. (1998) Prokaryotes: The unseen majority, *Proceedings of the National Academy of Sciences*, **95** (12), pp. 6578–6583.

Eric Steinhart

The Singularity

Beyond Philosophy of Mind

Abstract: *Thought about the singularity intersects the philosophy of mind in deep and important ways. However, thought about the singularity also intersects many other areas of philosophy, including the history of philosophy, metaphysics, the philosophy of science, and the philosophy of religion. I point to some of those intersections.*

According to Chalmers (this volume, p. 14), thought about the singularity is closely associated with many important philosophical topics. As expected, Chalmers is mainly interested in topics that fall in or near the philosophy of mind (e.g. consciousness, personal persistence). Those topics are of great importance. But the singularity is also connected with topics in the history of philosophy, metaphysics, philosophy of science, and philosophy of religion. And it is connected with topics in ethics and political philosophy. Chalmers is correctly disappointed by the dearth of philosophical interest in the singularity (*ibid.*, pp. 13–14). I hope to arouse further philosophical interest by writing in a deliberately provocative way about the links between the singularity and topics outside of the philosophy of mind.

As with everything in the future, the singularity has a past. For the historian of philosophy, the singularity may inspire new ways of looking at some old traditional themes. One such theme is the deep *metaphysics* of intelligence. This metaphysics goes back to the *nous* of Anaxagoras and the *logos* of Heraclitus; it was developed in fascinating ways by the Stoics. Of course, theists have long argued that intelligence plays a foundational role: our universe exists because it was designed by a divine intelligence. And there is a long tradition that portrays our universe as a progressive process in which intelligence is always rising to greater heights (Lovejoy, 1936, Chapter IX). For the Hegelian, the singularity looks much like the final self-realization of

Spirit in absolute knowing (Zimmerman, 2008). Peirce develops an evolutionary cosmology in which the universe makes endless progress towards a final singularity (Hausman, 1993). Adams used reasoning strikingly like that of Kurzweil to develop his *law of acceleration* (1904; 1909). After World War II, one leading figure in the philosophical history of the singularity is Teilhard de Chardin (1955/2002).

Also after World War II, much work related to the singularity is done in the context of *physical eschatology* (Cirkovic, 2003). Barrow and Tipler (1986) sketch a view of the future of the universe as converging towards an ultimate Omega Point. For them, the singularity is the final phase in a long evolutionary process. However, evolution is not random; on the contrary, it is *progressive*. It is governed by some hypothesized *law of optimality*. For Barrow and Tipler, this law is the *Final Anthropic Principle*: 'Intelligent information-processing must come into existence in the Universe, and, once it comes into existence, it will never die out' (*ibid.*, p. 23). Tipler later interprets this law as his *Eternal Life Postulate* (1995, p. 11). For Kurzweil, this law is the *Law of Accelerating Returns*: 'As order exponentially increases, time exponentially speeds up' (Kurzweil, 1999, p. 30; 2005, Chapters 1 and 2). Kurzweil thinks of evolution as a self-bootstrapping positive feedback loop in which *value* is always exponentially increasing (1999, p. 32).

According to their authors, these laws of optimality, like other natural laws, were inferred from empirical data by inductive arguments. For Barrow and Tipler, the induction is by inference to the best explanation. They say that some anthropic principle is the best explanation for the apparent fine-tuning of our universe for life. Ultimately, they derived their Final Anthropic Principle from an atheistic version of the design argument. For Kurzweil, the induction seems to be mostly empirical generalization (1999; 2005). Advocates of the singularity say that these laws can be used to predict the singularity much as the laws of Newtonian gravitation were used by Halley to predict his comet. Kurzweil is well-known for his detailed timelines (1999, pp. 261–80; 2005, Chapters 1 and 2). Of course, either the singularity will happen as predicted or it will not. If it does not, sceptics are entitled to reject the alleged laws as mere pseudoscience. However, singularitarians have an intriguing reply. The reply is to push these laws deeper: they should not have been formulated as laws about our universe. Correctly formulated, they work at much greater depths.

Perhaps these laws are related to the laws of optimality discussed by recent philosophers. Leslie argues that ethical requirements have creative power (1970; 1979). Rescher states the law of optimality like this: 'whatever possibility is for the best is *ipso facto* the possibility that is actualized' (2000, p. 815; 1984, Chapter 2). At least for *axiarchists* like Leslie and Rescher, the laws of optimality are deeper than natural laws. They are metaphysical principles that are not empirically falsifiable. Further reflection on the laws of optimality may lead the singularitarian to look at Leibniz's cosmological argument (1697/1988, pp. 84–5). Leibniz says that there must be some ultimate sufficient reason for existence. Of course, Leibniz also says that this reason must be grounded in God. But the singularitarian is more likely to prefer an atheistic account: why is there something rather than nothing? Because there is some law of optimality that eventually ensures the singularity.

Whether or not the singularity occurs as predicted in our universe, thinking about the singularity leads to new ways for the philosopher of religion to think about the old design arguments for God. Many writers say the design arguments lead to a regression of designers (e.g. Hume, 1779/1990, pp. 72–3; Doore, 1980, pp. 153–5; Dawkins, 2008, pp. 136, 146, 188). Our god designed our universe; but our god was itself designed by some previous god. There are three ways this regression can go: (1) each previous god is more divine than the next god; (2) each previous god is just as divine as the next god; and (3) each previous god is less divine than the next god. Since only the third option avoids some obvious fatal flaws, the regression of gods eventually bottoms out in some least god. When this regression is turned around, it presents itself as a progression of ever more divine gods. This progression is analogous to the progression of ever greater artificial intellects discussed by singularitarians. It is often said that the singularity will be achieved through the *recursive self-improvement* of artificial intellects (Good, 1965; Kurzweil, 2005, pp. 27–8; Schmidhuber, 2007; Chalmers, this volume, pp. 15–26). Anyone interested in the design argument is free to apply recursive self-improvement to the progression of gods.

This leads to an *evolutionary theogony*. Since evolution is self-bootstrapping, it brings some initial god into being out of nothingness. Adopting what Leslie calls extreme axiarchism (1970, p. 286; 1979, p. 6), one might argue that an abstract law of optimality brings this initial god into existence. The initial god exists because it ought to. The initial god has some minimal level of perfection (it is minimally benevolent, intelligent, and powerful). However small this minimal level may be, it is sufficient for the design and creation of the next more divine god. The result is a series of gods in which each previous god designs and creates the next god. The next god is always more divine than the previous god. It is interesting to note that this progression of gods is compatible with Aquinas's Third Way (*Summa Theologica*, Part 1, Q. 2, Art. 3). And yet, surprisingly, this evolutionary theogony is atheistic. Every god is surpassed by some greater god. Just as there is no biggest number, so there is no maximally perfect being — the god of theism does not appear in this sequence. This evolutionary theogony is an atheistic interpretation of the Leibnizian doctrine of the *striving possibles* (Leibniz, 1697/1988; Rescher, 1991, pp. 171–2, 188, 191; Blumenfeld, 1981). And these gods are just big computers. They are not the objects of revealed religion — they are the objects of pure computer science.

The evolutionary theogony entails the production of an endless series of ever better universes. Just as our local god has designed and created our universe, so every god designs and creates its own universe. Just as later gods are more divine, so later universes are better. Philosophers who have reflected on the ontological argument have described such sequences of greater gods making greater universes (Henle, 1961). The classical problem of evil vanishes: every god does the best it can; but no god is unsurpassable. For the singularitarian who is willing to embrace this evolutionary theogony, the singularity becomes much harder to refute. The evolution of intelligence is at work on a super-cosmic scale. Within every next universe, the internal evolutionary process runs farther. Universes can be classified on a Kurzweil scale (2005, pp. 14–21): universes in the n-th rank evolve only through Kurzweil's n-th epoch. Perhaps our universe is just too primitive for the singularity — it merely lies in the fifth rank of universes. Not to worry: the singularity will happen in some near descendent of our universe. It is reasonable to complain that this type of thinking is extremely speculative — but any philosopher has to accept *some* theory of the ultimate origination of concrete actuality. The only question is how well this evolutionary theogony compares with its rivals (which are primarily theistic).

Uploading is often associated with the singularity (Kurzweil, 2005, pp. 198–204; Chalmers, this volume, pp. 45–67). For Chalmers, and for other philosophers of mind, uploading is a lens that focuses thought on the nature of consciousness and problems of personal persistence. But uploading is connected with many topics outside of philosophy of mind. Uploading raises fascinating issues for political philosophers. If we are going to be uploaded into some cyber-world, surely it will be politically structured. Obviously, speculation about cyber-politics is analogous to classical utopian speculation. Human writers have designed many utopias (e.g. Plato's *Republic*, Bacon's

Bensalem, and so on). But what sorts of utopias will the great artilects design? Chalmers considers a Kantian link between intelligence and benevolence (pp. 40–1). Bostrom suggests that superhuman artilects will also be superhuman ethical reasoners (2003, p. 280). On this line of thought, the cyber-world will have heavenly laws. Can we even speculate about those laws? And how would they be enforced? One idea is that the artilects may alter the ethical natures of uploaded humans. Presumably, it will be easy for the artilects to edit our software brains. Perhaps they will get rid of our old sinful natures — editing out the lust and aggression. Perhaps we will also be *morally enhanced* (Douglas, 2008; Faust, 2008). It is easy to see that uploading raises many fascinating ethical and political questions.

Uploading is a secular model of the resurrection theory of John Hick (1976, Chapter 15). And many Christian writers have used uploading (or something very much like it) as an analogy for resurrection (Polkinghorne, 1985, pp. 180–1; 2002; Reichenbach, 1978, p. 27; Mackay, 1997, pp. 248–9). An adventurous thinker might work to link the evolutionary theogony with Hick's *pareschatology* (1976, Chapter 21). If the evolutionary theogony is not compatible with orthodox Christian theology, it may at least be compatible with certain radical expressions of Christian soteriology. Many writers associated with the singularity have also explicitly defended resurrection theories. Moravec talks about resurrection (1988, pp. 122–4). Tipler says we will be resurrected in the computers of the far future (1995, Chapters 9–13). And Kurzweil says that he will use advanced technology to resurrect his dead father (Kushner, 2009, p. 61). Since uploading is technically possible, the resurrection of the body is intelligible. Uploading allows anyone interested in bodily life after death to replace a religious mystery with techno-scientific intelligibility. Strikingly, the resurrection of the body becomes a sub-topic in pure (and perhaps applied) computer science.

Philosophers of science have long known that scientific theories often posit ideal objects. Physical theories discuss frictionless planes and ideal gasses. And pure computer science posits ideal computers (e.g. Turing machines). For the modal realist, these ideal objects exist in other possible universes. Although these ideal objects do not exist in our universe, they nevertheless remain legitimate objects of scientific discussion. They are included in the scopes of our best scientific theories. From the point of view of metaphysics and philosophy of religion, whether or not the singularity actually happens in the future of our universe is not very interesting. What is far more interesting is that singularitarian thought suggests that many of the objects and processes that once lay in the domain of revealed religion now lie in the domain of pure computer science.

References

Adams, H. (1904) A law of acceleration, in Adams, H. (1919) *The Education of Henry Adams*, chapter 34, New York: Houghton Mifflin.

Adams, H. (1909) The rule of phase applied to history, in Adams, H. & Adams, B. (1920) *The Degradation of the Democratic Dogma*, pp. 267–311, New York: Macmillan.

Barrow, J. & Tipler, F. (1986) *The Anthropic Cosmological Principle*, New York: Oxford University Press.

Blumenfeld, D. (1981) Leibniz's theory of the striving possibles, in Woolhouse, R.S. (ed.) *Leibniz: Metaphysics and Philosophy of Science*, pp. 77–88, New York: Oxford University Press.

Bostrom, N. (2003) Ethical issues in advanced artificial intelligence, in Schneider, S. (2009) *Science Fiction and Philosophy*, chapter 22, Malden, MA: Wiley-Blackwell.

Cirkovic. M. (2003) Resource letter: PEs-1: Physical eschatology, *American Journal of Physics*, **71** (2), pp. 122–133.

Dawkins, R. (2008) *The God Delusion*, New York: Houghton-Mifflin.

Doore, G. (1980) The argument from design: Some better reasons for agreeing with Hume, *Religious Studies*, **16** (2), pp. 145–161.

Douglas, T. (2008) Moral enhancement, *Journal of Applied Philosophy*, **25** (3), pp. 228–245.

Faust, H. (2008) Should we select for moral enhancement?, *Theoretical Medicine and Bioethics*, **29**, pp. 397–416.

Good, I. (1965) Speculations concerning the first ultraintelligent machine, in Alt, F. & Rubinoff, M. (eds.) *Advances in Computers*, vol. 6, New York: Academic Press.

Hausman, C. (1993) *Charles S. Peirce's Evolutionary Philosophy*, New York: Cambridge University Press.

Henle, P. (1961) Uses of the ontological argument, *The Philosophical Review*, **70** (1), pp. 102–109.

Hick, J. (1976) *Death and Eternal Life*, New York: Harper & Row.

Hume, D. (1779/1990) *Dialogues Concerning Natural Religion*, New York: Penguin.

Kurzweil, R. (1999) *The Age of Spiritual Machines*, New York: Penguin.

Kurzweil, R. (2005) *The Singularity is Near: When Humans Transcend Biology*, New York: Viking.

Kushner, D. (2009) When man and machine merge: Interview with Ray Kurzweil, *Rolling Stone*, (19 February 2009), pp. 57–61.

Leibniz, G.W. (1697/1988) On the ultimate origination of the universe, in Schrecker, P. & Schrecker, A. (eds.) *Leibniz: Monadology and Other Essays*, New York: Macmillan.

Leslie, J. (1970) The theory that the world exists because it should, *American Philosophical Quarterly*, **7** (4), pp. 286–298.

Leslie, J. (1979) *Value and Existence*, Totowa, NJ: Rowman & Littlefield.

Lovejoy, A. (1936) *The Great Chain of Being*, Cambridge, MA: Harvard University Press.

Mackay, D. (1997) Computer software and life after death, in Edwards, P. (ed.) *Immortality*, pp. 248–249, Amherst, NY: Prometheus Books.

Moravec, H. (1988) *Mind Children: The Future of Robot and Human Intelligence*, Cambridge, MA: Harvard University Press.

Polkinghorne, J.C. (1985) The scientific worldview and a destiny beyond death, in MacGregor, G. (ed.) *Immortality and Human Destiny: A Variety of Views*, pp. 180–183, New York: Paragon House.

Polkinghorne, J.C. (2002) Eschatological credibility, in Peters, T., Russell, R.J. & Welker, M. (eds.) *Resurrection: Theological and Scientific Assessments*, pp. 43–55, Grand Rapids, MI: Eerdmans Publishing.

Reichenbach, B. (1978) Monism and the possibility of life after death, *Religious Studies*, **14** (1), pp. 27–34.

Rescher, N. (1984) *The Riddle of Existence: An Essay in Idealistic Metaphysics*, New York: University Press of America.

Rescher, N. (1991) *G.W. Leibniz's Monadology: An Edition for Students*, Pittsburgh, PA: University of Pittsburgh Press.

Rescher, N. (2000) Optimalism and axiological metaphysics, *The Review of Metaphysics*, **53** (4), pp. 807–835.

Schmidhuber, J. (2007) Godel machines: Fully self-referential optimal universal self-improvers, in Goertzel, B. & Pennachin, C. (eds.) *Artificial General Intelligence*, pp. 199–226, Berlin: Springer-Verlag.

Teilhard de Chardin, P. (1955/2002) *The Phenomenon of Man*, Wall, B. (trans.), New York: Harper Collins. Originally written 1938–1940.

Tipler, F. (1995) *The Physics of Immortality: Modern Cosmology, God and the Resurrection of the Dead*, New York: Anchor Books.

Zimmerman, M. (2008) The singularity: A crucial phase in divine self-actualization?, *Cosmos and History: The Journal of Natural and Social Philosophy*, **4** (1–2), pp. 347–370.

Burton Voorhees

Parsing the Singularity[1]

The singularity, so the story goes, will take place when artificially
intelligent machines take over the design of the next generation of AI
machines, leading to an exponential speed-up in the development of
even more intelligent machines. The result: artificial intelligence that
exceeds human intelligence by a degree greater than that by which we
humans exceed ants. Humanity may end up, in singer Bruce Cockburn's
words, as the insect life of paradise.[2] If so, will we be the friendly bees
producing honey, or the roaches that get squashed when caught?

David Chalmers carries out an excellent analysis of the philosophical issues such a possibility raises and he makes the important distinction between an Intelligence Singularity and a Speed Singularity,
focusing attention on the first. There are issues, however, that were
neglected.

First off, there is the matter of a distinction between human intelligence and AI as it exists today. This is the question of getting from
here to there. Taking the game of chess: human chess players cannot
analyse the huge number of possibilities that arise when a chess position is followed out for more than a few moves into the future, something that is easy for Deep Blue. Human chess players rely on pattern
recognition, something at which current AIs are quite poor.

In the general setting of everyday life, with all its complexities,
human beings are capable of recognizing or establishing partial patterns of possibilities in the mind as an object of contemplation and
then allowing the final elements that complete the pattern to emerge,
either quickly or slowly, through a process akin to directed fluctuation
enhancement. In a real sense, the partial pattern poses a question, creating an attraction that calls forth an answer. The final step in this process is actually recognizing that the result of this process is, indeed, an

[1] Supported by NSERC Discovery Grant 0024817

[2] From the song 'That's What Friends Are For' on the album Breakfast in New Orleans.

answer that fits. When that happens we say something like 'Ah Ha!' The digital processing behind present day AIs cannot do this, relying instead on brute speed in crunching through all possibilities.

But suppose this is a divide that can be bridged. Assume we have built a generally intelligent machine with superhuman capacities. What could this mean? In particular, what might these capabilities be and how might they manifest? Would this simply be a machine that could answer technical questions, carry out logical calculations, generate and prove mathematical theorems better and faster than humans?

The mathematician Keith Devlin (2000) distinguishes four levels of cognitive intelligence. At the first level is the ability to extrapolate an existing physical situation into the immediate future as, for example, a cheetah chasing a baboon must anticipate its moves. At the second level is the ability to keep an object in mind after it is no longer physically present. A chimpanzee, seeing an experimenter put a banana into a basket, will later look into the basket to retrieve the banana. At the third level is the ability, only present in humans, to detach properties from objects and use them in combinatorial thought play. The chimpanzee can know that the basket contains a yellow banana, but it cannot imagine an orange banana with green stripes, or that while in the basket it has morphed into a red apple with a blue stem. Finally, at the fourth level is the ability for purely abstract symbolic thought.

In this picture, human cognition consists of an abstract Platonic world existing on, but independent of, a sensory-motor base. While the degree of independence is a matter of debate, there is an emerging consensus that most human thought takes place in sensory-motor terms and abstract thought schemata are metaphorically abstracted from this base (Lakoff and Johnson, 1980; 1999; Gibbs, Jr., 2008; Pereira, 2007; Fauconnier and Turner, 2002).

This raises another question: what would it be like to be an advanced AI? Would it develop self-consciousness, thinking of itself as an embodied entity in a non-self world? Would the fundamental structures and processes supporting this self be similar to those underlying our human sense of self?[3] Would it use metaphor in the way that we do, given that its physical base would be radically different? Would it be more interesting than being a bat? Would these super AI creations be purely abstract formal intelligences? What sort of sensory channels would we give them at first, and how would this influence the sort of intelligence they possessed? If such an AI were

---

[3] This is a significant question since we don't at present know exactly what underlies our human self-consciousness.

provided, for example, with electromagnetic radiation sensors covering the range from long wavelength radio to extreme gamma radiation, how many dimensions would its 'colour' space have? Would it even have a colour space? And what about providing sensitivity to magnetic fields, neutrino fluxes, gravitational waves?

Might not an advanced AI lust after new sensations and design its successors with even more extended sensory capacities, leading to a *Sensory Singularity* as the limits of the human sensorium are extended and new, humanly unimaginable dimensions of possible sensation are discovered.

Taking Intelligence, Speed, and Sensory Singularities together gives something approaching the *scientia dei* of the Scholastics (e.g. Heidegger, 1984). How might we conceive of a consciousness capable of grasping, essentially instantaneously, all possible patterns, their changes and potentialities, across all possible modes of sensation?

And, how might it conceive of us?

References

Devlin, K. (2000) *The Math Gene*, New York: Basic Books.

Fauconnier, G. & Turner, M. (2002) *The Way We Think: Conceptual Blending and the Mind's Hidden Complexities*, New York: Basic Books.

Heidegger, M. (1984) *The Metaphysical Foundations of Logic*, Heim, M. (trans.), Bloomington, IN: Indiana University Press.

Lakoff, G. & Johnson, M. (1980) *Metaphors We Live By*, Chicago, IL: University of Chicago Press.

Lakoff, G. & Johnson, M. (1999) *Philosophy in the Flesh*, New York: Basic Books.

Gibbs, Jr., R.W. (ed.) (2008) *The Cambridge Handbook of Metaphor*, Cambridge: Cambridge University Press.

Pereira, F.C. (2007) *Creativity and Artificial Intelligence*, Berlin: Monton De Gruyter.

David J. Chalmers

The Singularity:
A Reply to Commentators

1. Introduction

I would like to thank the authors of the 26 contributions to this symposium on my article 'The Singularity: A Philosophical Analysis'. I
learned a great deal from reading their commentaries. Some of the
commentaries engaged my article in detail, while others developed
ideas about the singularity in other directions. In this reply I will concentrate mainly on those in the first group, with occasional comments
on those in the second.

A singularity (or an intelligence explosion) is a rapid increase in
intelligence to super-intelligence (intelligence of far greater than
human levels), as each generation of intelligent systems creates more
intelligent systems in turn. The target article argues that we should
take the possibility of a singularity seriously, and argues that there will
be super-intelligent systems within centuries unless certain specific
defeating conditions obtain.

I first started thinking about the possibility of an intelligence explosion as a graduate student in Doug Hofstadter's AI lab at Indiana University in the early 1990s. Like many, I had the phenomenology of
having thought up the idea myself, though it is likely that in fact I was
influenced by others. I had certainly been exposed to Hans Moravec's
1988 book *Mind Children* in which the idea is discussed, for example.
I advocated the possibility vigorously in a discussion with the AI
researchers Rod Brooks and Doug Lenat and the journalist Maxine
McKew on the Australian TV show *Lateline* in 1996. I first discovered the term 'singularity' on Eliezer Yudkowsky's website in 1997,
where I also encountered the idea of a combined intelligence and
speed explosion for the first time. I was fascinated by the idea that all
of human history might converge to a single point, and took that idea

to be crucial to the singularity *per se*; I have been a little disappointed that this idea has receded in later discussions.

Since those early days I have always thought that the intelligence explosion is a topic that is both practically and philosophically important, and I was pleased to get a chance to develop these ideas in a talk at the 2009 Singularity Summit and then in this paper for *JCS*. Of course the main themes in the target article (the intelligence explosion, negotiating the singularity, uploading) have all been discussed at length before, but often in non-academic forums and often in non-rigorous ways. One of my aims in the target article was to put the discussion on a somewhat clearer and more rigorous analytic footing than had been done in previously published work. Another aim was to help bring the issues to an audience of academic philosophers and scientists who may well have much to contribute.

In that respect I am pleased with the diversity of the commentators. There are nine academic philosophers (Nick Bostrom, Selmer Bringsjord, Richard Brown, Joseph Corabi, Barry Dainton, Daniel Dennett, Jesse Prinz, Susan Schneider, Eric Steinhart) and nine AI researchers (Igor Aleksander, Ben Goertzel, Marcus Hutter, Ray Kurzweil, Drew McDermott, Jurgen Schmidhuber, Murray Shanahan, Roman Yampolskiy, and Bringsjord again). There are also representatives from cultural studies (Arkady Plotnitsky), cybernetics (Francis Heylighen), economics (Robin Hanson), mathematics (Burton Voorhees), neuroscience (Susan Greenfield), physics (Frank Tipler), psychiatry (Chris Nunn), and psychology (Susan Blackmore), along with two writers (Damien Broderick and Pamela McCorduck) and a researcher at the Singularity Institute (Carl Shulman).

Of the 26 articles, about four are wholeheartedly pro-singularity, in the sense of endorsing the claim that a singularity is likely: those by Hutter, Kurzweil, Schmidhuber, and Tipler. Another eleven or so seem to lean in that direction or at least discuss the possibility of a singularity sympathetically: Blackmore, Broderick, Corabi and Schneider, Dainton, Goertzel, McCorduck, Shanahan, Steinhart, Shulman and Bostrom, Voorhees, and Yampolskiy. Three come across as mildly sceptical, expressing a deflationary attitude toward the singularity without quite saying that it will not happen: Dennett, Hanson, and Prinz. And about seven express wholehearted scepticism: Aleksander, Bringsjord, Greenfield, Heylighen, McDermott, Nunn, and Plotnitsky.

About twelve of the articles focus mainly on whether there will or will not be a singularity or whether there will or will not be AI: the seven wholehearted sceptics along with McCorduck, Prinz, Schmidhuber, Shulman and Bostrom, and Tipler. Three articles focus mainly on how best to negotiate the singularity: Goertzel, Hanson, and Yampolskiy. Three focus mainly on the character and consequences of a singularity: Hutter, Shanahan, and Voorhees. Three focus mainly on consciousness: Brown, Dennett, and Kurzweil. Three focus mainly on personal identity: Blackmore, Corabi and Schneider, and Dainton. Two focus on connections to other fields: Broderick and Steinhart. Numerous other issues are discussed along the way: for example, uploading (Greenfield, Corabi and Schneider, Plotnitsky) and whether we are in a simulation (Dainton, Prinz, Shulman and Bostrom).

I will not say much about the connections to other fields: Broderick's connections to science fiction and Steinhart's connections to theology. These connections are fascinating, and it is clear that antecedents of many key ideas have been put forward long ago. Still, it is interesting to note that very few of the science fiction works discussed by Broderick (or the theological works discussed by Steinhart) focus on a singularity in the sense of a recursive intelligence explosion. Perhaps Campbell's short story 'The Last Evolution' comes closest here: here humans defend themselves from aliens by designing systems that design ever smarter systems that finally have the resources to win the war. There is an element of this sort of recursion in some works by Vernor Vinge (originator of the term 'singularity'), although that element is smaller than one might expect. Most of the other works discussed by Broderick focus simply on greater-than-human intelligence, an important topic that falls short of a full singularity as characterized above.

At least two of the articles say that it is a bad idea to think or talk much about the singularity as other topics are more important: environmental catastrophe followed by nuclear war (McDermott) and our dependence on the internet (Dennett). The potential fallacy here does not really need pointing out. That it is more important to talk about topic B than topic A does not entail that it is unimportant to talk about topic A. It is a big world, and there are a lot of important topics and a lot of people to think about them. If there is even a 1% chance that there will be a singularity in the next century, then it is pretty clearly a good idea for at least a hundred people (say) to be thinking hard about the possibility now. Perhaps this thinking will not significantly improve the outcome, but perhaps it will; we will not even be in a position to make a reasoned judgment about that question without doing a good bit of thinking first. That still leaves room for thousands to think about the internet and for millions to think about the environment, as is already happening.

This reply will largely follow the shape of the original article. After starting with general considerations, I will spend the most time on the argument for an intelligence explosion, addressing various objections and analyses. In later sections I discuss issues about negotiating the singularity, consciousness, uploading, and personal identity.

2. The Argument for an Intelligence Explosion

The target article set out an argument for the singularity as follows.

(1) There will be AI (before long, absent defeaters).
(2) If there is AI, there will be AI+ (soon after, absent defeaters).
(3) If there is AI+, there will be AI++ (soon after, absent defeaters).
_____
(4) There will be AI++ (before too long, absent defeaters).

Here AI is human-level artificial intelligence, AI+ is greater-than-human-level artificial intelligence, and AI++ is far-greater-than-human-level artificial intelligence (as far beyond smartest humans as humans are beyond a mouse). 'Before long' is roughly 'within centuries' and 'soon after' is 'within decades', though tighter readings are also possible. Defeaters are anything that prevents intelligent systems from manifesting their capacities to create intelligent systems, including situational defeaters (catastrophes and resource limitations) and motivational defeaters (disinterest or deciding not to create successor systems).

The first premise is an equivalence premise, the second premise is an extension premise, and the third premise is an amplification premise. The target article gave arguments for each premise: arguments from brain emulation and from evolution for the first, from extendible technologies for the second, and from a proportionality thesis for the third. The goal was to ensure that if someone rejects the claim that there is a singularity, they would have to be clear about which arguments and which premises they are rejecting.

This goal was partially successful. Of the wholehearted singularity sceptics, three (Bringsjord, McDermott, and Plotnitsky) engage these arguments in detail. The other four (Aleksander, Greenfield, Heylighen, and Nunn) express their scepticism without really engaging these arguments. The three mild sceptics (Dennett, Hanson, and Prinz) all engage the arguments at least a little.

Greenfield, Heylighen, and Nunn all suggest that intelligence is not just a matter of information processing, and focus on crucial factors in human intelligence that they fear will be omitted in AI: understanding, embodiment, and culture respectively. Here it is worth noting that nothing in the original argument turns on equating intelligence with information processing. For example, one can equate intelligence with understanding, and the argument will still go through. The emulation and evolution arguments still give reason to think that we can create AI with human-level understanding, the extendibility point gives reason to think that AI can go beyond that, and the explosion point gives reasons to think that systems with greater understanding will be able to create further systems with greater understanding still.

As for embodiment and culture, in so far as these are crucial to intelligence, AI can simply build them in. The arguments apply equally to embodied AI in a robotic body surrounded by other intelligent systems. Alternatively, one can apply the emulation argument not just to an isolated system but to an embedded system, simulating its physical and cultural environment. This may require more resources than simulating a brain alone, but otherwise the arguments go through as before. Heylighen makes the intriguing point that an absence of values may serve as a resource limitation that slows any purported intelligence explosion to a convergence; but the only reason he gives is that values require a rich environmental context, so this worry is not a worry for AI that exists in a rich environmental context.

Aleksander makes a different argument against the singularity. First, knowledge of existing AI suggests that it is far off, and second, designing a system with greater than human intelligence requires complete self-knowledge and a complete cognitive psychology. On the first point, the distance between current AI and human-level AI may cast doubt on claims about human-level AI within years and perhaps within decades, but it does not do much to cast doubt on my arguments for the premise of AI within centuries. On the second point, it is far from clear that complete self-knowledge is required here. Brute force physical emulation could produce human-level AI without much theoretical understanding; the theoretical understanding could come later once one can experiment easily with the emulation. And paths to AI such as artificial evolution and machine learning have the potential to take a route quite different from that of human intelligence, so that again self-knowledge is not required.

Dennett engages with my arguments just for a moment. He expresses a good deal of scepticism about the singularity, but his only concrete objection is that any measure of intelligence that humans devise will be so anthropocentric that it will distort what it contrives to measure. To which one can respond: an anthropocentric measure will at least capture something that we humans care about, so that if the argument is sound, the conclusion will still be one of enormous significance. One can also note that even if we use a less anthropocentric measure of intelligence that humans do not devise, the argument may still go through. Either way, Dennett does not give any reason to deny any of the argument's premises.

Hanson expresses a different deflationary attitude to the argument, saying that its conclusion is too weak to be significant (!). The reason he gives is that there exist other sources of intelligence growth in our environment — the Flynn effect (IQ scores increase by three points per decade) and cultural development — and left to their own devices these will themselves lead to arbitrary increases in intelligence and to AI++. Here I think Hanson ignores the crucial temporal element in the conclusion. Perhaps the Flynn effect might lead to AI++ given enough time, but within centuries the level of AI+ (90 IQ points in 300 years?) is the best we can hope for. Economic and cultural development is perhaps more powerful, but again there is little reason to think it can yield human:mouse increases in community intelligence over centuries. Hanson himself observes that we have seen that sort of increase over the last 100,000 years. He goes on to suggest that that faster growth combined with the Flynn effect can be expected to yield the same sort of increase over centuries, but he gives no argument for this enormous speed-up. *Prima facie*, these sources of growth can at best be expected to lead to AI+ levels of human intelligence within centuries, not to AI++ levels.

Prinz invokes Bostrom's simulation argument to suggest that if a singularity is likely, it has probably happened already: we are probably simulated beings ourselves (as there will be many more simulated beings than non-simulated beings), and these will be created by super-intelligent beings. He then suggests that our creators are likely to destroy us before we reach AI++ and threaten them — so it is predictable that one of my defeaters (catastrophe) will occur. I am not unsympathetic with Bostrom's argument, but I think there are some holes in Prinz's use of it. In particular, it is quite likely that the great majority of simulations in the history of the universe will be unobserved simulations. Just as one finds with existing simulations, it is reasonable to expect that super-intelligent beings will run millions of universe simulations at once (leaving them to run overnight, as it were) in order to gather statistics at the end and thereby do science. They may well be able to devise anti-leakage mechanisms that make unobserved simulations cause little danger to them (the greatest danger of leakage comes from observed simulations, as discussed in the target paper). If so, our future is unthreatened. Prinz also suggests that we may destroy ourselves for relatively ordinary reasons (disease, environmental destruction, weapons) before reaching a singularity. Perhaps so, perhaps not, but here in any case I am happy enough with the limited conclusion that *absent defeaters* there will be a singularity.

Plotnitsky suggests that complexity may undermine my arguments from emulation and evolution for premise (1). If the brain is sufficiently complex, we may not be able to emulate it before long. And if evolution requires sufficient complexity, artificial evolution may be impossible in the relevant time frame.

Here, again, I was counting on the centuries time frame to make the premises of the arguments more plausible. In a recent report grounded in neurobiological evidence, Sandberg and Bostrom suggest that brain emulation with the needed degree of complexity may be possible within decades. I am a little sceptical of that prediction, but I do not see much reason to think that the intelligent processes in the brain involve complexity that cannot be emulated within centuries. I think it is more than likely that by the end of this century we will be able to simulate individual cells, their connections, and their plasticity very well. Another century should be more than enough to capture any remaining crucial internal structure in neurons and remaining features of global architecture relevant to intelligent behaviour. Throw in another century for embodiment and environmental interactions, and I think we reach the target on conservative assumptions. Of course these seat-of-the-pants estimates are not science, but I think they are reasonable all the same.

The role of complexity in the evolutionary argument is a little trickier, due to the enormous timescales involved: hundreds of millions of years of evolution culminating in a brain certainly involve much more complexity than the brain itself! On the other hand, unlike the emulation argument, the evolutionary argument certainly does not require an emulation of the entire history of evolution. It just requires a process that is relevantly similar to that history in that it produces an intelligent system. But it can certainly do so via a quite different route. The question now is whether the evolution of intelligence *essentially* turns on complexity of a level that we cannot hope to replicate artificially within centuries. I do not have a knockdown argument that this is impossible (Shulman and Bostrom's discussion of evolutionary complexity is worth reading here), but I would be surprised. Artificial evolution already has some strong accomplishments within its first few decades.

Shulman and Bostrom focus on a different objection to the evolutionary argument. It may be that evolution is extraordinarily hard, so that intelligence evolves a very small number of times in the universe. Due to observer selection effects, we are among them. But we are extraordinarily lucky. Normally one can reasonably say that extraordinary luck is epistemically improbable, but not so in this case. Because of this, the inference from the fact that evolution produced intelligence to the claim that artificial evolution can be expected to produce intelligence is thrown into question.

I do not have much to add to Shulman and Bostrom's thorough discussion of this issue. I am fairly sympathetic with their 'self-indication assumption', which raises the rational probability that the evolution of intelligence is easy, on the grounds that there will be many more intelligent beings under the hypothesis that the evolution of intelligence is easy. I also agree with them that parallel evolution of intelligence (e.g. in octopuses) gives evidence for 'evolution is easy' that is not undercut by observer effects, although as they note, the issues are not cut and dried.

All in all, I think that the evolutionary argument is worth taking seriously. But the worries about complexity and about observer effects raise enough doubts about it that it is probably best to give it a secondary role, with the argument from emulation playing the primary role. After all, one sound argument for the equivalence premise is enough to ensure the truth of a conclusion.

Bringsjord appeals to computational theory to argue that my argument is fatally flawed. His main argument is that Turing-level systems ($\mathcal{M}_2$) can never create super-Turing-level systems ($\mathcal{M}_3$), so that starting from ordinary AI we can never reach AI++, and one of my premises must be false. A secondary argument is that humans are at level $\mathcal{M}_3$ and AI is restricted to level $\mathcal{M}_2$ so that AI can never reach human level. I think there are multiple problems with his arguments, two problems pertaining especially to the first argument and two problems pertaining especially to the second.

First: intelligence does not supervene on computational class. So if we assume humans are themselves in $\mathcal{M}_2$, it does not follow that systems in $\mathcal{M}_3$ are required for AI++. We know that there is an enormous range of intelligence (as ordinarily conceived) within class $\mathcal{M}_2$: from mice to apes to ordinary people to Einstein. Bringsjord gives no reason to think that any human is close to the upper level of $\mathcal{M}_2$. So there is plenty of room within $\mathcal{M}_2$ for AI+ and AI++, or at least Bringsjord's argument gives no reason to think not. His argument in effect assumes a conception of intelligence (as computational class) that is so far from our ordinary notion that it has no bearing on arguments involving relatively ordinary notions of intelligence.

Second: if level $\mathcal{M}_3$ processes are possible in our world, then it is far from obvious that level-$\mathcal{M}_2$ AI could not create it. Computational theory establishes only that a very limited sort of 'creation' is impossible: roughly, creation that exploits only the AI's internal recourses plus digital inputs and outputs. But if the AI has the ability to observe and manipulate non-computational processes in nature, then there is nothing to stop the AI from producing a series of Turing-computable outputs that themselves lead (via the mediation of external physical processes) to the assembly of a super-Turing machine.

Third: there is little reason to think that humans are at level $\mathcal{M}_3$. Bringsjord gives no argument for that claim here. Elsewhere he has appealed to Gödelian arguments to make the case. I think that Gödelian arguments fail; for an analysis, see Chalmers (1995).

Fourth: even if humans are at level $\mathcal{M}_3$, there is then little reason to think that AI must be restricted to level $\mathcal{M}_2$. If we are natural systems, then presumably there are non-computational processes in nature that undergird our intelligence. There is no obvious reason why we could not exploit those in an artificial system, thereby leading to AI at level $\mathcal{M}_3$. Bringsjord appeals to an unargued premise saying that our AI-creating resources are limited to level $\mathcal{M}_2$, but if we are at $\mathcal{M}_3$ there is little reason to believe this premise.

McDermott holds that my argument for premise (3) fails and that more generally there is little reason to think that AI++ is possible. He says first, 'the argument is unsound, because a series of increases from AI_n to AI_{n+1}, each exponentially smaller than the previous one, will reach a limit'. Here McDermott appears to overlook the definitions preceding the argument, which stipulate that there is a positive δ such that the difference in intelligence between AI_n and AI_{n+1} is at least δ for all n. Of course it is then possible to question the key premise saying that if there is AI_n there will be AI_{n+1} (itself a consequence of the proportionality theses), but McDermott's initial formal point gives no reason to do so.

McDermott goes on to question the key premise by saying that it relies on an appeal to extendible methods, and that no method is extendible without limits. Here he misconstrues the role of extendible methods in my argument. They play a key role in the case for premise (2), where extendibility is used to get from AI to AI+. Indefinite extendibility is not required here, however: small finite extendibility is enough. And in the case for premise (3), extendibility plays no role.

So McDermott's doubts about indefinite extendibility have no effect on my argument.

I certainly do not think that a single extendible method is likely to get us from AI all the way to AI++. It is likely that as systems get more intelligent, they will come up with new and better methods all the time, as a consequence of their greater intelligence. In the target article, McDermott's worries about convergence are discussed under the label 'diminishing returns' (a 'structural obstacle') in section 4. Here I argue that small differences in design capacity tend to lead to much greater differences in the systems designed, and that a 'hill-leaping' process through intelligence space can get us much further than mere hill-climbing (of which ordinary extendibility is an instance). McDermott does not address this discussion.

The issues here are certainly non-trivial. The key issue is the 'proportionality thesis' saying that among systems of certain class, an increase of δ in intelligence will yield an increase of δ in the intelligence of systems that these systems can design. The evaluation of that thesis requires careful reflection on the structure of intelligence space. It is a mild disappointment that none of the commentators focused on the proportionality thesis and the structure of intelligence space. I am inclined to think that the success of my arguments (and indeed the prospects for a full-blown singularity) may turn largely on those issues.

Because these issues are so difficult, I do not have a knockdown argument that AI++ is possible. But just as McDermott says that it would be surprising if the minimal level of intelligence required for civilization were also the maximal possible level, I think it would be surprising if it were close to the maximal possible level. Computational space is vast, and it would be extremely surprising if the meanderings of evolution had come close to exhausting its limits.

Furthermore: even if we are close to the algorithmic limits, speed and population explosions alone might produce a system very close to AI++. Consider a being that could achieve in a single second as much as the Manhattan project achieved with hundreds of geniuses over five years. Such a being would be many levels beyond us: if not human:mouse, than at least human:dog! If we assume hundreds of AI+ (which McDermott allows) rather than hundreds of geniuses, then the difference is all the greater. McDermott's reasons give little reason to doubt that this sort of system is possible. So even on McDermott's assumption, something quite close to a singularity is very much in prospect.

On the other side, two of the articles vigorously advocate a singularity using arguments different from mine. Tipler argues that physics (and in particular the correct theory of quantum gravity) makes a singularity inevitable. An indefinitely expanding universe leads to contradictions, as it requires black hole evaporations, which violate quantum-mechanical unitarity. So the universe must end in collapse. A collapsing universe contradicts the second law of thermodynamics unless event horizons are absent, which requires a spatially closed universe and an infinite series of 'Kasner crushings'. No unintelligent process could produce this series, and no carbon-based life could survive so close to the collapse. So artificial intelligence is required. Physics requires a singularity.

I do not have the expertise in physics to assess Tipler's argument. I take it that certain key claims will be questioned by most physicists, however: for example, the claim that black hole evaporation violates unitarity, the claim that the universe must end in collapse, and the claim that a universe with an initial singularity or final collapse must be spatially closed. I also note that the last two steps of the argument in the previous paragraph seem questionable. First, to produce the Kasner crushings, non-artificial but non-carbon-based intelligence would presumably serve as well as AI. Second, for all Tipler has said, an AI process at or below human-level intelligence will be able to bring about the crushings. The more limited conclusion that the laws of physics require non-carbon-based intelligence would still be a strong one, but it falls short of requiring a full-blown singularity.

Schmidhuber says that we should stop discussing the singularity in such an abstract way, because AI research is nearly there. In particular, his Gödel machines are well on the path to self-improving intelligence that will lead to an intelligence explosion. Again, I lack the expertise to fully assess the argument, but past experience suggests that a certain amount of caution about bold claims by AI researchers advocating their own frameworks is advisable. I am certainly given pause by the fact that implementation of Gödel machines lags so far behind the theory. I would be interested to see the evidence that the sort of implementation that may lead to a full-blown intelligence explosion is itself likely to be practically possible within the next century, say.

Before moving on, I will note one of the most interesting responses to the singularity argument I have come across. In a discussion at Berkeley, a teacher in the audience noted that what goes for the design of artificial systems may also apply to the teaching of human systems. That suggests the follow analogue (or parody?) of I.J. Good's argument for a singularity:

Let a super-teacher be defined as a teacher who we teach to surpass the teaching activities of any existing teacher however competent. Since the teaching of teachers is one of these teaching activities, a super-teacher could teach even better teachers. There would then unquestionably be a 'teaching explosion', and the results of current education would be left far behind.

It is an interesting exercise to evaluate this argument, consider possible flaws, and consider whether those apply to the argument for an intelligence explosion. I suppose that it is far from clear that we can teach a super-teacher. We can certainly train ordinary teachers, but it is not obvious that any simple extension of these methods will yield a super-teacher. It is also far from clear that a proportionality thesis applies to teaching: just because one teacher is 10% better than another, that does not mean that the former can teach their pupils to be 10activity) than the pupils of the latter. Even in ordinary cases it is arguable that this thesis fails, and as we move to extraordinary cases it may be that the capacity limits of the brain will inevitably lead to diminishing returns. Now, a proponent of the original intelligence explosion argument can argue that AI design differs from teaching in these respects. Still, the argument and the analogy here certainly repay reflection.

3. Negotiating the Singularity

Three of the articles (by Goertzel, Hanson, and Yampolskiy) concern how we should best negotiate the singularity, and three (by Hutter, Shanahan, and Voorhees) concern its character and consequences. Most of these do not engage my article in much depth (appropriately, as these were the areas in which my article had the least to say), so I will confine myself to a few comments on each.

Goertzel's very interesting article suggests that to maximize the chance of a favourable singularity, we should build an 'AI Nanny': an AI+ system with the function of monitoring and preventing further attempts to build AI+ until we better understand the processes and the risks. The idea is certainly worth thinking about, but I have a worry that differs from those that Goertzel considers. Even if building an AI Nanny is feasible, it is certainly much more difficult than building a regular AI system at the same level of intelligence. So by the time we have built an AI nanny at level-n, we can expect that there will exist a regular AI system at level-$n+1$, thereby rendering the nanny system obsolete. Perhaps an enormous amount of coordination would avoid the problem (a worldwide AI-nanny Manhattan project with resources far outstripping any other project?), but it is far from clear that such coordination is feasible and the risks are enormous. Just one rogue breakaway project before the AI nanny is complete could lead to consequences worse than might have happened without the nanny. Still, the idea deserves serious consideration.

Hanson says that the human-machine conflict is similar in kind to ordinary intergenerational conflicts (the old generation wants to maintain power in face of the new generation), and is best handled by familiar social mechanisms, such as legal contracts whereby older generations pay younger generations to preserve certain loyalties. Two obvious problems arise in the application to AI+. Both arise from the enormous differences in power between AI+ systems and humans (a disanalogy with the old/young case). First, it is far from clear that humans will have enough to offer AI+ systems in payment to offset the benefits to AI+ systems in taking another path. Second, it is far from clear that AI+ systems will have much incentive to respect the existing human legal system. At the very least, it is clear that these two crucial matters depend greatly on the values and motives of the AI systems. If the values and motives are enough like ours, then we may have something to offer them and they may have reason to respect legal stability (though even this much is far from clear, as interactions between indigenous and invading human societies tends to bring out). But if their values and motives are different from ours, there is little reason to think that reasoning based on evidence from human values will apply. So even if we take Hanson's route of using existing social mechanisms, it will be crucial to ensure that the values and motives of AI systems fall within an appropriate range.

Yampolskiy's excellent article gives a thorough analysis of issues pertaining to the 'leakproof singularity': confining an AI system, at least in the early stages, so that it cannot 'escape'. It is especially interesting to see the antecedents of this issue in Lampson's (1973) confinement problem in computer security. I do not have much to add to Yampolskiy's analysis. I am not sure that I agree with Yampolskiy's view that the AI should never be released, even if we have reason to believe it will be benign. As technology progresses, it is probably inevitable that someone will produce an unconfined AI+ system, and it is presumably better if the first unconfined AI+ is benign.

Likewise, I have little to add to Hutter's analysis of the character of an intelligence explosion. On his questions of whether it will be visible from the outside or the inside, I am somewhat more inclined to give positive answers. From the outside, even an 'inward' explosion is likely to have external effects, and an 'outward' explosion may at least involve a brief period of interaction with outsiders (or alternatively, sudden death). Hutter is probably right that observation of the entire explosion process by outsiders is unlikely, however. From the inside, a uniform speed explosion might not be detectable, but there are likely to be sources of non-uniformity. It is likely that the world will contain unaccelerated processes to compare to, for example. And algorithmic changes are likely to lead to qualitative differences that will show up even if there is a speed explosion. Hutter is certainly right that it is not easy to draw a boundary between speed increases and intelligence increases, as the earlier example of an instant Manhattan project suggests. But perhaps we can distinguish speed improvements from algorithmic improvements reasonably well, and then leave it as a matter for stipulation which counts as an increase in 'intelligence'. Hutter's discussion of the potential social consequences of AI and uploading is well-taken.

Shanahan and Voorhees focus on the different sorts of intelligence that AI+ and AI++ systems might have. Voorhees discusses differences in both cognitive and sensory intelligence. Shahahan suggests that a singularity may evolve an evolution from pre-reflective and reflective creatures (our current state) to post-reflective creatures, and that this evolution may involve a sea change in our attitude to philosophical questions. It is striking that Shanahan's super-intelligent post-reflective beings appear to hold the Wittgensteinian and Buddhist views with which Shanahan is most sympathetic. Speaking for myself, I might suggest (by parity of reasoning) that AI++ systems will certainly be dualists! More likely, systems that are so much more intelligent than us will have philosophical insights of a character that we simply have not anticipated. Perhaps this is the best hope for making real progress on eternal philosophical problems such as the problem of consciousness. Even if the problem is too hard for humans to solve, our super-intelligent descendants may be able to get somewhere with it.

4. Consciousness

Although the target article discussed consciousness only briefly, three of the commentators (Brown, Dennett, and Kurzweil) focus mainly on that issue, spurred by my earlier work on that topic or perhaps by the name of the journal (*Journal of Consciousness Studies*) that hosted the original symposium on which this book is based.

Kurzweil advocates a view of consciousness with which I am highly sympathetic. He holds that there is a hard problem of consciousness distinct from the easy problems of explaining various functions; there is a conceptual and epistemological gap between physical processes and consciousness (requiring a 'leap of faith' to ascribe consciousness to others); and that artificially intelligent machines that appear to be conscious will almost certainly be conscious. As such, he appears to hold the (epistemological) further-fact view of consciousness discussed briefly in the target article, combined with a functionalist (as opposed to biological) view of the physical correlates of consciousness.

Dennett reads the target article as a mystery story, one that superficially is about the singularity but fundamentally is about my views about consciousness. I think that is a misreading: my views about consciousness (and especially the further-fact view) play only a marginal role in the article. Still, Dennett is not the only one to find some puzzlement in how someone could at once be so sympathetic to AI and functionalism on one hand and to further-fact views about consciousness on the other. I thought I addressed this in the target article by noting that the two issues are orthogonal: one concerns whether the physical correlates of consciousness are biological or functional, while the second concerns the relationship between consciousness and those physical correlates. Dennett has nothing to say about that distinction. But even without getting into philosophical technicalities, it is striking that someone like Kurzweil has the same combination of views as me. Perhaps this is evidence that the combination is not entirely idiosyncratic.

Dennett likes my fading and dancing qualia arguments for functionalism, and wonders why they do not lead me to embrace type-A materialism: the view that there is no epistemological gap between physical processes and consciousness. The answer is twofold. First, while fading and dancing qualia are strange, they are not logically impossible. Second, even if we grant logical impossibility here, the arguments establish only that biological and non-biological systems (when functionally equivalent in the same world) are on a par with respect to consciousness: if one is conscious, the other is conscious. If there is no epistemic gap between biological processes and consciousness (as type-A materialism suggests), it would follow that there is no epistemic gap between non-biological processes and consciousness. But if there *is* an epistemic gap between biological processes and consciousness (as I think), the argument yields an equally large epistemic gap between non-biological processes and consciousness. So the argument simply has no bearing on whether there is an epistemic gap and on whether type-A materialism is true. As far as this argument is concerned, we might say: type-A materialism in, type-A materialism out; epistemic gap in, epistemic gap out.

Dennett says that mere logical possibilities are not to be taken seriously. As I see things, logical possibilities make the difference between type-A and type-B materialism. If zombies are so much as logically possible, for example, there is an epistemic gap of the relevant sort between physical processes and consciousness. Still, it is possible to put all this without invoking logical possibility (witness Kurzweil, who thinks that zombies are 'scientifically irrelevant' because unobservable but who nevertheless thinks there is an epistemic gap). The thought-experiment gives reason to think that non-biological systems can be conscious; but as with our reasons for thinking that other people are conscious, these are non-conclusive reasons that are compatible with an epistemic gap. And as before, the argument shows at best that the epistemic gap for non-biological systems is no larger than the epistemic gap for biological systems. It does nothing to show that the gap is non-existent.

I will not go into all the very familiar reasons for thinking there is an epistemic gap: apart from matters of logical possibility, there is Mary in her black-and-white room who does not know what it is like to see red, and most fundamentally the distinction between the hard and the easy problems of consciousness. Dennett quotes me as saying that there is no point presenting me with counterarguments as no argument could shake my intuition, but this is a misquotation. I very much like seeing and engaging with counterarguments, but non-question-begging arguments are required to get to first base. Like the arguments of Dennett (1995) that I responded to in Chalmers (1997), Dennett's arguments here work only if they presuppose their conclusion. So the arguments fail to meet this minimal standard of adequacy.

Toward the end of his article Dennett descends into psychoanalysis, offering seven purported reasons why I reject his type-A materialism: faith, fame, Freud, and so on. I am not against psychoanalysis in philosophy, but in this case the psychoanalyses are hopeless. Dennett must at least recognize that my reasons for holding that there is an epistemic gap are very widely shared, both inside and outside philosophy. Even such apparent arch-reductionists as Ray Kurzweil and Steven Pinker seem to share them. It is easy to see that Dennett's purported analyses do not have a hope of applying in these cases. Now, it is not out of question a deep psychoanalysis could reveal a very subtle mistake or illusion that huge numbers of intelligent people are subject to. In Dennett's more serious moments he has taken stabs at analyses of this sort. It would be nice to see him bring this seriousness to bear once more on the topic of consciousness. It would also be nice to see him bring it to bear on the topic of the singularity.[1]

Brown uses considerations about simulated worlds to raise problems for my view of consciousness. First, he cites my 1990 discussion piece 'How Cartesian Dualism Might Have Been True', in which I argued that creatures who live in simulated environments with separated simulated cognitive processes would endorse Cartesian dualism. The cognitive processes that drive their behaviour would be entirely distinct from the processes that govern their environment, and an investigation of the latter would reveal no sign of the former: they will not find brains inside their heads driving their behaviour, for example. Brown notes that the same could apply even if the creatures are zombies, so this sort of dualism does not essentially involve consciousness. I think this is right: we might call it process dualism, because it is a dualism of two distinct sorts of processes. If the cognitive processes essentially involve consciousness, then we have something akin to traditional Cartesian dualism; if not, then we have a different sort of interactive dualism.

Brown goes on to argue that simulated worlds show how one can reconcile biological materialism with the conceivability and possibility of zombies. If biological materialism is true, a perfect simulation of a biological conscious being will not be conscious. But if it is a perfect simulation in a world that perfectly simulates our physics, it will be a physical duplicate of the original. So it will be a physical duplicate without consciousness: a zombie.

I think Brown's argument goes wrong at the second step. A perfect simulation of a physical system is not a physical duplicate of that system. A perfect simulation of a brain on a computer is not made of neurons, for example; it is made of silicon. So the zombie in question is a merely functional duplicate of a conscious being, not a physical duplicate. And of course biological materialism is quite consistent with functional duplicates.

It is true that from the point of view of beings in the simulation, the simulated being will seem to have the same physical structure that the original being seems to us to have in our world. But this does not entail that it is a physical duplicate, any more than the watery stuff on Twin

---

[1] I do apologize to Dennett, though, for not citing his 1978 article 'Where am I?' in the context of uploading. It certainly had a significant influence on my discussion. Likewise, my question about a reconstruction of Einstein was intended as an obvious homage to Hofstadter's 'A Conversation with Einstein's Brain' (1981). These articles are such classics that it is easy to simply to take them as part of the common background context, just as one might take Turing's work as part of the background context in discussing artificial intelligence. But I acknowledge my debt here.

Earth that looks like water really is water. (See note 7 in 'The Matrix as Metaphysics' for more here.) To put matters technically (non-philosophers can skip!), if P is a physical specification of the original being in our world, the simulated being may satisfy the primary intension of P (relative to an inhabitant of the simulated world), but it will not satisfy the secondary intension of P. For zombies to be possible in the sense relevant to materialism, a being satisfying the secondary intension of P is required. At best, we can say that zombies are (primarily) conceivable and (primarily) possible — but this possibility mere reflects the (secondary) possibility of a microfunctional duplicate of a conscious being without consciousness, and not a full physical duplicate. In effect, on a biological view the intrinsic basis of the microphysical functions will make a difference to consciousness. To that extent the view might be seen as a variant of what is sometimes known as Russellian monism, on which the intrinsic nature of physical processes is what is key to consciousness (though unlike other versions of Russellian monism, this version need not be committed to an *a priori* entailment from the underlying processes to consciousness).

5. Uploading and Personal Identity

The last part of the target article focuses on uploading: transferring brain processes to a computer, either destructively, non-destructively, gradually, or reconstructively. Two of the commentaries (Greenfield and Plotnitsky) raise doubts about uploading, and three (Blackmore, Dainton, and Schneider and Corabi) focus on connected issues about personal identity.

Greenfield argues that uploading will be very difficult because of the dynamic plasticity of neurons: they are not fixed components, but adapt to new situations. However, there appears to be no objection in principle to simulation processes that simulate all the relevant complexity: not just the static behaviour but the dynamic adaptation of neurons. As Greenfield herself notes, this will require a simulation of all the chemical and biochemical machinery that makes its plasticity and sensitivity possible, but extra detail required here is a difference in degree rather than a difference in kind. Perhaps this sort of simulation is not realistic in the near term, but there is little reason to think that it will not be possible within a time frame of centuries.[2]

----

[2]   Greenfield says that I do not define 'singularity' in the target article, but it is defined in the first sentence of the article in very much the way that I define it in the second paragraph of this reply. A difference is that in the current article I speak of 'systems' rather than

Plotnitsky suggests that to create an exact copy of a human being we may need to repeat the history of the universe from the beginning. But of course uploading does not require an exact copy. The most important thing is to create simulations of the essential components of the cognitive system that at least approximate their patterns of functioning and their relations to other components, to within something like the range of background noise or other ordinary fluctuations. Then an upload can be expected to produce behaviour that we *might* have produced in a similar environment, even if it does not produce exactly the behaviour that we *would* have produced. That is a much easier task. Again it is certainly a non-trivial task and one that may not be accomplished within decades, but it is hard to see that there is an obstacle of principle here.

Corabi, Schneider, and Dainton discuss uploading in the context of personal identity. Their discussions presuppose a good amount of philosophical background, and consequently the remainder of this section is philosophically technical.

Corabi and Schneider are doubtful that we can survive uploading. They initially frame their arguments in terms of background metaphysical premises such as substrate and bundle views of individuals, and of three-dimensional and four-dimensional views of identity over time. I do not have firm views on these metaphysical questions and think that they may not have determinate answers, for reasons discussed in my article 'Ontological Anti-Realism'. However, as Corabi and Schneider themselves note, their central arguments against uploading do not depend on these premises, so I will consider them independently.

Corabi and Schneider argue against destructive uploading on the grounds that it can yield strong spatiotemporal discontinuity in the path of an individual. I might be in Australia at one moment (in a biological body) and then in the US the next moment (in uploaded form) without travelling through the points in between. They suggest that objects do not behave this way (at least if they are construed as substances). However, it is not hard to find objects that exhibit this sort of discontinuous behaviour.

In 1713 Yale University moved from Wethersfield to New Haven. I do not know the exact circumstances, but it is not hard to imagine that the move happened with the issuing of a decree. At that moment, the university moved from one place to another without passing through

---

'machines' in order to accommodate the possibility that the intelligence explosion takes place in humans by a process of enhancement.

the places in between. One could also imagine versions where it exists for a brief period at both locations, or in which there is a temporal gap during which it is located nowhere. I take it that universities are objects, so there is no general objection to objects behaving this way. There are also objects such as electronic databases that can quite clearly be destructively uploaded from one location to another. Whether we construe objects as substrates or as bundles, a plausible theory of objects should be able to accommodate phenomena such as this. So I do not think that a plausible theory of objects will rule out discontinuous motion of this sort.

Perhaps Corabi and Schneider hold that there is a disanalogy between universities and people. Perhaps people are fundamental entities where universities (and databases) are non-fundamental entities, for example, and perhaps the continuity constraint is more plausible where fundamental entities are concerned. They say explicitly that they intend their arguments to apply on a materialist view (on which people are not fundamental), however, so this cannot be what is going on. And if we assume a substance dualist view on which people are fundamental non-physical entities, there is not much reason to suppose that non-physical entities are subject to the same continuity constraints as fundamental physical entities.

Corabi and Schneider also argue against gradual (destructive) uploading. They say that it is subject to the same issues concerning spatiotemporal discontinuity, at the 'dramatic moment at which the data is assembled by the computer host and the isomorph is born'. Here I suspect that they are conceiving of gradual uploading in the wrong way. As I conceive of gradual uploading, there is no such dramatic moment. A functional isomorph of the original is present throughout. Its neurons are replaced one at a time by uploaded copies, leading from a 100% biological system to a 99%–1% system (biological–silicon, say), a 98%–2% system, and so on until there is a 100% isomorph of the original. In so far as the person changes location it will be a gradual change, one neuron at a time.

The same misconception may be at play in Corabi and Schneider's formal argument against gradual uploading, where they appeal to a scenario in which a single person is gradually uploaded to two locations. Their premise (A) says 'If [gradual] uploading preserves the continuity of consciousness, then the continuity of consciousness can be duplicated in multiple locations'. As I conceive of gradual uploading, this premise is much less obvious than Corabi and Schneider suggest. Gradually uploaded systems are not built up from nothing; rather they are connected to the original system throughout. At the initial stages, the 99% of the brain will be causally integrated with a single uploaded version of the remaining 1%. It is not easy to see how it could be causally integrated with two such systems. Perhaps one could be a back-up copy that does not causally affect the brain, but then it will not count as causally integrated. Perhaps two copies could both be integrated with the brain by a sort of causal overdetermination, but if so the combined system is most naturally treated as a single system.

Perhaps one might run a version of the latter scenario that eventually leads to two different independent systems, both of which will have a sort of continuity of consciousness with the original. This sort of case is best treated as a sort of fission case, perhaps akin to a case where one gradually splits my brain while I am still conscious. It is not easy to know what to say about those cases. Perhaps the most natural view of these cases holds that they involve a sort of survival that falls short of numerical identity but that nevertheless yields much of what we care about in numerical identity. In any case, the possibility of uploading does not seem to add any difficulties that do not already arise from the possibility of fission.

Dainton suggests that my claim that continuity of consciousness is our best guide to personal identity is in tension with the further-fact views and deflationary views of personal identity (both of which I have some sympathy for): if continuity secures identity, then there is no room for further facts and no room for deflation.

Considering first the further-fact view: here I meant only to be saying (as Dainton suggests) that physical facts and synchronic mental facts (facts about the states of a subject at specific times) can leave open questions about identity. In the article I suggested that once we specify continuity of consciousness over time in addition, there is no such open question. Still, on reflection I do not think the matter is cut and dried. One can consistently hold that continuity is the best guide that we have to consciousness without holding that it is an indefeasible guide. So the 'best guide' view is compatible with there being an epistemological further fact about identity over and above facts about continuity.

If we put things in terms of conceivability, the key issue is whether one can conceive of cases in which continuity takes a certain pattern and certain identity facts obtain, and cases with the same pattern in which other identity facts obtain. The case of gradual fission while conscious might be an example: perhaps I can conceive myself surviving as the left or the right hemisphere? It may even be that ordinary continuity of consciousness is epistemically compatible with identity and its absence. For example, it is not obviously inconceivable that a single stream of consciousness could involve different subjects at the start and at the finish (perhaps Cartesian egos could swap in and out of a stream of consciousness?) There are tricky issues here about how to characterize continuity without presupposing a single subject throughout, but I think one can appeal to a notion of q-continuity (analogous to Parfit's q-memory, and needed in any case to handle fission scenarios) that makes no such presupposition. Then there may at least be an epistemic gap between q-continuity and identity.

Another worry is provided by cases where continuity is absent, such as the gap between sleep and waking. Here it is arguable that one can conceive of both surviving this gap and failing to survive it. Dainton tries to remove the gap by extending his notion of continuity to C-continuity that obtains across this gap: roughly, the idea is that the pre-sleep and post-waking states are C-continuous iff, had the intermediate systems been conscious throughout, these states would have been part of a single stream of consciousness. Dainton has thought about this matter more than me, but my initial reaction is that even if continuity without identity is inconceivable, C-continuity without identity may be conceivable. The counterfactuals involved in C-continuity may be such as to change the facts about identity: for example, it seems quite consistent to say that a stream involves a single subject over time, but that if it had been interrupted then there would have been two subjects over time. But I may be wrong about this.

In so far as I take a further-fact view, I take the view that there is Edenic survival, with deep further facts about the identity of a self as there might have been in Eden. (Here, as in my 'Perception and the Fall from Eden' (2006), we can think of Eden as the world as presented by our experience and intuitions, and of Edenic survival as analogous to Edenic colours, the primitive colours that obtain in Eden.) I think Edenic survival involves primitive identity facts: facts about the identity of subjects over time that are not reducible to any other facts. If so, there will always be an epistemic gap between non-identity facts and identity facts. If continuity of consciousness is specified in a way that builds in identity (the same subject has a certain stream over time), then there need be no gap between continuity and identity, but if it is specified in a way that does not (as mere q-continuity, for example), then there will be such a gap. Some non-identity-involving condition such as q-continuity may serve as a sort of *criterion* for identity over time, but it will be a criterion that is merely contingently associated with identity (perhaps as a matter of natural necessity). From this perspective, a view (like Dainton's?) on which there is no epistemic gap between q-continuity and identity is already somewhat deflationary about identity and survival.

On Dainton's view, gradual uploading preserves identity (there is continuity of a stream of consciousness or the potential for it), while destructive uploading does not (there is no such potential). In so far as I hold an Edenic view, I think it is at least conceivable that someone survives destructive uploading; this once again brings out the epistemic gap between facts about C-continuity and facts about survival. I am far from sure that this is naturally possible, though: it is not out of the question that the sort of Parfit-style psychological relation found in cases of destructive uploading provides a contingent criterion for identity. On the Edenic view this is an issue for speculation that philosophy and science may have a hard time resolving. Still, in so far as continuity of consciousness provides a (contingent) sufficient condition for survival, then gradual uploading will serve at least as well to ensure survival as destructive uploading, and quite possibly better.

That said, I am a little sceptical about Edenic survival and somewhat more sympathetic to deflationary views. I think that once one gives up on Edenic survival, one should hold that there are no deep facts about survival. There are many relations connecting subjects over time, but none carry absolute weight. On a moderate version of the view (like Parfit's), there is some reason to care about each of these relations. On an extreme version of the view, there is no special reason to care about one rather than another (or at least no reason for the distinctive sort of caring that we typically associate with identity relations); the only thing that distinguishes them is that we do care about some rather than others. Either way, it is quite consistent to hold that continuity is crucial to the relation that we care about most or that we should care about most.

I am inclined to think the extreme view is more consistent, although it is also more counterintuitive. Moderate deflationary views do not seem to be able to tell a good story about *why* causal or continuity connections give us reason to care in an identity-like way. Our ordinary reasons for caring about these relations seem to stem from the fact that we take them to provide good criteria for something like Edenic survival. Once we have discarded Edenic survival, these reasons seem to have little residual force. Something similar may well apply to Dainton's view that reduces survival to a sort of continuity.

So in so far as I endorse a deflationary view, I incline toward an extreme deflationary view on which identity-based concern for the future is irrational or at best arational (after philosophical reflection).

Blackmore articulates this view nicely (Prinz and McDermott also appear to hold versions of it). As she notes, there may be other reasons to care about future selves. We can reasonably care that our current projects are fulfilled, for example, and it may be that caring about future selves (like caring about others in our community) makes things better for everyone. Speaking for myself, whether or not there are *reasons* for identity-based concern about the future, I find this sort of concern impossible to abandon. I think it will probably always function in our lives the way that other arational desires function. Still, all this means that if we come to care about our future uploaded selves in much the same way that we care about our future biological selves, then uploading will be on a par with ordinary biological survival.

Dainton goes on to consider the hypothesis that we are inhabiting a simulated universe. As well as considering Bostrom's simulation argument (which can be used to suggest that this hypothesis is quite likely) and some ethical and practical issues, he also engages my argument (in 'The Matrix as Metaphysics') that this hypothesis is not a sceptical hypothesis. That is: it is not correct to say that if we are inhabiting a simulated universe, tables and chairs (and so on) do not exist. Rather, ordinary external objects exist, and we should just revise our metaphysical views about their ultimate constitution. Here Dainton objects to my argument on the ground that the computational structure of a simulated universe does not suffice for it to yield a properly spatial universe. To yield a true space, the underlying processes in a simulation would need to be laid out in the right sort of spatial arrangement. Without that arrangement, the simulation scenario will indeed be a sceptical scenario.

I think this is the best way to resist the argument of 'The Matrix as Metaphysics', as I noted in that article. As with the case of survival and of colour, I think we may have a grip on a primitive concept of space — call it a concept of Edenic space — that requires certain special primitive relations to obtain, relations that may not obtain in a simulation. But, as with Edenic colour and Edenic survival, I think there are serious grounds for doubt about whether there is Edenic space in our world. Perhaps there might be Edenic space in a Newtonian world. But relativity theory and quantum mechanics both give grounds for doubt about whether space can be Edenic. Still, we are not inclined to say that there is no space in our world. In the case of colour, after we drop Edenic colour we tend to identify colours with whatever play the relevant roles, for example in producing our colour experience. Likewise, after dropping Edenic space, I think we identify space with whatever plays the relevant roles, both in scientific theory and in producing our experience. But this functionalist conception of space then opens the door for there to be space in a Matrix scenario: we identify space with whatever plays the relevant role in that scenario, just as we do in quantum mechanics and relativity (and even more so in more speculative theories that postulate more fundamental levels underneath the level of space). Of course there is much more to say here, for example about the choice between functionalist and primitivist conceptions of space; I try to say some of it in my forthcoming book *Constructing the World*.

6. Conclusion

The commentaries have reinforced my sense that the topic of the singularity is one that cannot be easily dismissed. The crucial question of whether there will be a singularity has produced many interesting thoughts, and most of the arguments for a negative answer seem to have straightforward replies. The question of negotiating the singularity has produced some rich and ingenious proposals. The issues about uploading, consciousness, and personal identity have produced some very interesting philosophy. The overall effect is to reinforce my sense that there is an area where fascinating philosophical questions and vital practical questions intersect. I hope and expect that these issues will continue to attract serious attention in the years to come.[3]

References

Bostrom, N. (2003) Are we living in a simulation?, *Philosophical Quarterly*, **53** (211), pp. 243–255.

Chalmers, D.J. (1990) How Cartesian dualism might have been true, [Online], http://consc.net/notes/dualism.html

Chalmers, D.J. (1995) Minds, machines, and mathematics, *Psyche*, **2**, pp. 11–20.

Chalmers, D.J. (1997) Moving forward on the problem of consciousness, *Journal of Consciousness Studies*, **4** (1), pp. 3–46.

Chalmers, D.J. (2005) The Matrix as metaphysics, in Grau, C. (ed.) *Philosophers Explore the Matrix*, Oxford: Oxford University Press.

Chalmers, D.J. (2006) Perception and the fall from Eden, in Gendler, T. & Hawthorne, J. (eds.) *Perceptual Experience*, Oxford: Oxford University Press.

Dennett, D.C. (1978) Where am I?, in Dennett, D.C., *Brainstorms*, Cambridge, MA: MIT Press.

Dennett, D.C. (1995) Facing backward on the problem of consciousness, *Journal of Consciousness Studies*, **3** (1), pp. 4–6.

Hofstadter, D.R. (1981) A conversation with Einstein's brain, in Hofstadter, D.R. & Dennett, D.C. (eds.) *The Mind's I*, New York: Basic Books.

Lampson, B.W. (1973) A note on the confinement problem, *Communications of the ACM*, **16**, pp. 613–615.

---

[3]   Thanks to Uziel Awret for all his work in putting together this volume and for his comments on this reply.

Moravec, H. (1988) *Mind Children: The Future of Robot and Human Intelligence*, Cambridge, MA: Harvard University Press.

Sandberg, A. & Bostrom, N. (2008) Whole brain emulation: A roadmap, *Technical Report 2008–3*, Future for Humanity Institute, Oxford University, [Online], http://www.fhi.ox.ac.uk/Reports/2008-3.pdf